Foundations of Radio for Scientists and Technologists

CHRISTOPHER JOHN COLEMAN
University of Bath and University of Adelaide

CAMBRIDGE
UNIVERSITY PRESS

CAMBRIDGE
UNIVERSITY PRESS

Shaftesbury Road, Cambridge CB2 8EA, United Kingdom

One Liberty Plaza, 20th Floor, New York, NY 10006, USA

477 Williamstown Road, Port Melbourne, VIC 3207, Australia

314–321, 3rd Floor, Plot 3, Splendor Forum, Jasola District Centre, New Delhi – 110025, India

103 Penang Road, #05–06/07, Visioncrest Commercial, Singapore 238467

Cambridge University Press is part of Cambridge University Press & Assessment, a department of the University of Cambridge.

We share the University's mission to contribute to society through the pursuit of education, learning and research at the highest international levels of excellence.

www.cambridge.org
Information on this title: www.cambridge.org/9781108470940

DOI: 10.1017/9781108684514

First published 2018

A catalogue record for this publication is available from the British Library

Library of Congress Cataloging-in-Publication data
Names: Coleman, Christopher, 1950– author.
Title: Foundations of radio for scientists and technologists /
Christopher Coleman (Department of Electrical and Electronic Engineering, Adelaide University).
Description: Cambridge : Cambridge University Press, 2018. |
Includes bibliographical references and index.
Identifiers: LCCN 2018015101 | ISBN 9781108470940 (hardback)
Subjects: LCSH: Radio–History. | Radio waves. | Radio wave propagation. |
Radio–Equipment and supplies.
Classification: LCC TK6550.C675 2018 | DDC 621.384–dc23
LC record available at https://lccn.loc.gov/2018015101

ISBN 978-1-108-47094-0 Hardback

Additional resources for this title available at www.cambridge.org/coleman2

Contents

Preface

Radio is one of the most pervasive technologies of the twentieth century and is a major element in all of our lives. Besides being the major technology that makes the broadcast and telecommunication industries possible, it is an important supporting technology for the transport industry, the military and the emergency services. In the sphere of scientific research, it is an important element in data gathering and, through radar, provides a primary research tool for geology, meteorology and environmental science. Furthermore, through radio telescopes, it is a major tool for astronomers. It is clear that many scientists will need to understand, and even design, radio systems of considerable sophistication. Whilst such scientists might not need to design the electronics, they will certainly need to understand its capability and performance limitations. However, such knowledge is usually the preserve of the professional RF engineer; knowledge he will have gleaned from a series of courses in diverse areas that include electronics, signal processing, communications, electromagnetism, antennas and propagation. Material from these topics constitutes the subject of radio. Many scientists, and technologists, do not have the luxury of being able to study such a range of material in detail. Consequently, it is the aim of the presented text to provide such readers with a basic understanding of radio, both its theory and practice.

This is not a book about the design of radio frequency circuits, but rather a book about the phenomenon of radio and how it works. Consequently, there is an even balance between the physics of radio and the technology that has made it possible. The primary aim in writing this book has been to make some quite advanced topics in radio accessible to a more general audience. However, due to the mathematical nature of most radio theory, there is a danger that such a book can end up as just a list of facts and formulas. To avoid this, the book uses a more physics-based approach to the complex theory of radio. This not only aids the reader's understanding but also avoids the requirement of a large prerequisite knowledge in advanced mathematics. Indeed, the reader only needs some basic knowledge of vectors, calculus and complex numbers. In particular, the book does not require knowledge of vector calculus. As a consequence, the book should be accessible to a large range of scientists, engineers and technicians.

The book develops the theory of electromagnetism in a historical fashion, from early ideas concerning electrostatics to the prediction and discovery of radio waves. It then looks at the fundamental technological developments that have made modern radio possible. The book intertwines the history of radio with the theory, hence giving the reader an idea of how, and why, certain technologies were developed. Importantly, the book discusses issues that can affect the performance of radio systems. The book develops, in some detail, the important topics of transmission lines, antennas and propagation. Additionally, it looks at some important modern radio technologies such as spread spectrum, cellular radio, MIMO and radar. Whilst the book does not aim to train the reader in radio electronics, the book will provide a sufficient background for the reader to progress onto relatively advanced texts in this area. Much of the material in this book grew out of courses given to non-specialists with a need to understand radio in more detail than is offered by most introductory texts. The students on these courses were mainly users of radio (ionospheric physicists and radar scientists for example) with a need to understand radio in greater depth. Such students are the intended readership of this book. However, the book might also be of interest to those with an electrical and electronic engineering (EEE) background. The coverage of radio in a typical EEE degree tends to be disjointed and it is possible that this book might prove useful to those who wish to fill gaps in their knowledge of radio, or simply to revise their knowledge. For those readers interested in the practical application of the material in this book, the online resources include numerous examples and exercises.

I would like to thank my wife, Marilyn, for her invaluable support and help in preparing this book.

1 Electromagnetism

Radio is a technology that is based upon electromagnetic phenomena and an understanding of electromagnetic theory is crucial to the understanding of radio. Ideas of electricity and magnetism have been in existence for many millennia, but the theory of electromagnetism was the result of a surge in activity over the last four centuries. The development of electromagnetic theory culminated in the Maxwell equations, equations that are crucial to our understanding of radio waves. Radio is an example of the triumph of theoretical science in that it was predicted through theory rather than being discovered by accident. It is the aim of the current chapter to describe electromagnetic theory through its historical development. The chapter has been written for those with very little knowledge in the area and so can be skipped by those who already possess a good knowledge of the subject. However, it is expected that some readers will be a little rusty on the theory and so this chapter will serve as revision for them.

1.1 Electricity

The first recorded observations of electrical effects go back to the Greeks. In the sixth century BC, Thales of Miletus observed that amber, when rubbed, would attract light objects. This phenomenon is exemplified by the old schoolboy trick of rubbing a comb on your trousers and then seeing it lift small scraps of paper. Today we know that matter is made up of atoms which contain particles with positive electric charge (protons), negative electric charge (electrons) and no charge (neutrons). Further, that like charges repel each other and that unlike charges attract. A simple model of a single atom consists of a number of electrons that orbit around a nucleus consisting of the same number of protons and possibly some neutrons (see Figure 1.1). The electrons are arranged in shells around the nucleus, each shell containing electrons of approximately the same energy (the energy increases with radius) and are designated, in order of energy, as K, L, M, N, O, P and Q (it should be noted that the energy gap between these shells is much larger than the range of energies within a shell). Due to quantum mechanical effects, the shells contain only a limited number of electrons (the K shell can contain a maximum of 2 electrons, the L shell 8 electrons, the M shell 18 electrons, the N shell 32, etc.). Matter will consist of a large collection of such atoms which, under normal circumstances, will be in overall electrical neutrality (the numbers of electrons and protons are equal). Under some circumstances, however, it is possible to increase, or decrease, the number

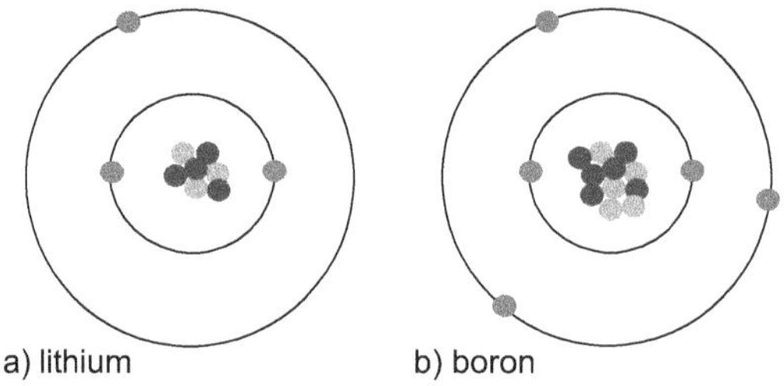

Fig. 1.1 Atomic structure consists of electrons orbiting an equal number of protons and possibly some neutrons.

of electrons and the material will become electrically charged. This is what is achieved in the above rubbing process, sometimes known as the triboelectric effect. The essential condition for the effect to exist is that the materials being rubbed together have different strengths of the force that bind their electrons to the nucleus (glass has a far stronger bond than rubber for example). When the materials are brought together, electrons in the material with the weaker force will be attracted to the material with the stronger force. When the materials are then separated, some of the transferred electrons will remain on the material with the stronger force and both materials will be charged, one positively (the one with the weaker force) and one negatively (the one with the stronger force).

Real matter can be quite complex in structure, with many materials composed of molecules that are complex combinations of different kinds of atoms. The heavier atoms (those with a large number of protons) can have many layers of electrons surrounding the nucleus and this means that the bond of the outer electrons can be relatively low. This can lead to high electron mobility in materials composed of such atoms. Materials for which the electrons are highly mobile, relative to the protons, are known as conductors and are exemplified by metals such as copper, silver and gold. Materials where the electrons are relatively immobile are known as insulators (glass and rubber being important examples). Insulators and conductors turned out to be of great importance in the development of electricity.

The seventeenth and eighteenth centuries were a period of great advances in our knowledge of electrical effects, much of it made possible by increasingly sophisticated machines for developing charged materials through the triboelectric effect. Figure 1.2 shows the basic mechanism of such machines. The rubber belt rolls over the glass cylinder and this causes electrical charge to build up on these components through the triboelectric effect. When the components separate, the belt will be negatively charged and the cylinder positively charged. The negative charges on the belt will eventually reach a conducting brush that sweeps them up onto a conducting metal wire along which they travel until reaching a conducting sphere on which they accumulate. In a similar fashion, the positive charge travels with the cylinder until it reaches a conducting brush. At this brush, the positive charge is neutralised by negative charge that has been drawn from the lower sphere along the conducting wire. In this fashion, positive charge accumulates on the lower sphere. As shown in Figure 1.2, the charge accumulates on opposing faces of the spheres. This occurs due to the mobility of electrons on conductors and the fact that

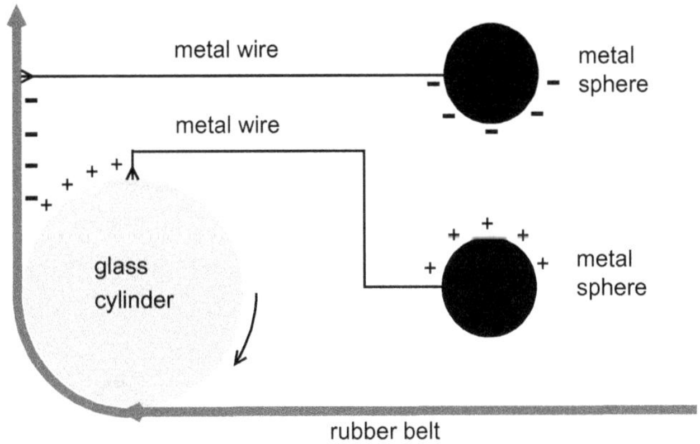

Fig. 1.2 A basic machine for creating positive and negative charge by the triboelectric effect.

opposing charges attract. The medium between the spheres is composed of air and this will tend to act as an insulator and so the charge will just accumulate on the spheres. Furthermore, the charges on the opposing spheres will balance each other out.

If a charged particle is placed between the spheres, it will be drawn towards the sphere with the opposing charge and repulsed by the sphere with the same charge. Consequently, if we want to increase the amount of negative charge on the upper sphere by directly moving positive charge to the lower sphere, this will require an external agency to do some work. This brings us to the important concept of *potential difference*. The potential difference between two points is defined to be the work done by an external force in moving positive charge between these points and is measured in terms of volts (1 volt is 1 joule per coulomb). In order to quantify this, we need to be able to calculate the force that one charge imposes upon another. The force F imposed on charge q by charge Q is given by *Coulomb's law*

$$F = \frac{1}{4\pi\epsilon_0}\frac{qQ}{r^2}, \tag{1.1}$$

where r is the distance that separates the charges and ϵ_0 is known as the *permittivity* of free space (i.e. space that is devoid of matter). This force is repulsive if the charges have the same sign and attractive if the sign is different. The law was formulated by Charles Augustin de Coulomb in 1784 as the result of much experimental work. The units of charge are known as coulombs, with a proton having a charge 1.60219×10^{-19} coulombs and an electron minus that amount. If distances are measured in metres and the force in newtons, $\epsilon_0 = 8.85 \times 10^{-12}$.

Force is *vector* in nature, i.e. it has both magnitude and direction. Consequently, we need some understanding of vector quantities. Pictorially, we can represent a vector as an arrow that points in the direction of the vector with its length equal to the magnitude (Figure 1.3). Vectors are not only useful for describing quantities such as force, but can also be used for describing the geometrical concept of position. The position of a point can be described by the vector that joins some arbitrary origin to this point, the magnitude being the distance from the origin to the point. An important concept in vectors is that of

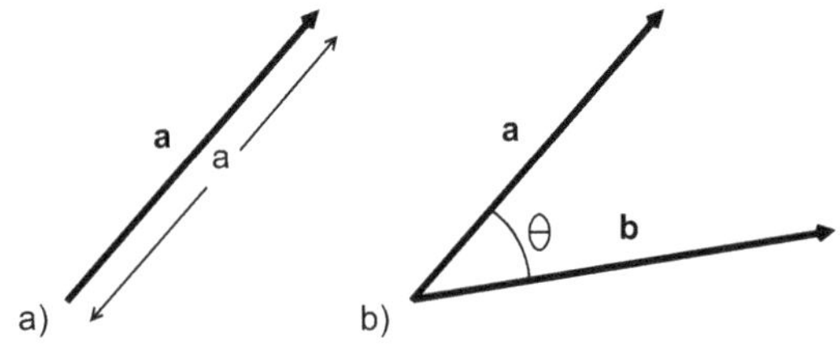

Fig. 1.3 a) Vector represented graphically as an arrow and b) angle between vectors for the vector dot product.

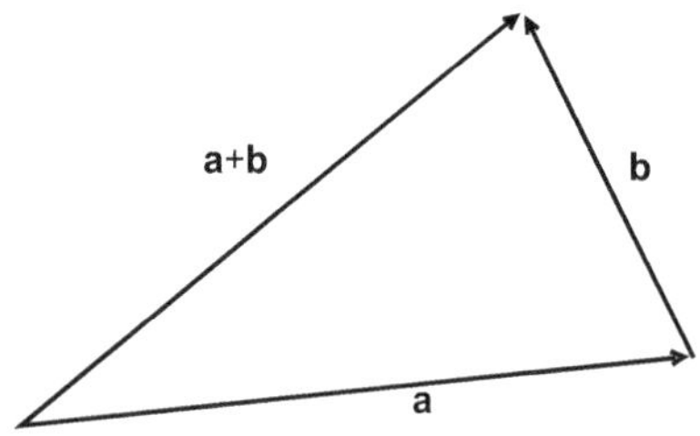

Fig. 1.4 The addition of vectors.

the *dot product* of two vectors **a** and **b**, written as $\mathbf{a}\cdot\mathbf{b}$. If the two vectors have magnitudes a and b, respectively, the dot product is defined to be $ab\cos\theta$ where θ is the angle between these vectors (see Figure 1.3). It can now be seen that $a=\sqrt{\mathbf{a}\cdot\mathbf{a}}$ and $b=\sqrt{\mathbf{b}\cdot\mathbf{b}}$. (Note that we often use $|\mathbf{x}|$ as mathematical shorthand for magnitude $x=\sqrt{\mathbf{x}\cdot\mathbf{x}}$ of the vector.) The dot product can be used to find the component of a force **F** in a particular direction. Let $\hat{\mathbf{t}}$ be a unit vector ($|\hat{\mathbf{t}}|=1$) in the direction of interest, then $\hat{\mathbf{t}}\cdot\mathbf{F}$ is the component of force in that direction.

An important operation we can perform on a vector **p** is to multiply it by a scalar s to get a new vector $s\mathbf{p}$ that points in the same direction as **p** but now has the magnitude sp. Another important operation when we have multiple vectors is their addition. For the vectors **a** and **b**, if we join the tip of the arrow representing **a** to the base of the arrow representing **b**, the sum $\mathbf{a}+\mathbf{b}$ is represented by the arrow from the base of the arrow representing **a** to the tip of the arrow representing **b** (see Figure 1.4).

In terms of vectors, Coulomb's law can be rewritten as

$$\mathbf{F}=\frac{1}{4\pi\epsilon_0}\frac{qQ}{r^2}\hat{\mathbf{r}}, \tag{1.2}$$

where $\hat{\mathbf{r}}$ is a unit vector ($\hat{\mathbf{r}}\cdot\hat{\mathbf{r}}=1$) in the direction from Q to q. An alternative way of looking at this is to regard charge Q as creating an *electric field* (sometimes known as the *electric intensity*)

$$\mathcal{E}=\frac{1}{4\pi\epsilon_0}\frac{Q}{r^2}\hat{\mathbf{r}} \tag{1.3}$$

that pervades space. When a charge q is placed in this field, it is acted upon by a force $q\mathcal{E}$ where $\mathcal{E}$ is the value of the field at the position of charge q ($\mathcal{E}$ will have units of volts per metre). The concept of a field that exists at all points of space was a revolution in thinking and was an extremely important step in the development of electromagnetism.

One can now ask what the field will be when there are charges at a variety of locations. Fortunately, it turns out that this field will simply consist of the sum of the fields due to the individual charges. Consequently, at a position $\mathbf{r}$, a system of N charges has the electric field

$$\mathcal{E}(\mathbf{r}) = \frac{1}{4\pi\epsilon_0}\sum_{i=1}^{N}\frac{Q_i}{|\mathbf{r}-\mathbf{r}_i|^3}(\mathbf{r}-\mathbf{r}_i), \tag{1.4}$$

where $\mathbf{r}_i$ is the position of the ith charge Q_i and $|\mathbf{r}-\mathbf{r}_i|$ is the distance from $\mathbf{r}_i$ to $\mathbf{r}$.

We now return to the question of the potential difference between points $\mathbf{r}_A$ and $\mathbf{r}_B$. This is the work done in moving a unit charge from a point $\mathbf{r}_A$ to a point $\mathbf{r}_B$. If there is a constant electric field, the work done in moving from point $\mathbf{r}_A$ to $\mathbf{r}_B$ is $-(\mathbf{r}_B-\mathbf{r}_A)\cdot\mathcal{E}$ (i.e. minus the field in the direction of $\mathbf{r}_B$ from $\mathbf{r}_A$ multiplied by the distance in that direction). When moving through the field produced by a finite number of charges, however, the force will vary from point to point. Consequently, we will need to split the path over which the unit charge moves into a number of short segments on each of which the electric field can be regarded as constant (see Figure 1.5). The potential difference will now be approximated by

$$V = -\sum_{i=1}^{M}\mathcal{E}(\mathbf{r}_i)\cdot(\mathbf{r}_i-\mathbf{r}_{i-1}), \tag{1.5}$$

where M is the number of segments. Taking the limit where the segment lengths tend to zero, the above sum becomes the mathematical operation of integration along a line, that is

$$V = -\int_{\mathbf{r}_A}^{\mathbf{r}_B}\mathcal{E}(\mathbf{r})\cdot d\mathbf{r}. \tag{1.6}$$

In the case of our finite system of charge, we will define the *potential* V of the system to be the potential difference when point $\mathbf{r}_A$ is a point at infinity and $\mathbf{r}_B$ is the test point $\mathbf{r}$, then

$$V(\mathbf{r}) = \frac{1}{4\pi\epsilon_0}\sum_{i=1}^{N}\frac{Q_i}{|\mathbf{r}-\mathbf{r}_i|}. \tag{1.7}$$

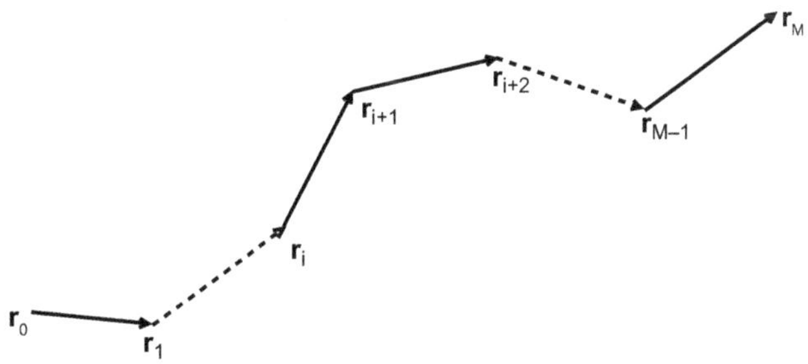

Fig. 1.5 Path for calculating work done when divided into segments ($\mathbf{r}_0 = \mathbf{r}_A$ and $\mathbf{r}_M = \mathbf{r}_B$).

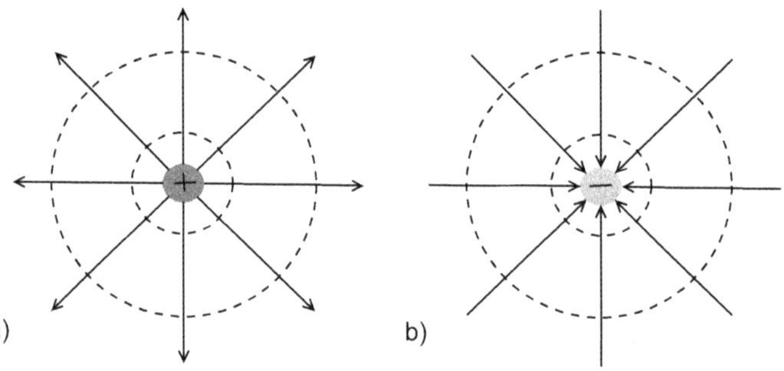

Fig. 1.6 Field lines and lines of constant potential for positive and negative charges.

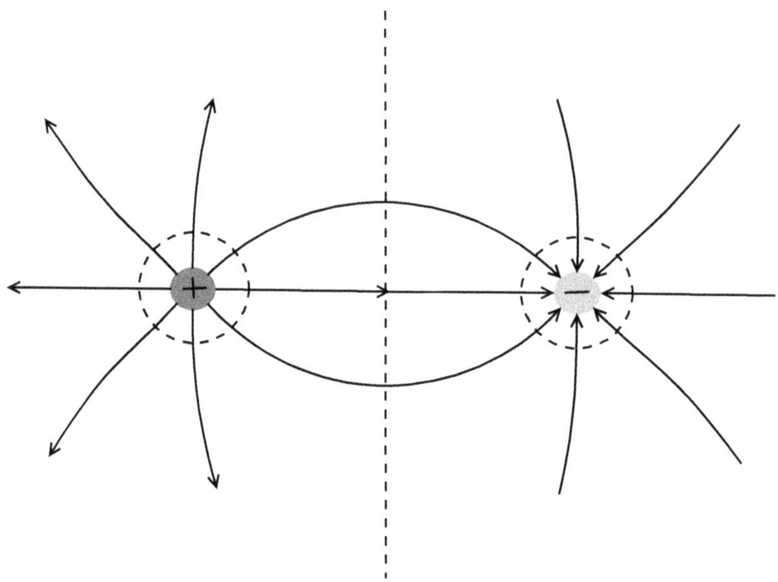

Fig. 1.7 Field lines and lines of constant potential for a dipole.

We can visualise a field in terms of what are known as *field lines*. Such lines have the property that, at any point, their tangent is in the direction of the field at that point. Figure 1.6 shows the field lines for positive and negative charges, the fields run in the radial direction (outwards and inwards respectively). It will be noted that the field lines spread out as we move away from the sources and so the density of field lines at any point is an indication of the strength of the field at that point. Also shown are the surfaces of constant potential (spherical surfaces around the charge that are depicted as broken lines). Figure 1.7 shows the field lines for positive and negative charges of equal magnitude that are separated by a finite distance d. This combination is often known as a dipole and is important in the development of radio theory. At great distances from the dipole the effects of the charges will almost balance out and so the field will be much weaker than

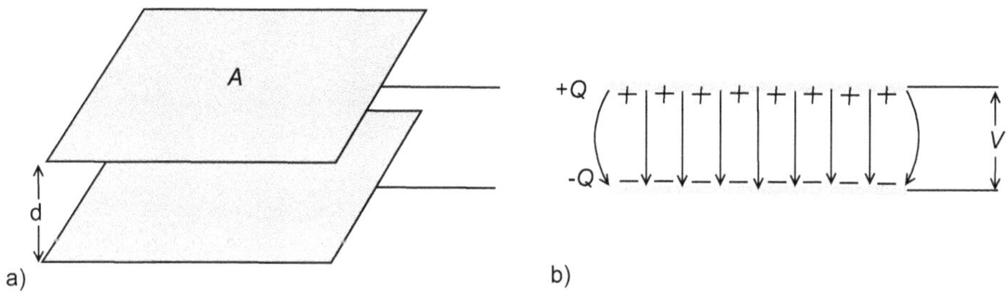

Fig. 1.8 a) Geometry of a parallel plate capacitor and b) field lines in a charged capacitor.

that of a single charge. At great distances, the field will have the form

$$\mathcal{E} = \frac{1}{4\pi\epsilon_0 r^3}(3\hat{\mathbf{r}}\mathbf{p}\cdot\hat{\mathbf{r}} - \mathbf{p}), \tag{1.8}$$

where $\mathbf{p} = Q(\mathbf{r}_+ - \mathbf{r}_-)$ is known as the dipole moment with $\mathbf{r}_+$ and $\mathbf{r}_-$ the positions of the positive and negative charges respectively.

We now return to the configuration of Figure 1.2 and note that the machine causes the accumulation equal numbers of opposite-signed charges, positive on the lower sphere and negative on the upper sphere. The spheres essentially store charge and are an example of an electrical device known as a *capacitor*. It will be noted that the potential on each sphere must be constant. This property follows from the fact that charges can move freely on a conducting sphere and so no further work is needed to move them around on the sphere. It turns out that the charge Q on the lower sphere is proportional to the potential difference $V = V_{+Q} - V_{-Q}$ between the spheres. The constant of proportionality C is known as the *capacitance* ($Q = CV$) and is measured in farads (coulombs per volt). Spheres are not the only capacitors and an important form of capacitor is known as the parallel plate capacitor (see Figure 1.8a). In this device the charge is accumulated on opposing faces of two parallel plates. The field between the plates is mainly constant (magnitude $\mathcal{E} = Q/\epsilon A$), except at the edges, where it adjusts to the zero field outside the capacitor. If the plates are distance d apart and have surface area A, the capacitance will be $C = \epsilon_0 A/d$. This value can be enhanced by inserting an insulating material between the plates. The capacitance will now given by $C = \epsilon A/d$ where ϵ is known as the *permittivity* of the insulator. When an insulator is added (see Figure 1.9a), the molecules become polarised (electrons are drawn towards the positive plate and protons towards the negative plate). The material will then consist of a collection of dipoles that are orientated along the original field line and this causes an additional field that partially counters the original field. The reduced field inside the dielectric will then result in an increased capacitance. The capacitor is an important component in electronic circuits and is represented by the symbol shown in Figure 1.9b.

If we connect the two sides of a capacitor by a conductor, electrons will flow from the negative side to the positive side until all the charge has been neutralised. For a perfect conductor, this will happen instantaneously. In reality, however, conductors are imperfect and there will be some resistance to the flow due to collisions on the molecular scale. The flow through an imperfect conductor is described by Ohm's law, according to which

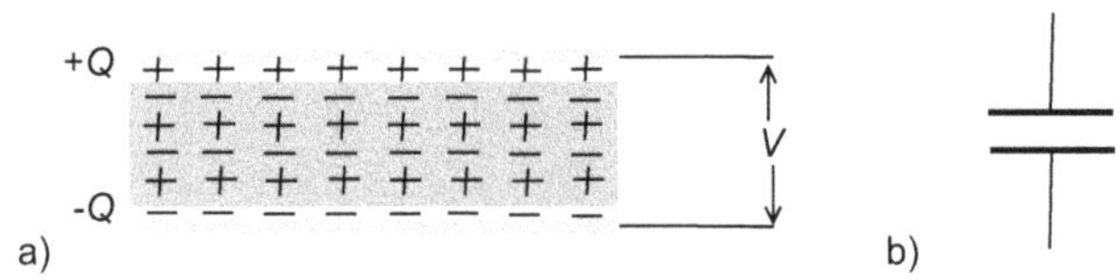

Fig. 1.9 a) Parallel capacitor with dielectric and b) symbol for capacitor.

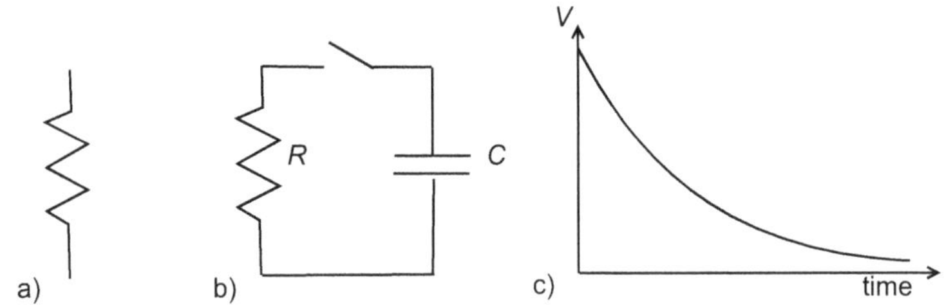

Fig. 1.10 Resistor and a capacitor drained by a resistor.

the potential drop V across the conductor is proportional to the current $\mathcal{I}$ through the conductor. Current is the rate at which charge flows in a conductor and is measured in amperes (1 ampere is 1 coulomb per second). Somewhat confusingly, current has always been taken to be flow of positive charge from higher to lower potential (the opposite direction to the reality of electron flow) and so is the rate of decrease of charge Q on the capacitor plate ($\mathcal{I} = -\mathrm{d}Q/\mathrm{d}t$ in the language of calculus). The constant of proportionality in Ohm's law is known as the resistance R ($V = R\mathcal{I}$) and has units of ohms (1 ohm is 1 amp per volt). George Ohm proposed his famous law in 1827 and it is an important relation in circuit theory. In the case of a wire of length L and cross-sectional area A, the resistance is given by $R = L/A\sigma$ where σ is a material property known as its conductivity.

An imperfect conductor is known as a resistor and is an important component in electronic circuits. A resistor is a lossy device and dissipates energy as heat at a rate $R\mathcal{I}^2$ (this is known as *Ohmic loss*). Figure 1.10b shows a simple circuit consisting of a capacitor and a resistor that dissipates the energy stored in the capacitor (Figure 1.10a shows the symbol used to represent the resistor). When the switch is thrown, a current $\mathcal{I}$ will flow through the resistor and the voltage drop across the capacitor will be given by $\mathcal{V} = R\mathcal{I}$. As the resistor drains the capacitor, the voltage across the capacitor will drop since the charge will be steadily depleted (see Figure 1.10c). Since $Q = C\mathcal{V}$ we will have $\mathcal{I} = -C\mathrm{d}\mathcal{V}/\mathrm{d}t$ and hence $\mathcal{V} = -RC\mathrm{d}\mathcal{V}/\mathrm{d}t$. This is an ordinary differential equation that has the solution $\mathcal{V} = V_0 \exp(-t/RC)$ where V_0 is the initial voltage difference between the capacitor plates and t is the time after switch on.

Much of the early development of the science of electricity was hindered by the need to use machines, such as that shown in Figure 1.2, to generate electric charge. In 1794, however, this process was revolutionised through the invention of the *battery* by Alessandro Volta, a device that creates charge through a chemical process rather than a mechanical process. Figure 1.11 shows a single-cell version of Volta's battery

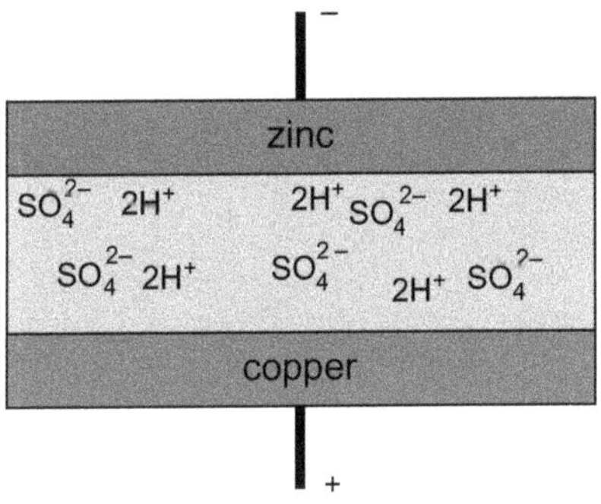

Fig. 1.11 Volta's battery.

(Volta in fact made a stack of these in order to produce large potential differences). It consists of a layer of copper (the anode), a layer of felt that is soaked in a mixture of water and sulphuric acid (the electrolyte) and a layer of zinc (the cathode). Within the electrolyte, the sulphuric acid will disassociate into SO_4^{2-} and H^{2+} ions. At the copper plate electrons are drawn into the electrolyte to combine with hydrogen ions and form hydrogen gas, hence causing an accumulation of positive charge. Meanwhile, at the zinc plate, this is counterbalanced by zinc ions dissolving into the electrolyte, hence causing an accumulation of negative charge. The chemistry can be summarised as

$$\mathrm{Cu} + \mathrm{Zn} + 3\mathrm{H_2SO_4} = \mathrm{Zn}^{2+} + 2\mathrm{SO_4^{2-}} + \mathrm{Cu}^{2+} + \mathrm{SO_2} + \mathrm{H_2} + 2\mathrm{H_2O}. \tag{1.9}$$

An important concept in electromagnetic theory (and many other field theories) is the concept of *flux*. Consider a flat surface with area A and unit normal $\mathbf{n}$. If $\mathbf{G}$ is a constant vector field, it will have a flux $\mathbf{n} \cdot \mathbf{G}A$ across the surface (i.e. the normal component of the field multiplied by the area of the surface). A good illustration of the notion of flux comes from the study of fluid flow. Such a medium is usually described in terms of its velocity field, a vector field that gives the magnitude and direction of the fluid velocity at a given point. The flux is then the total volume of fluid crossing the surface in a unit time. For a general surface surface S with unit normal $\mathbf{n}$, the *flux* through S is defined by the integral over the surface of the normal component of the vector field, i.e. $\int_S \mathbf{G}(\mathbf{r}) \cdot \mathbf{n} dS$. The surface integral is a calculus concept that can be understood by approximating the surface by a set of small flat surface elements on each of which $\mathbf{n}$ and $\mathbf{G}$ can be approximated by constant values. If the ith element has area ΔS_i, we approximate $\mathbf{n}$ by a constant vector $\mathbf{n}_i$ and $\mathbf{G}$ by a constant vector $\mathbf{G}_i$. The total flux through S is then approximated by the sum of the fluxes $\mathbf{G}(\mathbf{r}_i) \cdot \mathbf{n}_i \Delta S_i$ through these smaller elements, i.e.

$$\text{total flux through } S \approx \sum_{i=1}^{N} \mathbf{G}_i \cdot \mathbf{n}_i \Delta S_i. \tag{1.10}$$

In the limit of this sum as the areas of the surface elements tend to zero, the above sum then becomes the surface integral $\int_S \mathbf{G}(\mathbf{r}) \cdot \mathbf{n} dS$.

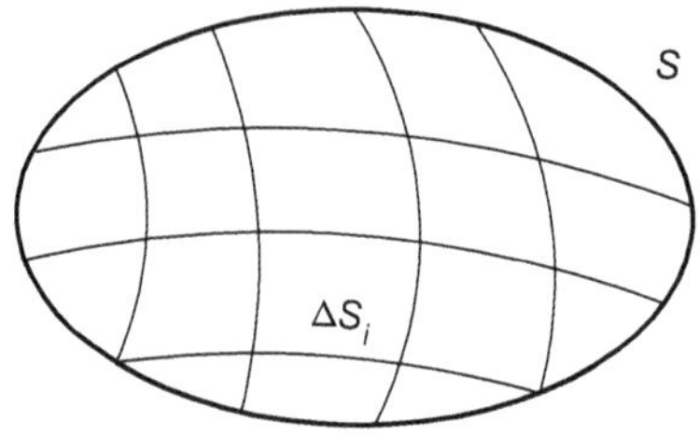

Fig. 1.12 Flux surface integral.

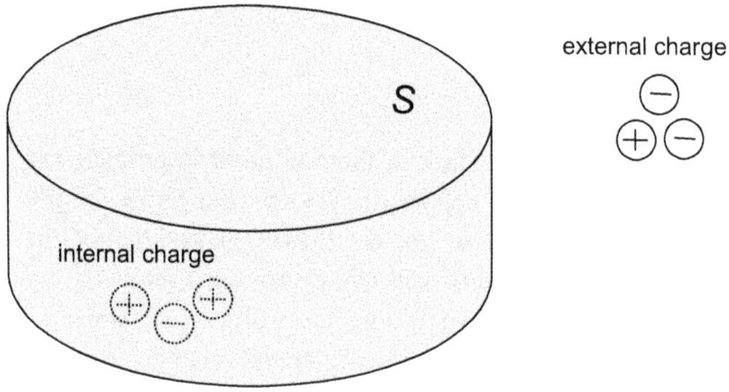

Fig. 1.13 Gauss' law.

An important property of the electric field is that the total flux through a closed surface S is proportional to the charge contained within that surface. This is known as *Gauss' Law* which, in mathematical terms, is given by

$$\int_S \epsilon \boldsymbol{\mathcal{E}}(\mathbf{r}) \cdot \mathbf{n} dS = \text{total charge within } S, \tag{1.11}$$

where S is an arbitrary closed surface in space and $\mathbf{n}$ is unit normal on this surface. Gauss' law is one of the fundamental laws of electromagnetism. A simple example is given by a single charge located at the origin and a surface S that consists of a sphere of radius a with centre at the charge. The field is given by Eq. 1.3 and from which $\boldsymbol{\mathcal{E}} \cdot \mathbf{n} = Q/4\pi\epsilon_0 a^2$ since $\mathbf{n}$ is a unit vector in the radial direction (i.e. the field direction). Since $\boldsymbol{\mathcal{E}} \cdot \mathbf{n}$ is constant, we simply multiply by the area of the sphere ($4\pi a^2$) to get the integral over the sphere. As a consequence $\int_S \boldsymbol{\mathcal{E}}(\mathbf{r}) \cdot \mathbf{n} dS = q/\epsilon_0$, which is Gauss' law.

1.2 Magnetism

At the time of their discovery of electrostatic attraction, the Greeks were also aware that the mineral magnetite (the oxide of iron Fe_3O_4) could attract pieces of non-oxide iron. Further, that the iron itself could be magnetised by stoking with the magnetite. The Chinese were also aware that magnetite (also known as lodestone) was a naturally occurring *magnet* that could attract iron. Indeed, the Chinese also discovered the effect of a magnet orientating itself with respect to Earth. By the twelfth century, both the Chinese and Europeans were using compasses in the form of lodestones for navigation. However, it took until 1600 for the Earth itself to be recognised as having the property of a magnet. This was recognised by William Gilbert in his book 'De Magnete', one of the first works on magnetism. The recognition of the Earth's magnetic properties led to the designation of the two ends of a magnet as North and South. However, unlike electric sources where positive and negative charge can have separate existence, the sources of magnetic fields are always found in North/South pairs. Because of this, the flux of a magnetic field through a closed surface S is zero, i.e.

$$\int_S \mathcal{B}(\mathbf{r}) \cdot \mathbf{n} dS = 0, \tag{1.12}$$

where $\mathcal{B}$ is the magnetic field (sometimes known as the *magnetic flux density*).

Since the magnetic poles always appear in North and South pairs, the basic source of magnetism is the magnetic dipole. This has a field

$$\mathcal{B} = \frac{\mu_0}{4\pi r^3}(3\hat{\mathbf{r}} \cdot \mathbf{M}\hat{\mathbf{r}} - \mathbf{M}), \tag{1.13}$$

where $\mathbf{M}$ is known as the dipole moment and μ_0 is a constant that is known as the permeability of free space. The dipole will have the field lines shown in Figure 1.14 (also shown is the field lines of Earth's magnetic field for which north is at 79° latitude).

If the basic magnetic source is the dipole, how do we interpret the dipole moment? It turns out that, if we suspend a magnetic dipole of moment $\mathbf{m}$ in field $\mathcal{B}$, the dipole will experience a torque

$$\tau = \mathbf{m} \times \mathcal{B}. \tag{1.14}$$

This is a more complex behaviour than the interaction of an electric charge with an electric field. In particular, it involves a *vector product*, defined by $\mathbf{a} \times \mathbf{b} = ab\sin(\theta)\hat{\mathbf{n}}$ where θ is the angle between the vectors and $\hat{\mathbf{n}}$ is a unit vector that is perpendicular to

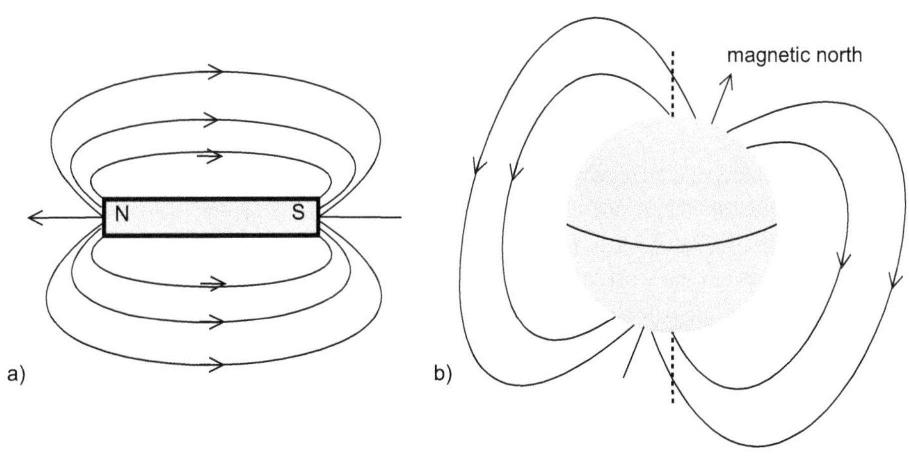

Fig. 1.14 Magnetic field lines.

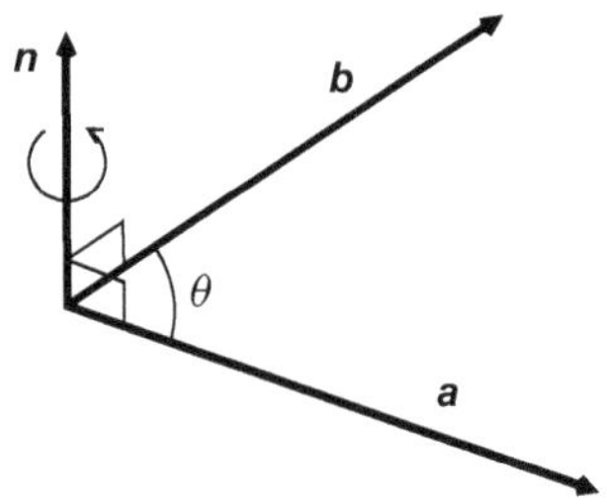

Fig. 1.15 Vector product.

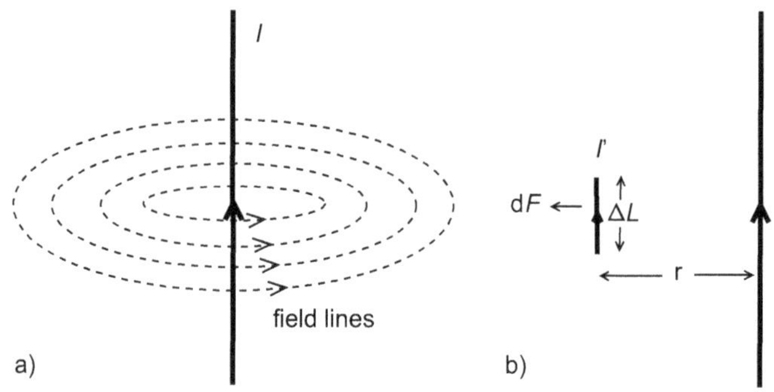

Fig. 1.16 The magnetic field of a current-carrying wire.

both **a** and **b** (direction defined by the right-hand screw rule as shown in Figure 1.15). An important consequence of 1.14 is that a magnet, freely suspended in the magnetic field of Earth, will rotate until it aligns with Earth's field lines (i.e. until the torque becomes zero), an effect that is used in navigation in the guise of a compass.

In 1820, Hans Christian Oersted discovered the magnetic effect of current. By observing the deflections of a compass, he showed that a long straight wire carrying a steady current $\mathcal{I}$ caused a magnetic field that had circular field lines centred on the wire (see Figure 1.16a) and a magnitude $\mathcal{B}$ that depended upon the distance r from the wire

$$\mathcal{B} = \frac{\mu_0 \mathcal{I}}{2\pi r}, \tag{1.15}$$

where $\mathcal{I}$ is the current in the wire. The unit for the magnetic field is usually the tesla, a quantity that is one newton per ampere per metre. Further, in such units, the permeability has the value $4\pi \times 10^{-7}$.

From the work of Oersted, it became clear that the moving charge had the ability to cause a magnetic field. Further, that a current carrying wire could experience the force of a magnetic field. According to Oersted, a wire element of length ΔL carrying a current $\mathcal{I}'$ will suffer a force

$$\Delta \mathbf{F} = \mathcal{I}' \Delta L \mathbf{t} \times \boldsymbol{\mathcal{B}}, \tag{1.16}$$

where **t** is a unit vector in the direction of the current.

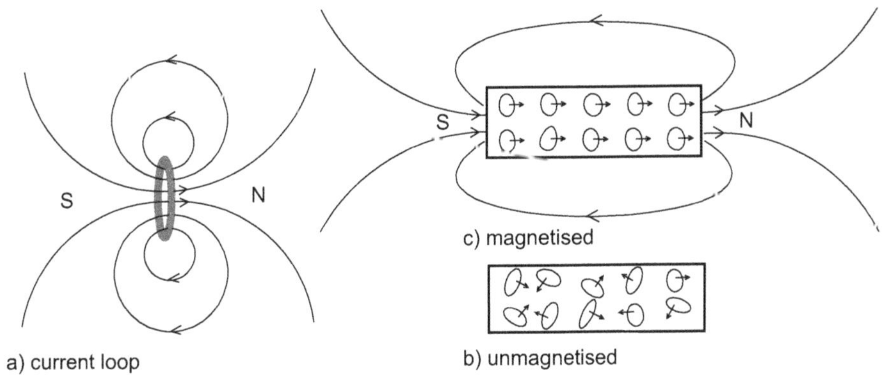

Fig. 1.17 The current loop and the current loop model of magnetism.

The fact that moving charge could produce a magnetic field led to further illumination of the concept of a magnetic dipole. Around 1820, the formula of Oersted was further generalised to allow for an arbitrary circuit C by the work of Jean-Baptiste Biot and Felix Savart. The Biot–Savart formula is

$$\mathcal{B}(\mathbf{r}) = \frac{\mu_0 I}{4\pi r}\int_C \frac{\mathrm{d}\mathbf{r}' \times (\mathbf{r}-\mathbf{r}')}{|\mathbf{r}-\mathbf{r}'|^3}. \tag{1.17}$$

For points at a large distance from a planar current loop, this expression reduces to (1.13) with $\mathbf{M} = \mathcal{I}A\mathbf{n}$ where $\mathbf{n}$ is a unit vector perpendicular to the plane of the loop and A is the area of the loop. The magnetic dipole can thus be pictured as a loop of current. In fact, at a molecular level, we can interpret this as orbiting electrons or spinning charge. All matter will consist of many such dipoles, but in most matter these will be in a random configuration and hence have no net effect. However, for materials such as magnetite, these dipoles are aligned with each other and hence the material will exhibit magnetic properties. Materials such as iron can be magnetised when external fields align their dipoles and materials such as steel can retain this magnetism.

Current can be regarded as a stream of charge travelling down a wire and the above considerations suggest that the force $\mathbf{F}$ that acts upon a charge q will be

$$\mathbf{F} = q\mathcal{E} + q\mathbf{v} \times \mathcal{B}, \tag{1.18}$$

where $\mathbf{v}$ is the velocity vector of the charge. This force is often known as the *Lorentz force* and is important for understanding the interaction of matter with the electromagnetic field.

A general law connecting magnetic fields and current was discovered by Andre-Marie Ampère in 1823. Consider a surface S through which current passes and which is bounded by a curve C. *Ampère's law*, in its mathematical form, then states that

$$\int_C \mathcal{B}(\mathbf{r}) \cdot \mathrm{d}\mathbf{r} = \mu_0 \mathcal{I}, \tag{1.19}$$

where $\mathcal{I}$ is the total current passing through the surface S. If we consider the case of a long straight wire carrying current $\mathcal{I}$, we could take the curve C to be a circle of radius a that is centred on the wire. In this case, the magnetic field B will be constant on C

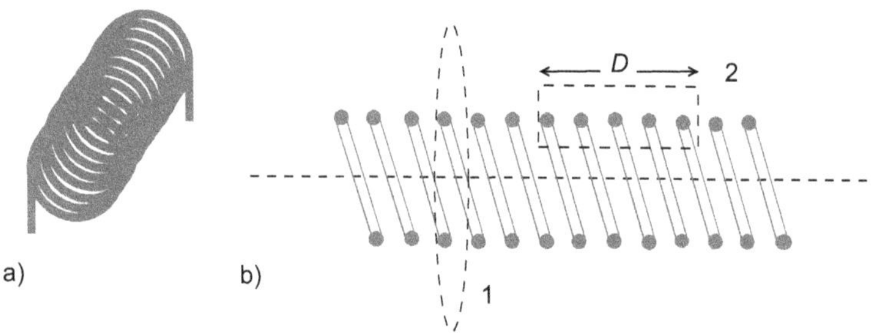

Fig. 1.18 The magnetic field of a long solenoid.

and we will have $\int_C \boldsymbol{\mathcal{B}}(\mathbf{r})\cdot d\mathbf{r} = 2\pi\mu_0 a\mathcal{B}$. Substituting from (1.15), the right-hand side becomes $\mu_0\mathcal{I}$, i.e. we have Ampère's law.

Ampère's law is a useful result for determining complex magnetic fields. Consider the example of an infinitely long solenoid (a good approximation to a long solenoid, as in Figure 1.18a). By symmetry, the only dependence of the magnetic field is the radial distance r from the axis. Further, by analogy with the limiting case of an infinite wire, the field lines outside the wire will be circular and centred on the solenoid axis, i.e. the field will point in the rotational direction. We first apply Ampère's law on a circular curve C with radius r and centred on the axis (the curve 1 in Figure 1.18b). On curve C the magnetic field will be constant and so $\int_C \boldsymbol{\mathcal{B}}(\mathbf{r})\cdot d\mathbf{r} = 2\pi r\mathcal{B}$, where $\mathcal{B}$ is the component magnetic field that is tangent to C. Consequently, if current $\mathcal{I}$ flows through the solenoid, Ampère's law will imply $\mathcal{B} = \mu_0\mathcal{I}/2\pi r$. Inside the solenoid, Ampère's law will imply that the rotational component of the field is zero and so, as a consequence, we take the field to be parallel to the solenoid axis. If we apply Ampère's law on the rectangular curve in Figure 1.18 (curve 2), we find that $\int_C \boldsymbol{\mathcal{B}}(\mathbf{r})\cdot d\mathbf{r} = D\mathcal{B}$ where $\mathcal{B}$ is the magnitude of the magnetic field parallel to the axis. The current through the loop will be $nD\mathcal{I}$ where n is the number of turns per unit length on the solenoid. As a consequence, Ampère's law will imply that $\mathcal{B} = \mu_0 n\mathcal{I}$ inside the solenoid. Up to now, we have implicitly assumed that the radial component of the magnetic field is zero, but we can verify this using (1.12). We take the surface S to be a cylinder of radius r, and length D, with the same axis as the solenoid. The magnetic flux through the cylinder ends will cancel, but the contribution from the curved surface will be $2\pi rD\mathcal{B}$, where $\mathcal{B}$ is now the radial component of the magnetic field. Equation 1.12 will then imply that this radial component is zero.

What emerges from our considerations is that the various integral results, such as Ampère's and Gauss' laws, constitute a powerful and self-contained description of electromagnetism. In order to complete this description, however, we need to introduce one further integral law and this is the subject of the next section.

1.3 Electromagnetism

We now consider the consequences of the variation of fields with time. This brings us to Faraday's law, one of the key discoveries in the development of electromagnetic theory.

In 1830, Michael Faraday discovered magnetic induction when he noted that, by moving a loop of wire in and out of a magnetic field, he could cause a current to flow in the loop. This was an important discovery as, hitherto, the battery and the charged capacitor had been the only means of driving a current through an electrical circuit. Somewhat confusingly, the effective potential of this new sort of generator came to be known as the *electromotive force* (or EMF for short). Faraday concluded that the EMF induced in a circuit was proportional to the rate of change of magnetic flux through that circuit, a result that is known as *Faraday's law*.. For a surface S with bounding curve C (see Figure 1.19), the flux is given by $\Phi = \int_S \mathcal{B}(\mathbf{r},t)\cdot\mathbf{n}dS$ and the law has the mathematical form

$$\int_C \mathcal{E}(\mathbf{r},t)\cdot d\mathbf{r} = -\frac{d}{dt}\int_S \mathcal{B}(\mathbf{r},t)\cdot\mathbf{n}dS. \tag{1.20}$$

Figure 1.20 shows two different ways in which magnetic flux can vary in a circuit. If we consider the loop in Figure 1.20a to be rotating at angular speed ω, the flux Φ through the loop will be $\Phi = A\mathcal{B}\sin(\omega t)$ where A is the area of the loop and $\mathcal{B}$ is the magnetic field (we assume the plane of the loop is parallel to the field when $t = 0$). As a consequence, an EMF $A\mathcal{B}\cos(\omega t)$ will be generated and this causes an *alternating current* to flow in the load. In Figure 1.20b a conducting bar with load moves over a rectangular circuit at speed v and so causes the total area of the circuit to change at rate vd. As a consequence, the magnetic flux will increase at a rate $vd\mathcal{B}$ and so, by Faraday's law, an EMF of $-vdB$ will be generated in the circuit ($\mathcal{B}$ is a magnetic field orthogonal to the loop). We can view this last example from the viewpoint of the Lorentz force. A unit charge, located on the conducting bar, will suffer a Lorentz force $v\mathcal{B}$ in the clockwise direction due to

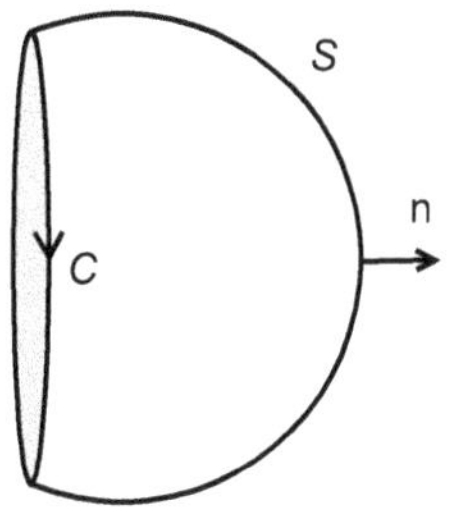

Fig. 1.19 Geometry for Faraday's law.

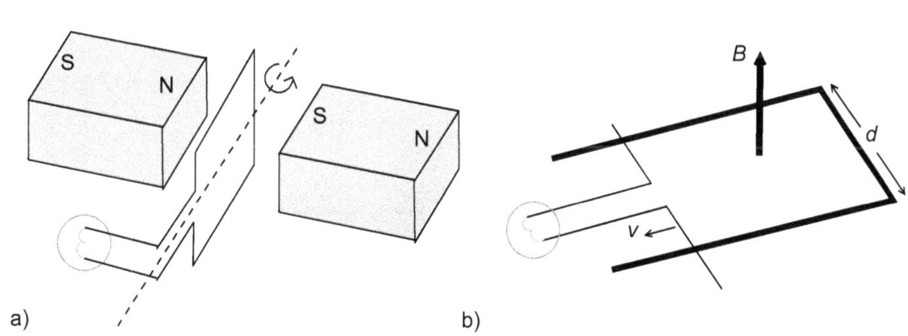

Fig. 1.20 Magnetic induction.

the imposed motion transverse to the bar. Consequently, integrating along the bar, we obtain an EMF of $-vd\mathcal{B}$.

A time-varying current brings us to the concept of mutual impedance. Consider a solenoid with a wire loop wrapped around it. If we now drive the solenoid by a alternating current $\mathcal{I}_1(t)$, there will be a magnetic field $\mathcal{B}(t) = \mu_0 n\mathcal{I}_1(t)$ through the loop and hence a flux $\Phi = A\mu_0 n\mathcal{I}_1(t)$ where A is the area of the loop. According to Faraday's law, this will generate an EMF of

$$\mathcal{E}_2 = -L_{21}\frac{\mathrm{d}\mathcal{I}_1}{\mathrm{d}t} \tag{1.21}$$

in the wire, where $L_{21} = A\mu_0 n$ is known as the *mutual inductance* of the wire loop and solenoid. If the solenoid has a finite length l with N_1 turns then $n = N_1/l$ and the mutual inductance will be given by $L_{12} = \mu_0 N_1 A/l$. Further, and if the loop has N_2 turns, the mutual inductance will now be given by $L_{12} = \mu_0 N_1 N_2 A/l$. A device with mutually interacting windings is known as a *transformer* and is represented by the symbol shown in Figure 1.21b. The longer solenoid is often known as the primary and the loop winding as the secondary. If current flows through the secondary, it is clear the secondary itself will cause additional flux and so

$$\mathcal{E}_2 = -L_{21}\frac{\mathrm{d}\mathcal{I}_1}{\mathrm{d}t} - L_{22}\frac{\mathrm{d}\mathcal{I}_2}{\mathrm{d}t}, \tag{1.22}$$

where L_{22} is known as the self inductance of the secondary winding. It is clear that the primary will also experience self inductance and that the EMF generated in the primary will take the form

$$\mathcal{E}_1 = -L_{11}\frac{\mathrm{d}\mathcal{I}_1}{\mathrm{d}t} - L_{12}\frac{\mathrm{d}\mathcal{I}_2}{\mathrm{d}t}, \tag{1.23}$$

where it should be noted that $L_{12} = L_{21}$. For the solenoid, it is obvious that it will induce a flux $\Phi = A\mu_0\mathcal{I}_1(t)A\mu_0 N_1/l$ in itself and so it will have a self inductance $L_{11} = \mu_0 N_1^2 A/l$. Likewise, the secondary will have a self inductance $L_{22} = \mu_0 N_2^2 A/l$. The unit of inductance is the henry and is named after Joseph Henry who discovered magnetic induction independently of Faraday and at about the same time. An inductance of 1 henry will result in an EMF of 1 volt in a closed loop for a change of 1 amp in the current over a period of 1 second.

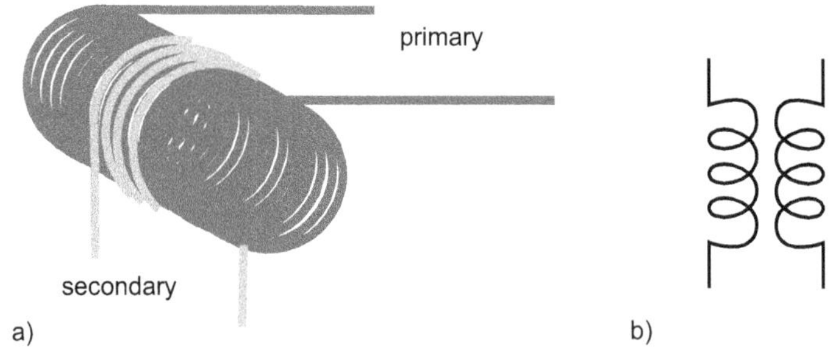

Fig. 1.21 Transformer and transformer symbol.

From the above considerations, it will be noted that a solenoid that carries a current $\mathcal{I}$ will always have a self-induced EMF, even if the secondary winding does not exist. For this reason, such a device is known as an *inductor* and is an important component in radio technology (it is represented by the symbol shown in Figure 1.21a). For such a component we will have the self-induced EMF

$$\mathcal{E} = -L\frac{\mathrm{d}\mathcal{I}}{\mathrm{d}t}, \tag{1.24}$$

where $\mathcal{I}$ is the current in the inductor and L is the self inductance ($L = \mu_0 N^2 A/l$ for a solenoid with length l and cross-sectional area A). We can enhance the inductance of the solenoid by winding it around a core made up of ferromagnetic material (iron and cobalt for example). As can be seen in Figure 1.22, the solenoid field causes the current loops within the core to align and this causes an increase in the magnetic flux density. Consequently, the inductance in the solenoid will now become $L = \mu N^2 A/l$, where μ is known as the *permeability* of the core. Permeability is a material property of the solenoid core and has units of henries per metre.

We now consider the circuit shown in Figure 1.23b, consisting of a series capacitor, resistor, inductor and switch. Before the switch is closed we assume the capacitor to be charged to a voltage V_0. After the switch is closed, however, the capacitor discharges and the potential difference across the capacitor will decay.

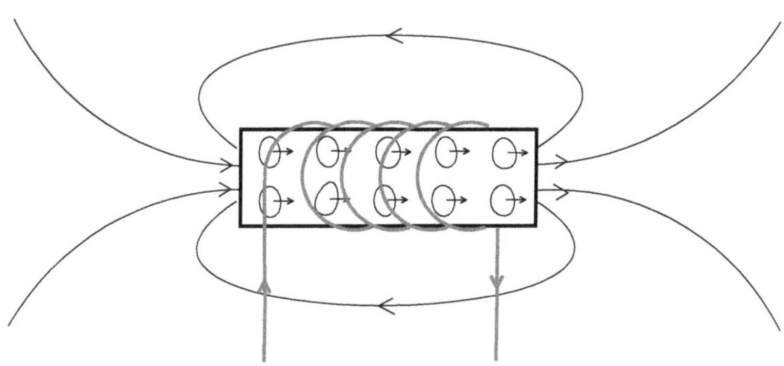

Fig. 1.22 Magnetic flux enhanced by a ferromagnetic material.

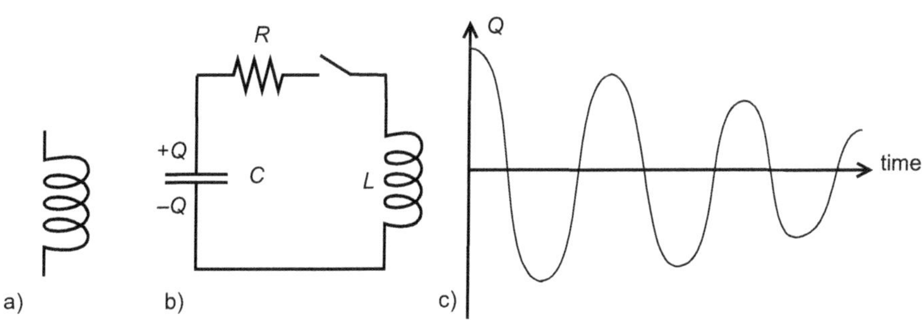

Fig. 1.23 Inductor and oscillating circuit.

We can analyse the circuit by noting the famous Kirchhoff circuit laws:

1. The total current into any junction is equal to the total current out.
2. The total voltage drop around any circuit loop is zero.

(Since they both have units of volts, voltage is a terminology often used for both potential difference and EMF.) From the first law we obtain that the same current $\mathcal{I}$ flows in and out of all components and from the second law we obtain that

$$-L\frac{d\mathcal{I}}{dt} - \mathcal{I}R + \frac{Q}{C} = 0. \tag{1.25}$$

In terms of the charge Q on the upper capacitor plate, we have the ordinary differential equation (ODE)

$$L\frac{d^2Q}{dt^2} + R\frac{dQ}{dt} + \frac{Q}{C} = 0, \tag{1.26}$$

which can be solved to yield

$$Q = V_0 C \exp(-\zeta\omega_0 t)\cos(\sqrt{1-\zeta^2}\omega_0 t), \tag{1.27}$$

where $\omega_0 = 1/\sqrt{LC}$ and $\zeta = R\sqrt{C/L}/2$. It will be noted that when the capacitor is discharged through an inductor, it will *ring*, i.e. there will be oscillations in the circuit at an angular frequency of $\sqrt{1-\zeta^2}\omega_0$. Further, the oscillations will decay at a rate that is dependent upon the amount of resistance R in the circuit.

The frequency ω at which a circuit rings is of great important to us in radio. Consider the parallel combination of a capacitor C, an inductor L and a load resistance R with the inductor driven by harmonic voltage source $V_S\cos(\omega t)$ (see Figure 1.24). By the Kirchhoff current law, we have that

$$\mathcal{I} = C\frac{d\mathcal{V}}{dt} + \frac{\mathcal{V}}{R} \tag{1.28}$$

and from the Kirchoff voltage law

$$\mathcal{V} = V_S\cos(\omega t) - L\frac{d\mathcal{I}}{dt}. \tag{1.29}$$

As a consequence

$$V_S\cos(\omega t) - \mathcal{V} = LC\frac{d^2\mathcal{V}}{dt^2} + \frac{L}{R}\frac{d\mathcal{V}}{dt} \tag{1.30}$$

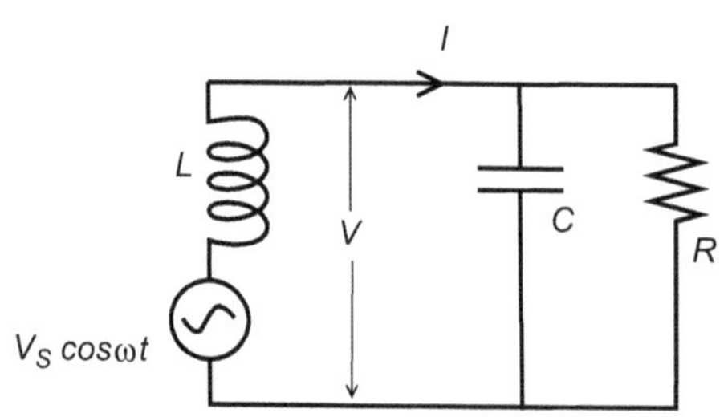

Fig. 1.24 A sinusoidally driven circuit.

and, dividing by LC, we obtain

$$\frac{d^2\mathcal{V}}{dt^2} + \frac{1}{RC}\frac{d\mathcal{V}}{dt} + \frac{\mathcal{V}}{LC} = \frac{V_S}{LC}\cos(\omega t). \tag{1.31}$$

After the source is switched on, the solution will settle down to the steady state

$$\mathcal{V}(t) = \frac{V_S}{LC}\frac{(\omega_0^2-\omega^2)\cos(\omega t) + \frac{\omega}{RC}\sin(\omega t)}{(\omega_0^2-\omega^2)^2 + \frac{\omega^2}{R^2C^2}}, \tag{1.32}$$

where $\omega_0 = 1/\sqrt{LC}$, i.e. the voltage in the circuit oscillates at the forcing frequency ω. It will be noted, however, that as frequency ω approaches ω_0, the amplitude of oscillations will increase, reaching a peak value of QV_S where $Q = R/\omega L$. The circuit is said to *resonate* at frequency ω_0 and Q is a measure of the strength of this resonance.

1.4 Maxwell's Equations

Until about 1860, Eqs. (1.11), (1.12), (1.19) and (1.20) were presumed to correctly reflect the content of electromagnetic theory. Whilst these equations imply that time-varying magnetic flux will cause an electric field, they do not imply that time-varying electric flux will cause a magnetic field. Around 1860 the physicist James Clerk Maxwell became convinced that time-varying electric flux should cause a magnetic field. Indeed, there are good reasons for believing that Ampère's law needs some form of modification. Consider Ampère's law in form

$$\int_C \boldsymbol{\mathcal{B}}(\mathbf{r})\cdot d\mathbf{r} = \mu_0\mathcal{I}. \tag{1.33}$$

We consider the case of two charged spheres that are made to discharge by connecting them through a conducting wire that, as a result, carries a current $\mathcal{I}$. Referring to Figure 1.25, if we apply Eq. (1.33) using surface S_1 we obtain that $\int_{C_1}\boldsymbol{\mathcal{B}}(\mathbf{r})\cdot d\mathbf{r} = \mu_0\mathcal{I}$ and using surface S_2 we obtain that $\int_{C_2}\boldsymbol{\mathcal{B}}(\mathbf{r})\cdot d\mathbf{r} = 0$. If we add these two results, the path integrals will cancel (the curves are identical but the integrals are evaluated in opposite directions) and this will imply that $\mathcal{I} = 0$. This clearly poses a problem for electromagnetic theory. Maxwell's solution was to add another term to Eq. (1.33), which he called the *displacement current*. The resulting equation is

$$\int_C \boldsymbol{\mathcal{B}}(\mathbf{r},t)\cdot d\mathbf{r} = \mu_0\frac{d}{dt}\int_S \epsilon\boldsymbol{\mathcal{E}}(\mathbf{r},t)\cdot \mathbf{n}dS + \mu_0\mathcal{I}. \tag{1.34}$$

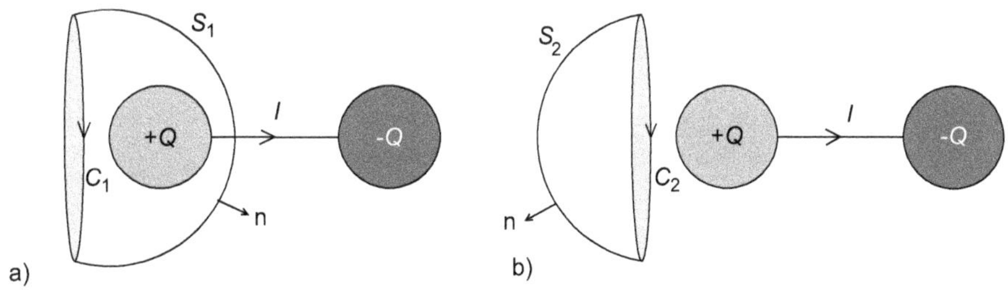

Fig. 1.25 Configuration for testing Ampère's law.

If we now consider this modified Ampère's law on the surfaces S_1 and S_2 we now obtain $\int_{C_1} \mathcal{B}(\mathbf{r}) \cdot \mathrm{d}\mathbf{r} = \mu_0 \frac{\mathrm{d}}{\mathrm{d}t} \int_{S_1} \epsilon \mathcal{E}(\mathbf{r}) \cdot \mathbf{n} \mathrm{d}S + \mu_0 \mathcal{I}$ and $\int_{C_2} \mathcal{B}(\mathbf{r}) \cdot \mathrm{d}\mathbf{r} = \mu_0 \frac{\mathrm{d}}{\mathrm{d}t} \int_{S_2} \epsilon \mathcal{E}(\mathbf{r}) \cdot \mathbf{n} \mathrm{d}S$. If we now add these two results, we obtain that $\mu_0 \frac{\mathrm{d}}{\mathrm{d}t} \int_S \epsilon \mathcal{E}(\mathbf{r}) \cdot \mathbf{n} \mathrm{d}S + \mu_0 \mathcal{I}$ where $S = S_1 + S_2$. Since S will be a closed surface, Gauss' law will therefore imply that $\mathrm{d}Q/\mathrm{d}t + \mathcal{I} = 0$, the definition of the current, i.e. we now have consistency.

We can now state the integral form of what are known as *Maxwell's equations*, equations that fully describe an electromagnetic (EM) field. The first pair of these four vector equations are:

$$\int_C \frac{\mathcal{B}}{\mu} \cdot \mathrm{d}\mathbf{r} = \frac{\mathrm{d}}{\mathrm{d}t} \int_S \epsilon \mathcal{E} \cdot \mathbf{n} \mathrm{d}S + \mathcal{I} \tag{1.35}$$

and

$$\int_C \mathcal{E} \cdot \mathrm{d}\mathbf{r} = -\frac{\mathrm{d}}{\mathrm{d}t} \int_S \mathcal{B} \cdot \mathbf{n} \mathrm{d}S, \tag{1.36}$$

where S is a surface through which a total current $\mathcal{I}$ flows and C is a contour that is the boundary of S (see Figure 1.19). The second pair are

$$\int_S \mathcal{B} \cdot \mathbf{n} \mathrm{d}S = 0 \tag{1.37}$$

and

$$\int_S \epsilon \mathcal{E} \cdot \mathbf{n} \mathrm{d}S = Q, \tag{1.38}$$

where S is now a closed surface that contains an amount of charge Q. It should be noted that we have included the possibility of a magnetic medium (i.e. $\mu \neq \mu_0$) and, for convenience, have written $\mathcal{E}(\mathbf{r},t)$ as $\mathcal{E}$ and $\mathcal{B}(\mathbf{r},t)$ as $\mathcal{B}$.

1.5 Conclusion

In the present chapter we have developed electromagnetic theory up to the integral form of Maxwell's equations. Sources apart (current and charge), these equations possess a remarkable symmetry, something we have come to expect from the laws of physics. Indeed, Maxwell's equations are the only laws of physics to emerge unscathed from the relativity revolution that Einstein unleashed in 1905. It is fair to say that Maxwell's equations were the first truly relativistic theory of physics.

Maxwell showed that his equations predicted wavelike fields and, around 1886, this prediction led the German physicist Heinrich Hertz to construct apparatus that could transmit and receive electromagnetic waves. The wavelengths involved were those that we now commonly associate with radio and so it was with Hertz that radio was born. In the next chapter we will develop the basic theory behind radio and describe the experiments that led to its practical realisation. Further, we will describe some of the early work that helped turn this scientific discovery into one of the most dynamic technologies of the twentieth century.

2 Radio Waves

Around 1864, Maxwell was able to show that his equations implied that electromagnetic fields satisfied a wave equation. Remarkably, the speed of these waves turned out to be exactly that of light, suggesting that light is an electromagnetic wave phenomenon. It turns out that light is a very high-frequency form of electromagnetic energy and it was not long before physicists started looking for electromagnetic energy at lower frequencies, those we now associate with radio waves. This led to the classic work of Heinrich Hertz around 1886 in which he demonstrated the generation and detection of radio waves. In this chapter we will show that Maxwell's equations predict fields with wavelike behaviour and show that accelerating charge is the source of such waves. The chapter then concludes with a description of the experiments that led to the discovery of radio waves and the early technological developments that turned radio from a scientific curio into a technology.

2.1 Waves

We will start our discussion of waves by considering those waves that appear as ripples on the surface of water, something with which most of us are familiar. If we have initially calm water, we can cause a disturbance in the water through the forced motion of a platform at the surface of the water (see Figure 2.1) and this disturbance will then travel outwards from the object at a finite speed (c say). The disturbances will be in sympathy with the motion of the object, but a point at distance x from the object will not receive the effect of the motion until a time x/c later. If the motion of the body starts at time 0, at time T later the water will still be calm at distances greater than cT from the object. If the wave is confined to a channel, the wave will maintain its shape as it travels out, i.e. we have a wave height of form $\psi_+(x+ct)$ leftwards and $\psi_-(x-ct)$ rightwards. A plane with $x-ct=$ constant, or $x+ct=$ constant, is known as a *wavefront* (the boundary between still and disturbed water being an important example of a wavefront). A further property of water waves is that they transfer energy from their source (the platform) to a detector at some distance from the source. This can be seen from the fact that when a water wave passes a previously motionless platform it will be set into motion in sympathy with the wave. Such a process is utilised these days in the form of a wave power generator.

Water waves exhibit all of the important features of waves. These include the ability to transport energy over large distances, the energy being transported at a finite speed c. Further, that the source can create an arbitrarily shaped wave that maintains its shape as

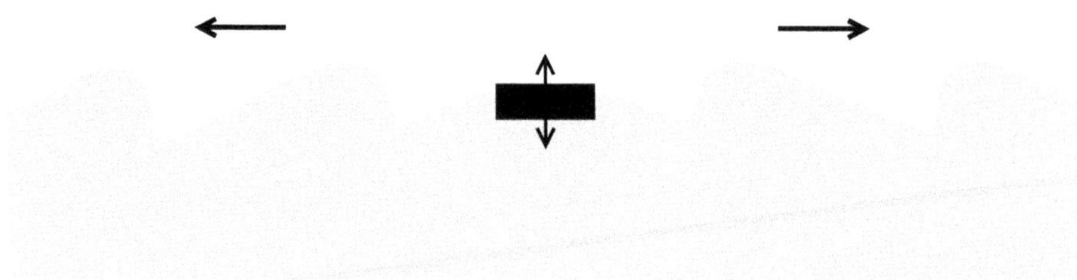

Fig. 2.1 Water waves generated by an oscillating platform.

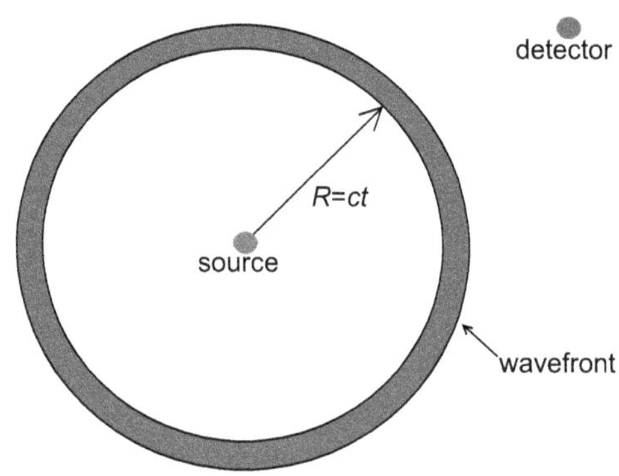

Fig. 2.2 Pulse travelling out from a bounded source.

it travels outwards and, as a consequence, can transfer arbitrary information over large distances. Water waves will satisfy a *wave equation* of the form

$$\frac{\partial^2\psi}{\partial x^2} - \frac{1}{c^2}\frac{\partial\psi^2}{\partial t^2} = 0. \tag{2.1}$$

In the case of the rightwards wave ψ_-, this can be seen from the following argument. It will be noted that ψ_- is a function of a single variable $\phi = x - ct$. If f is a function of ϕ, an important result of calculus tells us that $\partial f/\partial x = (\partial\phi/\partial x)\mathrm{d}f/\mathrm{d}\phi = \mathrm{d}f/\mathrm{d}\phi$ and that $\partial f/\partial t = (\partial\phi/\partial t)\mathrm{d}f/\mathrm{d}\phi = -c\mathrm{d}f/\mathrm{d}\phi$. As a consequence, $\partial^2\psi_-/\partial x^2 = \mathrm{d}\psi_-^2/\mathrm{d}\phi^2$ and $\partial^2\psi_-/\partial t^2 = c^2\mathrm{d}\psi_-^2/\mathrm{d}\phi^2$, from which it can be seen that (2.1) will be satisfied for arbitrary $\psi_-(x-ct)$. In a similar fashion, (2.1) will also be satisfied by arbitrary $\psi_+(x+ct)$.

A more realistic wave is one that is generated by a bounded source (see Figure 2.2). The waves will travel outwards from such a source and so the wave field ψ will be a function of $R - ct$ where R is the distance from the source (the wavefront is spherical in this case). For most physical fields, the energy in the field is proportional to the field squared (the field is the wave height in the case of a water wave). Consequently, as the wave propagates it will spread out and the rate of spread will need to conserve energy. In the case of water waves, these are two-dimensional (they only exist on the surface of the water) and so the field will need to fall off as $1/\sqrt{R}$. Many fields (including sound and electromagnetic fields) will exist in three-dimensional space and so, in order to conserve energy, the field will need to fall off as $1/R$.

2.2 Electromagnetic Waves

We want to show that Maxwell's equations allow solutions that behave as waves. However, we first need to introduce some further ideas concerning vectors. The space we live is said to be three-dimensional and this means that we can specify any point through three numbers, its coordinates. There are many different sorts of coordinate system, but we will concentrate on what are known as Cartesian coordinates. These are arguably the simplest coordinates and were invented in the seventeenth century by the French mathematician René Descartes. In Cartesian coordinates we define an origin O for which the coordinates are $(0,0,0)$ and then the position of any point $\mathbf{r}$ is given by the vector $x\hat{\mathbf{x}}+y\hat{\mathbf{y}}+z\hat{\mathbf{z}}$, where $\hat{\mathbf{x}}$, $\hat{\mathbf{y}}$ and $\hat{\mathbf{z}}$ are unit length vectors ($|\hat{\mathbf{x}}|=|\hat{\mathbf{y}}|=|\hat{\mathbf{z}}|=1$) that are mutually orthogonal ($\hat{\mathbf{x}}\cdot\hat{\mathbf{y}}=\hat{\mathbf{y}}\cdot\hat{\mathbf{z}}=\hat{\mathbf{x}}\cdot\hat{\mathbf{z}}=0$). The vectors $\hat{\mathbf{x}}$, $\hat{\mathbf{y}}$ and $\hat{\mathbf{z}}$ are known as *basis vectors*. Essentially, the coordinates are the distances that we must travel in the direction of the associated basis vectors in order to travel from the origin O to position $\mathbf{r}$. Any vector field $\mathbf{F}(\mathbf{r})$ can also be represented in terms of the above basis vectors as $F_x(\mathbf{r})\hat{\mathbf{x}}+F_y(\mathbf{r})\hat{\mathbf{y}}+F_z(\mathbf{r})\hat{\mathbf{z}}$ where $F_x(\mathbf{r})$, $F_y(\mathbf{r})$ and $F_z(\mathbf{r})$ are known as the components of the vector field (often abbreviated to just F_x, F_y and F_z). The important thing about the Cartesian coordinate system is that basis vectors are the same vectors at all points in space (not so for more exotic coordinate systems such as those of the polar variety). Concerning vectors expressed in component form, there are two important results that the reader should note

$$\mathbf{a}\cdot\mathbf{b}=a_xb_x+a_yb_y+a_zb_z \tag{2.2}$$

and

$$\mathbf{a}\times\mathbf{b}=(a_yb_z-a_zb_y)\hat{\mathbf{x}}+(a_zb_x-a_xb_z)\hat{\mathbf{y}}+(a_xb_y-a_yb_x)\hat{\mathbf{z}}. \tag{2.3}$$

We will assume that there is an EM wave travelling in the $\hat{\mathbf{x}}$ direction. Consequently, the fields will depend on x and t alone (i.e. $\mathcal{E}(\mathbf{r})=\mathcal{E}(x,t)$ and $\mathcal{B}(\mathbf{r})=\mathcal{B}(x,t)$) and then only through the combination $\phi(x,t)=x-ct$ where c is some, as yet to be determined, propagation speed (we assume a right-moving wave for the present). We will now study the implication of these assumptions for the Maxwell equations. We first consider the equations

$$\int_S \mathcal{B}\cdot\mathbf{n}dS=0 \text{ and } \int_S \epsilon\mathcal{E}\cdot\mathbf{n}dS=Q \tag{2.4}$$

and apply these equations over a small cube that does not contain any charge ($Q=0$), the cube having centre at the general point $\mathbf{r}$ and sides of length Δ.

There will be contributions to the integral from the six faces of the cube, the integral on each being approximated as the area of the face multiplied by the normal component of the field at the centre of each face. Consequently,

$$\begin{aligned}\int_S \mathcal{B}\cdot\mathbf{n}dS \approx&\mathcal{B}_x(x+\Delta/2,t)\Delta^2-\mathcal{B}_x(x-\Delta/2,t)\Delta^2\\&+\mathcal{B}_y(x,t)\Delta^2-\mathcal{B}_y(x,t)\Delta^2\\&+\mathcal{B}_z(x,t)\Delta^2-\mathcal{B}_z(x,t)\Delta^2,\end{aligned} \tag{2.5}$$

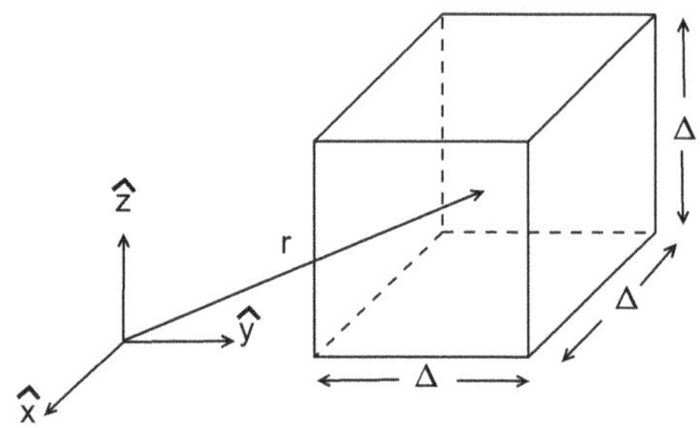

Fig. 2.3 Cube for evaluating Maxwell's equations.

where the contributions on the right-hand side are from, in order, the sides with normals $\hat{\mathbf{x}}$, $-\hat{\mathbf{x}}$, $\hat{\mathbf{y}}$, $-\hat{\mathbf{y}}$, $\hat{\mathbf{z}}$ and $-\hat{\mathbf{z}}$. The equation $\int_S \boldsymbol{\mathcal{B}} \cdot \mathbf{n} dS = 0$ will then imply that

$$\mathcal{B}_x(x+\Delta/2,t)\Delta^2 - \mathcal{B}_x(x-\Delta/2,t)\Delta^2 = 0. \tag{2.6}$$

From (2.6) we will have that $(\mathcal{B}_x(x+\Delta/2,t) - \mathcal{B}_x(x-\Delta/2,t))/\Delta = 0$ and so, in the limit $\Delta \to 0$, this will imply that $\partial\mathcal{B}_x/\partial x = 0$. The only way that this can hold is if $\mathcal{B}_x$ is itself constant and so we will take $\mathcal{B}_x = 0$ since a constant non-zero field throughout space is not the wavelike solution we are looking for. In a similar fashion, we will have $\mathcal{E}_x = 0$.

We now consider Eqs. (1.35) and (1.36) when applied over small rectangular circuits that are not threaded by any current ($\mathcal{I} = 0$). The circuits are static and so we can bring the time derivative under the integral to obtain

$$\int_C \frac{\boldsymbol{\mathcal{B}}}{\mu} \cdot d\mathbf{r} = \int_S \epsilon \frac{\partial \boldsymbol{\mathcal{E}}}{\partial t} \cdot \mathbf{n} dS \tag{2.7}$$

and

$$\int_C \boldsymbol{\mathcal{E}} \cdot d\mathbf{r} = -\int_S \frac{\partial \boldsymbol{\mathcal{B}}}{\partial t} \cdot \mathbf{n} dS. \tag{2.8}$$

Since the rectangles are small, the surface integral can be approximated by $\mathbf{n} \cdot \partial\boldsymbol{\mathcal{B}}/\partial t$ when evaluated at the centre of the rectangle and multiplied by the area of the rectangle. On each side of the rectangle, the line integral can be approximated by $\mathbf{t} \cdot \boldsymbol{\mathcal{B}}/\mu$ evaluated at the mid point of the side ($\mathbf{t}$ is a unit vector along the side) multiplied by the length of the side. Consider a rectangle in the xy plane (see Figure 2.4a), then (2.7) will become

$$-\frac{\mathcal{B}_x(x,t)}{\mu}\Delta - \frac{\mathcal{B}_y(x-\Delta/2,t)}{\mu}\Delta + \frac{\mathcal{B}_x(x,t)}{\mu}\Delta + \frac{\mathcal{B}_y(x+\Delta/2,t)}{\mu}\Delta = \epsilon\frac{\partial\mathcal{E}_z}{\partial t}\Delta^2, \tag{2.9}$$

where the terms on the left hand side are, in order, the contributions from the sides with directions $-\hat{\mathbf{x}}$, $-\hat{\mathbf{y}}$, $\hat{\mathbf{x}}$ and $\hat{\mathbf{y}}$. As a consequence,

$$\frac{1}{\Delta}\left(\frac{\mathcal{B}_y(x+\Delta/2,t)}{\mu} - \frac{\mathcal{B}_y(x-\Delta/2,t)}{\mu}\right) = \epsilon\frac{\partial\mathcal{E}_z}{\partial t} \tag{2.10}$$

and then, in the limit $\Delta \to 0$,

$$\frac{1}{\mu}\frac{\partial\mathcal{B}_y}{\partial x} = \epsilon\frac{\partial\mathcal{E}_z}{\partial t}. \tag{2.11}$$

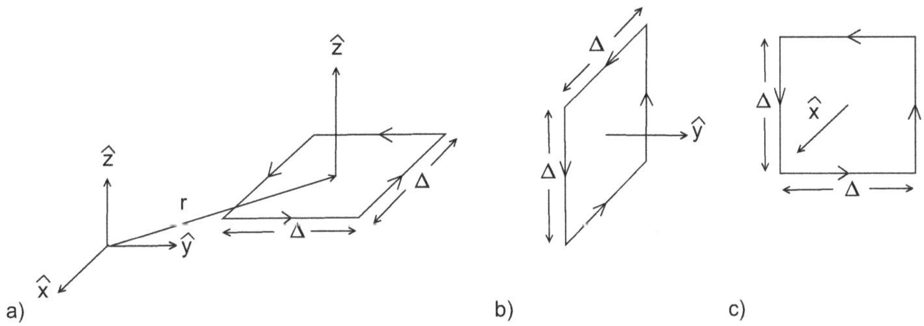

Fig. 2.4 Rectangular circuit for evaluating Maxwell's equations.

In a similar fashion, 2.8 will imply that

$$\frac{\partial \mathcal{E}_y}{\partial x} = -\frac{\partial \mathcal{B}_z}{\partial t}. \tag{2.12}$$

If we now consider a rectangle in the xz plane (see Figure 2.4b), similar arguments imply that

$$\frac{\partial \mathcal{E}_z}{\partial x} = \frac{\partial \mathcal{B}_y}{\partial t} \tag{2.13}$$

and

$$-\frac{1}{\mu}\frac{\partial \mathcal{B}_z}{\partial x} = \epsilon\frac{\partial \mathcal{E}_y}{\partial t}. \tag{2.14}$$

For a rectangle in the zy plane (see Figure 2.4c), the equations are identically satisfied. Bringing all of this together, we have $\mathcal{B}_x = \mathcal{E}_x = 0$ with the four remaining field components ($\mathcal{E}_y$, $\mathcal{E}_z$, $\mathcal{B}_y$ and $\mathcal{B}_z$) satisfying Eqs. (2.11) to (2.14).

If we take the t derivative of (2.11) and the x derivative of (2.13) we obtain

$$\frac{1}{\mu}\frac{\partial^2 \mathcal{B}_y}{\partial x \partial t} = \epsilon\frac{\partial \mathcal{E}_z^2}{\partial t^2} \tag{2.15}$$

and

$$\frac{\partial^2 \mathcal{E}_z}{\partial x^2} = \frac{\partial^2 \mathcal{B}_y}{\partial t \partial x}. \tag{2.16}$$

Then, eliminating $\partial^2 \mathcal{B}_y/\partial t \partial x$ between these two equations, we obtain the wave equation

$$\frac{\partial^2 \mathcal{E}_z}{\partial x^2} - \mu\epsilon\frac{\partial^2 \mathcal{E}_z}{\partial t^2} = 0, \tag{2.17}$$

where $c = 1/\sqrt{\epsilon\mu}$ is the propagation speed of the wave ($c = 3 \times 10^8$ m/s for an electromagnetic wave in a vacuum and this is often denoted by c_0). In a similar fashion

$$\frac{\partial^2 \mathcal{E}_y}{\partial x^2} - \mu\epsilon\frac{\partial^2 \mathcal{E}_y}{\partial t^2} = 0, \quad \frac{\partial^2 \mathcal{B}_z}{\partial x^2} - \mu\epsilon\frac{\partial^2 \mathcal{B}_z}{\partial t^2} = 0 \quad \text{and} \quad \frac{\partial^2 \mathcal{B}_y}{\partial x^2} - \mu\epsilon\frac{\partial^2 \mathcal{B}_y}{\partial t^2} = 0, \tag{2.18}$$

i.e. all field components satisfy the same wave equation.

Although the above equations imply that the field components $\mathcal{E}_y$, $\mathcal{E}_z$, $\mathcal{B}_y$ and $\mathcal{B}_z$ all satisfy wave equations, they are not independent of each other. For simplicity, we will assume a right-travelling wave in the $\hat{\mathbf{x}}$ direction. All of the field components will then be functions of $\phi = x - ct$ and, by similar arguments to the previous section, we find from (2.11) that

$$\frac{d\mathcal{B}_y}{d\phi} = -c\epsilon\mu\frac{d\mathcal{E}_z}{d\phi}. \tag{2.19}$$

Integrating with respect to ϕ, and noting that $\epsilon\mu = 1/c^2$, we obtain

$$\mathcal{B}_y = -\frac{1}{c}\mathcal{E}_z. \tag{2.20}$$

In a similar fashion, from (2.12) to (2.14),

$$\mathcal{E}_y = c\mathcal{B}_z\ ,\ \mathcal{E}_z = -c\mathcal{B}_y \text{ and } \mathcal{B}_z = \frac{1}{c}\mathcal{E}_y. \tag{2.21}$$

Since $\mathcal{E}_x = \mathcal{B}_x = 0$, it is clear that the propagation direction (the $\hat{\mathbf{x}}$ direction) will be orthogonal to both $\boldsymbol{\mathcal{E}}$ and $\boldsymbol{\mathcal{B}}$. Further, noting the relations in Eqs. (2.20) and (2.21), we obtain that $\boldsymbol{\mathcal{E}} \cdot \boldsymbol{\mathcal{B}} = \mathcal{E}_y\mathcal{B}_y + \mathcal{E}_z\mathcal{B}_z = c\mathcal{B}_z\mathcal{B}_y - c\mathcal{B}_y\mathcal{B}_z = 0$. As a consequence, we have that the propagation direction, the magnetic field $\boldsymbol{\mathcal{B}}$ and the electric field $\boldsymbol{\mathcal{E}}$ are all mutually orthogonal, a basic property of EM waves. Noting the relations (2.20) and (2.21), we also have the relation

$$\boldsymbol{\mathcal{B}} = -\frac{1}{c}\hat{\mathbf{x}} \times \boldsymbol{\mathcal{E}}, \tag{2.22}$$

where $\hat{\mathbf{x}}$ is a unit vector in the propagation direction. It is clear that an electric wave will always have a corresponding magnetic field. However, providing it is orthogonal to the propagation direction $\hat{\mathbf{x}}$, the components of the electric field $\boldsymbol{\mathcal{E}}$ can be arbitrary functions of $x - ct$, i.e. $\boldsymbol{\mathcal{E}} = \boldsymbol{\mathcal{E}}(x - ct)$. Such waves are known as *plane waves* due to the fact that the fields are constant across planes that are orthogonal to the propagation direction. Although plane waves might seem like a mathematical idealisation, they are found to give a good representation of real waves at large distances from their sources. Further, the orthogonality properties, and the relation (2.22), are retained by real waves. Figure 2.5 shows an example of a plane wave pulse that propagates in the $\hat{\mathbf{x}}$ direction and for which $\boldsymbol{\mathcal{E}}$ has the form $\mathcal{E}_y\hat{\mathbf{y}}$. Then, from (2.21), we then find that $\boldsymbol{\mathcal{B}} = (\mathcal{E}_y/c)\hat{\mathbf{z}}$. In general, a plane wave has a propagation direction $\mathbf{p}$ (a unit vector) with the electric field orthogonal to this direction. The electric field $\boldsymbol{\mathcal{E}}$ can be an arbitrary function of $\mathbf{p} \cdot \mathbf{r} - ct$ and the magnetic flux density is given by

$$\boldsymbol{\mathcal{B}} = -\frac{1}{c}\hat{\mathbf{p}} \times \boldsymbol{\mathcal{E}}. \tag{2.23}$$

Before we proceed any further, we need to address the issue of the energy content of an EM field. This is important since one of the major properties of waves is that they transport energy. In the case of the electric field, energy is most simply understood by considering the example of a capacitor. We consider the parallel plate capacitor of Figure 2.6a and calculate the work required to move a charge Q between the capacitor plates starting from a condition of no charge. This will be the energy of the field inside

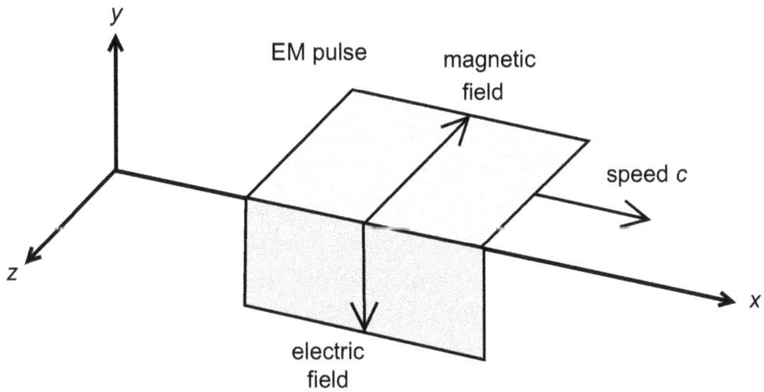

Fig. 2.5 Plane electromagnetic wave consisting of a pulse.

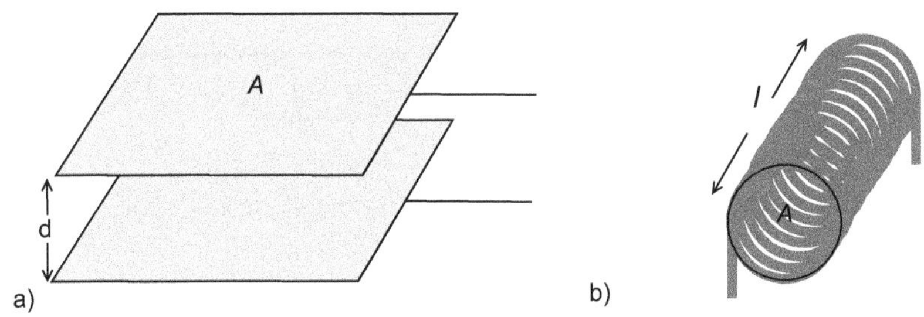

Fig. 2.6 Parallel plate capacitor and long solenoid.

the capacitor. If there are charges q and $-q$ on the plates of the capacitor, there will be a field of strength $\mathcal{E} = q/\epsilon A$ between the plates (A is the area of the capacitor plates). If we move a small amount of charge $\mathrm{d}q$ from the bottom plate to the top plate, this will require an amount of work $\mathrm{d}W = \text{force} \times \text{distance} = \mathrm{d}q\mathcal{E}d = \mathrm{d}q(qd/\epsilon_0 A)$ to be performed (d is the distance between the capacitor plates). If we build up the charge from 0 to Q in $\mathrm{d}q$ stages, the total work done will be sum of these stages, i.e. the integral

$$W_e = \int_0^Q \frac{qd}{\epsilon_0 A}\mathrm{d}q = Q^2 \frac{d}{2\epsilon_0 A} \tag{2.24}$$

and, since $\mathcal{E} = Q/\epsilon_0 A$,

$$W_e = \frac{1}{2}\epsilon_0 \mathcal{E}^2 V, \tag{2.25}$$

where V is the volume over which the capacitor field exists. For a general electric field $\boldsymbol{\mathcal{E}}$, the energy of the field contained inside a volume V will be

$$W_e = \frac{1}{2}\int_V \epsilon \boldsymbol{\mathcal{E}} \cdot \boldsymbol{\mathcal{E}} \mathrm{d}V. \tag{2.26}$$

The above volume integral can be understood by dividing V into N smaller volumes δV_i which are small enough for the electric field to be represented by a constant value $\boldsymbol{\mathcal{E}}_i$. Over ith volume, the energy field can be approximated by $\epsilon \boldsymbol{\mathcal{E}}_i \cdot \boldsymbol{\mathcal{E}}_i \delta V_i/2$ and then the total energy will be the sum of these contributions. The integral is then the limit as the size of the volumes tends to zero.

As we have seen above, the electric field cannot exist without the magnetic field in the case of an EM wave. Consequently, we also need to discuss the energy content of the magnetic field. This is most simply understood by considering the example of an inductor. We consider the long solenoid shown in Figure 2.6b and need to calculate the work required to establish a current flow of magnitude $\mathcal{I}$ at time T, starting from zero current at time 0. This will then be the energy W_h of the field inside the inductor. If a current i is to be driven through the inductor, the source will need to have an EMF equal to $-\mathcal{E}$ where $\mathcal{E}$ is the EMF that is generated in the inductor by the current flow. If we consider a time interval $\mathrm{d}t$, a charge $i\mathrm{d}t$ will flow into the inductor and hence the source will supply the inductor with energy $-i\mathcal{E}\mathrm{d}t$. Since $\mathcal{E} = -L\mathrm{d}i/\mathrm{d}t$ for an inductor with inductance L, the source will do the work $\mathrm{d}W_h = iL(\mathrm{d}i/\mathrm{d}t)\mathrm{d}t$. Summing the work done from time 0 to time T (i.e. integrating with respect to time)

$$W_h = \int_0^t Li\frac{\mathrm{d}i}{\mathrm{d}t}\mathrm{d}t = \int_0^I Li\mathrm{d}i = \frac{1}{2}L\mathcal{I}^2. \tag{2.27}$$

For a solenoid of length l, and cross-sectional area A, $L = \mu_0 NA/l$ and the field inside the solenoid is given by $\mathcal{B} = \mu_0 n\mathcal{I}$. As a consequence,

$$W_h = \frac{1}{2}\frac{B^2}{\mu_0}V, \tag{2.28}$$

where $V(= lA)$ is the volume occupied by the solenoid. For a general magnetic field $\boldsymbol{\mathcal{B}}$, the energy of the field contained inside a volume V will be

$$W_h = \frac{1}{2}\int_V \frac{\boldsymbol{\mathcal{B}}\cdot\boldsymbol{\mathcal{B}}}{\mu}\mathrm{d}V. \tag{2.29}$$

The total EM field energy can now be stated as

$$W = W_e + W_h = \frac{1}{2}\int_V \left(\epsilon\boldsymbol{\mathcal{E}}\cdot\boldsymbol{\mathcal{E}} + \frac{\boldsymbol{\mathcal{B}}\cdot\boldsymbol{\mathcal{B}}}{\mu}\right)\mathrm{d}V. \tag{2.30}$$

We now consider the energy flow of an EM wave. On noting the vector identity $(\mathbf{a}\times\mathbf{b})\cdot(\mathbf{a}\times\mathbf{b}) = (\mathbf{a}\cdot\mathbf{a})(\mathbf{b}\cdot\mathbf{b}) - (\mathbf{a}\cdot\mathbf{b})^2$, we obtain from (2.23) that $\boldsymbol{\mathcal{B}}\cdot\boldsymbol{\mathcal{B}} = \boldsymbol{\mathcal{E}}\cdot\boldsymbol{\mathcal{E}}/c^2$. As a consequence, for an EM wave,

$$W = \int_V \epsilon\boldsymbol{\mathcal{E}}\cdot\boldsymbol{\mathcal{E}}\mathrm{d}V. \tag{2.31}$$

Consider the rate at which energy is transported by an EM wave, i.e. the power. Let A be an area that is orthogonal to the propagation direction, then an amount of energy

$$dW = \epsilon\boldsymbol{\mathcal{E}}\cdot\boldsymbol{\mathcal{E}}Acdt \tag{2.32}$$

will be transported across the area A in the time interval $\mathrm{d}t$. Consequently the power density, the rate of transport of energy across a unit area, will be given by

$$\mathcal{P} = \frac{1}{\eta}\boldsymbol{\mathcal{E}}\cdot\boldsymbol{\mathcal{E}}, \tag{2.33}$$

where $\eta = \sqrt{\frac{\mu}{\epsilon}}$ is known as the *impedance* of the medium (in a space that is free of matter $\eta = \eta_0 = \sqrt{\frac{\mu_0}{\epsilon_0}} \approx 120\pi$ ohms). The use of the terminology 'impedance' can be justified on dimensional grounds. Firstly, the electric field $\boldsymbol{\mathcal{E}}$ has dimensions V/L where V is the dimension of voltage and L is the dimension of length. From the Maxwell equation (1.35) we have that $\boldsymbol{\mathcal{B}}/\mu$ has the dimensions of current I/L where I is the dimension of current. Consequently, noting relation (2.22), $\eta = c\mu$ must have the dimensions of V/I, i.e. the dimensions of impedance.

2.3 The Field of an Accelerating Charge

Consider a charge q that is initially resting at some origin O, in which case it will have an electric field $\boldsymbol{\mathcal{E}} = (q/4\pi\epsilon_0 r^3)\mathbf{r}$. At time 0 we accelerate the charge to a velocity $\Delta\mathbf{v}$ over a time period Δt, after which the charge continues at a constant speed $\Delta\mathbf{v}$. At time t after the acceleration period, the field lines will be those shown in Figure 2.7 (Purcell, 2013). Outside a sphere of radius $c(t+\Delta t)$, the field will still be that of a charge fixed at the origin O since the information about the changes to the source has not yet had time to arrive. Inside a sphere of radius ct about the charge, however, the field will have settled down to that of a charge moving with constant velocity Δv and hence will have the form $\boldsymbol{\mathcal{E}} = (q/4\pi\epsilon_0\tilde{r}^3)\tilde{\mathbf{r}}$ where $\tilde{\mathbf{r}} = \mathbf{r} - t\Delta\mathbf{v}$. A fundamental requirement of EM fields is that field lines be continuous, except at sources (i.e. current and charge). As a result, we obtain the field lines for the accelerating period by joining up the corresponding field lines on these two spheres. It will now be noted that the field lines in the period of acceleration are not completely radial and have a component $\mathcal{E}_t$ that is transverse to the radial direction. We will now assume that t is very much greater than Δt, i.e. we will consider the effect of the acceleration a long time after it has happened. Referring

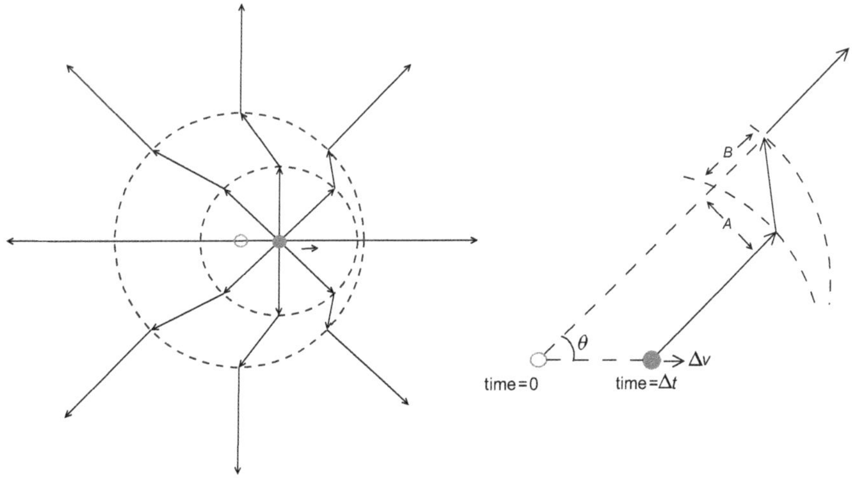

Fig. 2.7 Radiation by an accelerating charge.

to Figure 2.7, we have that $A \approx t\Delta v \sin\theta$ and $B \approx c\Delta t$. It therefore follows that

$$\frac{\mathcal{E}_t}{\mathcal{E}_r} = \frac{A}{B} \approx \frac{t\Delta v \sin\theta}{c\Delta t}, \tag{2.34}$$

where $\mathcal{E}_r$ is the radial component of the electric field. We further assume that Δv is very much smaller than the speed of light c and therefore $\tilde{r} \approx ct$. As a consequence, $\mathcal{E}_r \approx q/4\pi\epsilon_0 c^2 t^2$ for points between the spheres. From (2.34) we will now have that

$$\mathcal{E}_t = f\frac{q\sin\theta}{4\pi\epsilon_0 c^2 \tilde{r}}, \tag{2.35}$$

where $f = \Delta v/\Delta t$ is the acceleration of the charge. This expression also holds for a particle that is continuously accelerating, but we must now remember that a point at a distance $|\tilde{\mathbf{r}}|$ is responding to an acceleration that happened at time $t - |\tilde{\mathbf{r}}|/c$. As a consequence

$$\mathcal{E}_t = f(t - |\tilde{\mathbf{r}}|/c)\frac{q\mu_0 \sin\theta}{4\pi\tilde{r}}, \tag{2.36}$$

where the acceleration is now a function $f(t)$ of time. The important thing to note is that the field of an accelerating charge falls away as $1/|\tilde{\mathbf{r}}|$ as we move far away from the charge, whereas that of a charge in uniform motion falls away as $1/|\tilde{\mathbf{r}}|^2$. By accelerating a charge, we have extended its influence way beyond that of a static charge.

At large distances from the accelerating charge, the $\mathcal{E}_t$ term of the electric field will dominate and this will behave like a plane wave travelling radially outwards from the charge. Importantly, the wave can be given an arbitrary shape by controlling the behaviour of the charge acceleration. We now show that this wave carries energy away from the charge. In order to find the power P travelling outwards from the charge, we will need to integrate the power density $\mathcal{P}$ over a large sphere of radius R that surrounds the charge, i.e.

$$P = \frac{1}{\eta}\int_S \boldsymbol{\mathcal{E}} \cdot \boldsymbol{\mathcal{E}} \mathrm{d}S. \tag{2.37}$$

The sphere can be considered as a series of rings of area $2\pi R^2 \sin\theta d\theta$ that are parameterised by angle θ (see Figure 2.8). The power travelling through this ring will be $(\mathcal{E}_t^2/\eta)R^2 \sin\theta \mathrm{d}\theta$ and adding up the contributions from all rings, we obtain

$$P = \frac{1}{\eta}\int_0^\pi \mathcal{E}_t^2 2\pi R^2 \sin\theta \mathrm{d}\theta, \tag{2.38}$$

where $\mathcal{E}_t = f(t - R/c)q\mu_0 \sin\theta/4\pi R$. Then, noting that $\int_0^\pi \sin^3\theta \mathrm{d}\theta = 4/3$, we obtain the Larmor formula

$$P = \frac{f_{\text{ret}}^2 q^2 \mu_0^2}{6\eta\pi}, \tag{2.39}$$

where $f_{\text{ret}} = f(t - R/c)$ denotes the acceleration at a retarded time $t - R/c$. It will be noted that the power carried away by waves is proportional to the square of the acceleration.

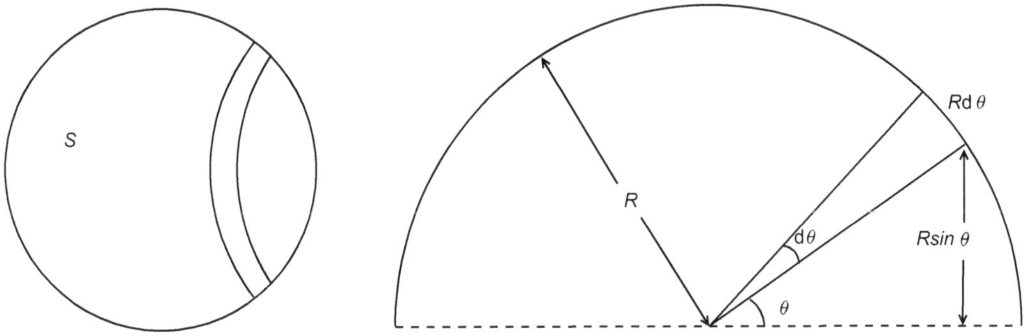

Fig. 2.8 Geometry for calculating power.

2.4 The Field of an Oscillating Charge

As we have seen, an accelerating charge will cause radio waves. For practical radio, however, we will need to produce accelerations in a sustainable fashion. It turns out that we can do this through an oscillating charge and, furthermore, this will have properties that are beneficial to radio. Consider a charge that performs harmonic oscillations over a distance $2l$ along a line that is parallel to unit vector $\hat{\mathbf{t}}$, i.e

$$\mathbf{r}_s = l\sin(\omega t + \phi)\hat{\mathbf{t}}, \tag{2.40}$$

where ω is the *oscillation frequency* in radians per second and ϕ is an arbitrary phase factor. The acceleration of the charge will be $f(t) = -\omega^2 l \sin(\omega t + \phi)$ and so, from (2.36), its field will have the transverse component

$$\mathcal{E}_t = -\omega^2 l \sin(\omega t - 2\pi R/\lambda + \phi)\frac{q\mu_0 \sin\theta}{4\pi R}, \tag{2.41}$$

where R is the distance from the charge and $\lambda = 2\pi c/\omega$. Furthermore, at large distances (many wavelengths away), this will be the dominant component. Figure 2.9 shows the field lines for a charge oscillating at a frequency of 3×10^7 Hz. (Note that Hz, short for *hertz*, is one cycle per second. Further, frequency f in hertz is related to angular frequency ω through $\omega = 2\pi f$.) The field is shown at equally spaced times over a full period of oscillation and it will be noted that the field lines exhibit a wavelike structure that moves outwards with time. The waves have a wavelength (distance between crests) $\lambda = 2\pi c/\omega$ which is 10 m in this case (the scales on the Figures are in metres).

In practical terms we can produce the effect of an oscillating charge by driving an oscillating current into a metal structure that is known as an antenna. The current will cause electrons to move backwards and forwards on the structure in a similar fashion to the single charge above. Consequently, at large distances, this will result in a field that is similar in nature to that of an oscillating charge. On the antenna structure, it will be noted that there will be a counterbalancing positive charge for each electron. However, since these charges are static, they have no effect upon the far field. The simplest practical antenna consists of a metallic rod with a generator driving oscillating current into its centre (see Figure 2.11). The electrons will move coherently and so the current can be

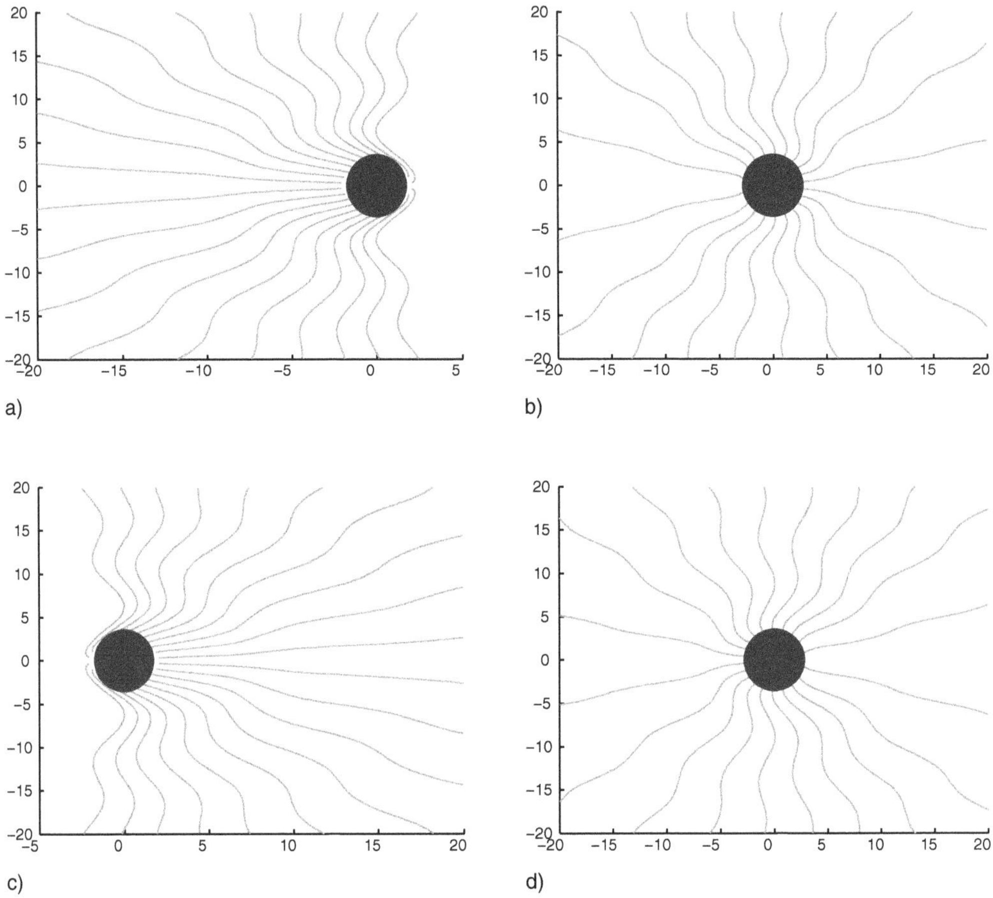

Fig. 2.9 Electric field lines of an oscillating charge over one period.

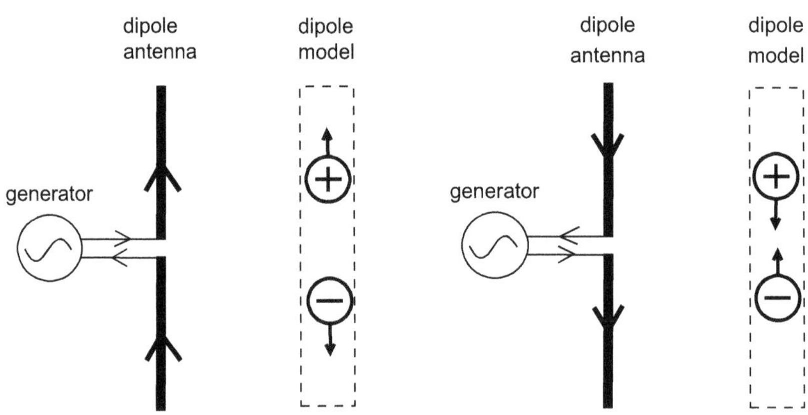

Fig. 2.10 Dipole antenna and model.

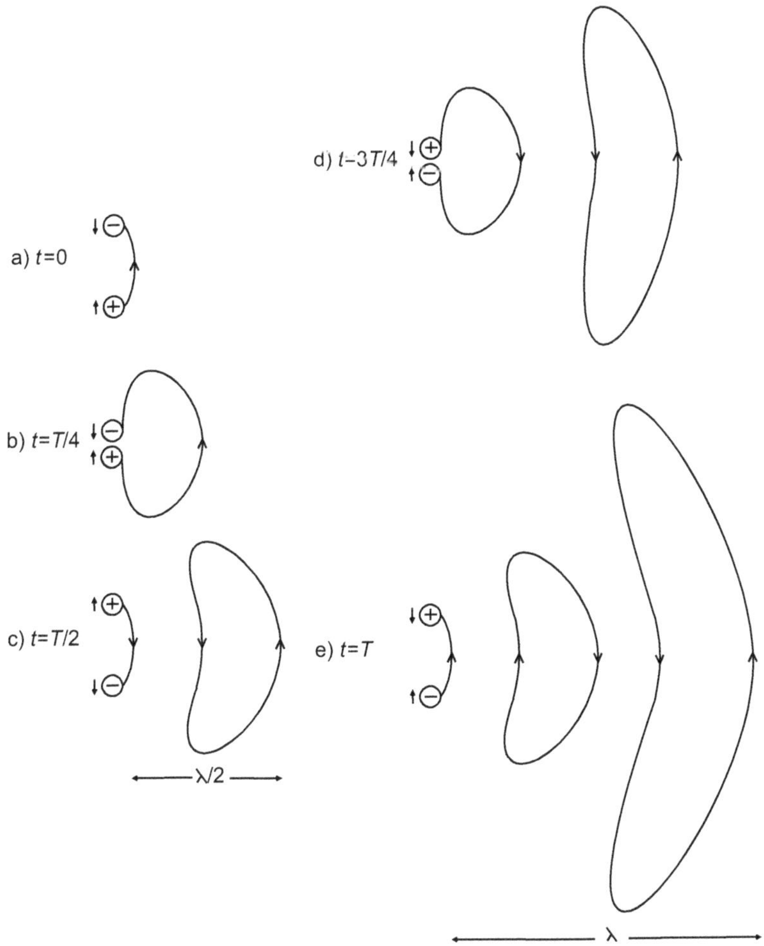

Fig. 2.11 Development of the dipole field over a period T of oscillation.

modelled as a pair of opposing charges that oscillate backwards and forwards along the antenna axis (i.e. an oscillating dipole). These charges move in opposite directions (they are 180° out of phase) and so will represent, at any time, a current in the same direction. For a dipole, the charges will oscillate backwards and forwards between the ends of the dipole, a distance $2l$, and the effective current will be $I = I_0 \cos(\omega t + \phi)$, where $I_0 = q\omega$. Consequently, at large distances, the transverse component of electric field will be

$$\mathcal{E}_t = -\sin(\omega t - 2\pi R/\lambda + \phi)\frac{\omega \mu_0 I_0 l \sin\theta}{2\pi R} \tag{2.42}$$

(this is the sum of the fields for positive and negative charge). Figure 2.11 shows the development of the dipole field through a full period of the charge oscillations. Field lines cannot cross and so, as the charges themselves cross, the field lines break off from the charges and the ends join together to form closed field lines. These closed field lines will then move away at the speed of light to make room for the new field lines that join the charges. Because of its similarity with a simple dipole of two charges, a driven metal rod is known as a *dipole antenna*.

The use of a sinusoidal generator will produce a radio wave at a single frequency and this can be advantageous when many radio users must share the radio spectrum. In this case, a receiver of radio signals can be designed to *filter* out all but the frequency of the desired signal. A sinusoidal signal, however, carries no information and must be modulated in some fashion if it is to convey information. We can achieve modulation by varying a suitable combination of I_0, ω and ϕ. If we vary I_0 alone we have *amplitude modulation*, ω alone we have *frequency modulation* and ϕ alone we have *phase modulation*. Unfortunately, the modulation will cause the radio signal to spread out around the *carrier frequency* ω and the width of the spread is known as the *bandwidth* of the signal.

2.5 The First Radio Systems

As we have seen in the previous two sections, we need to have accelerating charge in order to generate radio waves and charge oscillations are a convenient means of producing the necessary acceleration in a sustainable fashion. Around the year 1886, German physicist Heinrich Hertz demonstrated a practical device for producing radio waves. Using this device he was able to verify the existence of radio waves. The generator that Hertz invented is what we now call a *spark transmitter* and this was the mainstay of radio for the next thirty years. The essence of a spark transmitter and Hertz's receiving system is shown in Figure 2.12. In this Figure, when the key closes, the capacitor charges up until it reaches a point where the voltage across the spark gap is great enough for the insulation afforded by the air to break down and current to flow (the spark). The circuit, consisting of a series capacitor and inductor, is now closed and, as we have seen in Chapter 1, there is a damped oscillation that drives oscillatory current into a dipole antenna through a secondary winding on the inductor. When the oscillations in the circuit have dropped below a certain level, the sparking stops and the circuit breaks. The charge now builds up again on the capacitor until the break down point in the spark gap is once again reached and the oscillation starts again. As a consequence, the transmitted radio signal will consist of a sequence of damped oscillations. The radio wave generates an EMF in a receiving loop that is suitably orientated and this can cause a spark across the gap in

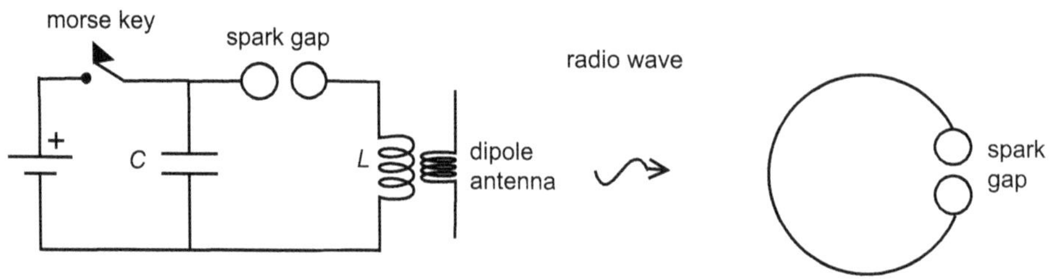

Fig. 2.12 A basic spark transmitter and a detector circuit.

the loop (Hertz needed to use a microscope to see this very small spark). According to Faraday's law, the EMF in the loop will be given by $\mathcal{E} = -A\partial\mathcal{B}_n/\partial t$ where $\mathcal{B}_n$ is the component of magnetic field that is normal to the plane of the loop and A is the area of the loop.

Whilst the transmitter shown in Figure 2.12 could, in theory, work for a small enough spark gap, the strength of the radio waves would have been far too small for the experiments of Hertz. To overcome this problem, Hertz used an induction coil to produce a high voltage. (This is the way that your car produces a spark to ignite the petrol in its engine.) Figure 2.13 illustrates the way in which this is achieved. Initially a current flows through a primary winding with a low number of turns. This causes a magnetic field in an iron core that attracts the lever on the right, which breaks the circuit. This sudden change in current produces a voltage pulse in the secondary winding. The secondary will have a far greater number of turns than the primary and so the pulse at the output will have a much larger voltage than that of the battery. While the key is closed, this process is repeated continuously resulting in a sequence of high-voltage pulses. In the Hertz transmitter, the battery and key of Figure 2.12 are replaced by the induction system of Figure 2.13. Spark transmitters were the predominant form of radio transmitter until well into the twentieth century and for this reason a radio operator was often known as a 'sparks', the terminology remaining into the current era.

Although Hertz verified the existence of radio waves, his apparatus was not practical as a piece of technology, mainly due to the inadequacy of the detector. Indeed, Hertz expressed the view that his discoveries would never amount to anything practical. Unfortunately, he did not live long enough to see how wrong he was on this score (he died in 1894). The first demonstration of the practical possibilities of radio were performed by the Indian physicist Sir Jagadish Chandra Bose. In November of 1894, Bose gave a public demonstration of radio-wave transmission over a distance of 23 metres, using his transmissions to set off a bell and some gunpowder. Also in that year, the British physicist Sir James Lodge gave a practical demonstration of the transmission of telegraph signals using radio. It was, however, left to the Italian inventor Guglielmo Marconi, to

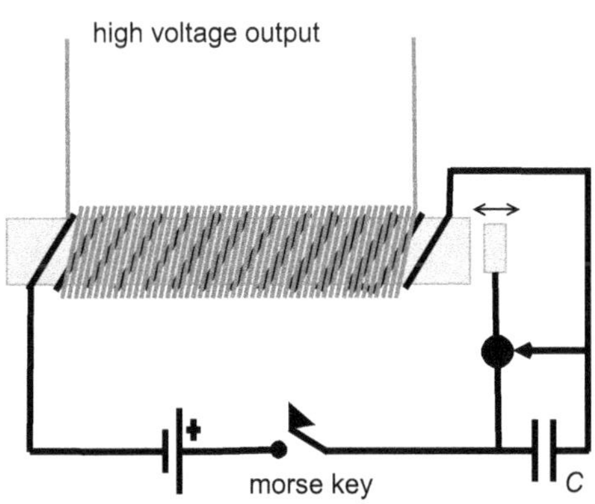

Fig. 2.13 Induction coil for providing high voltage pulses.

turn Hertz's remarkable discovery of radio into a commercially viable technology. In 1895 Marconi successfully transmitted radio signals over distances several kilometres (Hertz had only managed a few metres). The transmitting equipment of Marconi was much the same as that used by Hertz, but the radio receiver used a detection device known as a *coherer*. The coherer was a device invented by Edouard Branly in 1890. This device consisted of two metal electrodes with the space between them filled with loose iron fillings. Initially the device exhibits a high resistance, but when a radio signal appears across its electrodes the metal fillings cling together (i.e. they 'cohere') and the device becomes highly conductive. Figure 2.14 shows a typical radio receiver that is based on such a device (note that a dipole antenna can be used to receive a well as transmit a radio signal). An incoming radio wave will generate a time-varying current in the antenna and this, in turn, will generate a time-varying voltage in the secondary of the transformer. The situation will now resemble that of Figure 1.24 of Chapter 1 with the load R_L representing the detection system (i.e. the coherer and subsequent circuitry). When a strong enough radio signal is received at the resonant frequency of L and $C1$, the coherer becomes conducting and the DC (direct current) circuit involving the battery is completed. This DC pulse will then be detected by the radio operator as a click in the speaker. Unfortunately, once a radio signal has set a coherer into its conducting state, the fillings will remain cohered. Consequently, after a detection, the coherer needs to be agitated to loosen the fillings before the next detection can take place.

The problem with a coherer-based radio receiver is its lack of sensitivity and this continued to hold back the development of radio for almost a decade. The solution to this problem eventually came from an effect that was discovered by Ferdinand Braun in 1874. Braun found that galena crystals (a mineral form of lead sulphide) could conduct electricity asymmetrically (as much as a 2:1 asymmetry between forward and backward current is possible). Figure 2.15a shows a device, known as a *crystal detector*, that is based on this property (we now call such a device a *point contact diode*). A device of this form was first patented by Jagadish Bose in 1904, but a similar device using carborundum (silicon carbide) was also patented by Henry Dunwoody in 1907 and proved to be more reliable. In the Figure 2.15a, one connection to the detector is made through a mercury bath in which the galena sits and the other through a thin wire, known as a 'cat's whisker', that must be carefully adjusted to make contact at a point on the crystal that exhibits a suitable amount of asymmetry. This device made possible the receiver architecture

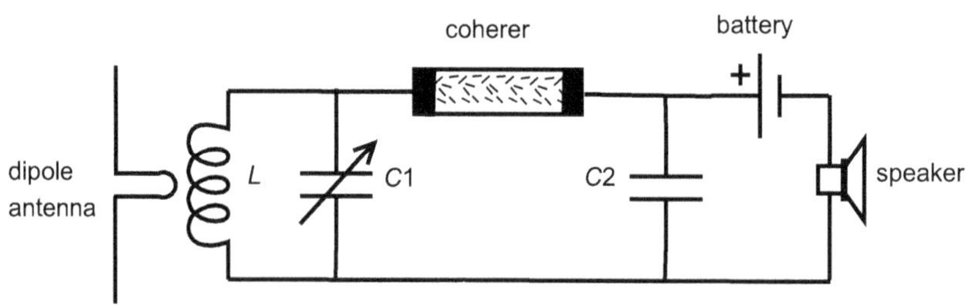

Fig. 2.14 Radio receiver based on a coherer.

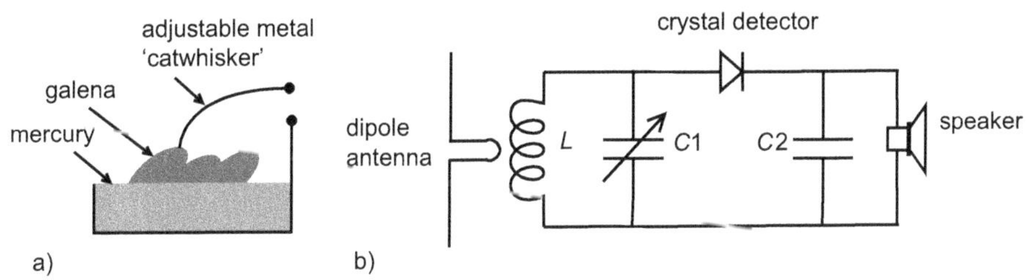

Fig. 2.15 A point contact diode and a receiver based on this device.

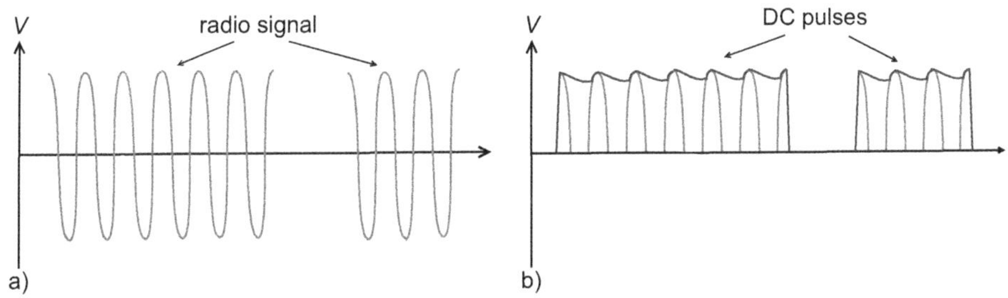

Fig. 2.16 Demodulation of radio frequency pulses by a crystal detector.

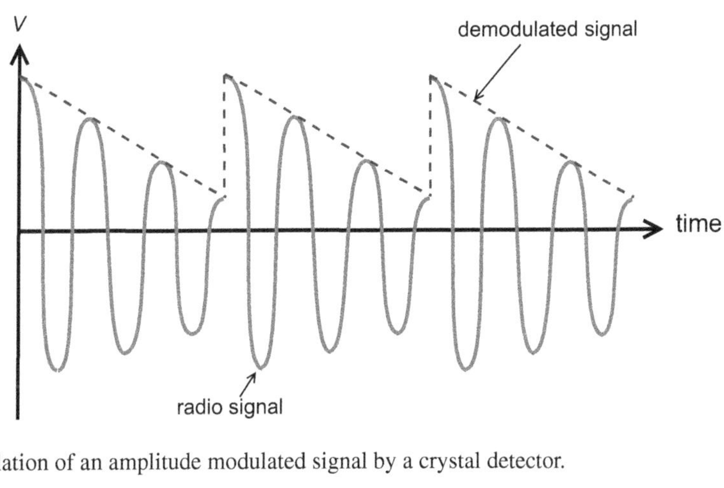

Fig. 2.17 Demodulation of an amplitude modulated signal by a crystal detector.

shown in Figure 2.15b. The field of the radio wave causes a current to flow in the antenna and hence an EMF in the circuit consisting of inductor L and capacitor $C1$ (set to resonate at the frequency of the desired incoming radio wave). The crystal detector only allows current flow in one direction and so, after smoothing by capacitor $C2$, the radio signal presents itself as a pulse at the speaker (see Figure 2.16). This DC pulse will then be detected by the radio operator as a click in the speaker.

The introduction of the crystal detector proved to be a boon for radio since it not only increased sensitivity, but also introduced a capability to demodulate signals with amplitude modulation. Because the output of a spark transmitter consists of a sequence of damped oscillations, it turns out to be already amplitude-modulated. The solid curve in Figure 2.17 shows typical behaviour of the amplitude of a signal that was produced by a spark transmitter and the broken curve shows the demodulated signal that presents itself to the speaker. Fortuitously, the modulation occurs at audio frequencies and this means that the output at the speaker will be a tone. Consequently, the signal from a keyed spark transmitter will consist of a sequence of tone pulses after crystal detection. This is far easier on the radio operator than a sequence of clicks.

2.6 Conclusion

In the current chapter we have developed the basic theory of radio waves, their generation, propagation and detection. Further, we have looked at some of the early developments in radio technology (see Garratt (1994) and Lee (1985) for further historical detail). Crucial to this technology is the idea of restricting the radio signal to a narrow band of frequencies in order that many users can be accommodated within the radio spectrum. In early radio, this was achieved through simple tuned circuits that were resonant at the desired frequency. However, the subsequent development of radio has required frequency selection at a much higher fidelity and so we digress in the next chapter to consider how this can be achieved.

3 Tuned Circuits

We have already seen that an ability to select frequencies is crucial to the effective development of radio technology. In early radio, the selection could be adequately achieved by very simple tuned circuits. However, the modern crowded airwaves have made the frequency selection requirement extremely stringent and simple tuned circuits are no longer adequate. Consequently, in the following chapter we introduce some more sophisticated tuned circuits that make higher-fidelity frequency selection possible. Additionally, we study how such circuits can also ensure the optimum transfer of power between the various stages of a radio. Power transfer and filtering are intimately connected since a filter works by transferring power well on its pass frequencies and badly on its blocking frequencies. In order to study tuned circuits we first develop some techniques for the harmonic analysis of circuits, techniques that we will use frequently throughout the rest of this book.

3.1 Time-Harmonic Circuits

In order that many users can utilise the available radio frequencies (often called the *radio spectrum*), we will need radio circuits that can isolate slices of frequency and we achieve this through what are known as *tuned circuits*. We will consider circuits in which the radio signals are sinusoidally varying with time at a specific frequency ω (*time-harmonic* circuits). Consequently, we can assume voltage and current to take the form

$$\mathcal{V}(t) = V_0 + V_1 \cos(\omega t + \phi_v) \text{ and } \mathcal{I}(t) = I_0 + I_1 \cos(\omega t + \phi_i). \tag{3.1}$$

We will assume that all other variations in time (such as modulation) take place over timescales that are very much greater than that of an oscillation period ($T = 2\pi/\omega$). Consequently, we can treat V_0, V_1, I_0 and I_1 as constant. In a linear circuit analysis, we can consider the DC (V_0 and I_0) aspects of the circuit separately from the RF (radio frequency) aspects. In analysing the DC aspects, we simply replace all capacitors by an open circuit and all inductors by a short circuit. To study the RF aspects, we now simply set I_0 and V_0 to be zero and instead study voltages and currents given by

$$v(t) = V_1 \cos(\omega t + \phi_v) \text{ and } i(t) = I_1 \cos(\omega t + \phi_i) \tag{3.2}$$

(note that we now use lower case letters to signify the RF part of the signal). At this point we will find it simpler to study our circuits in terms of *complex numbers*! A complex

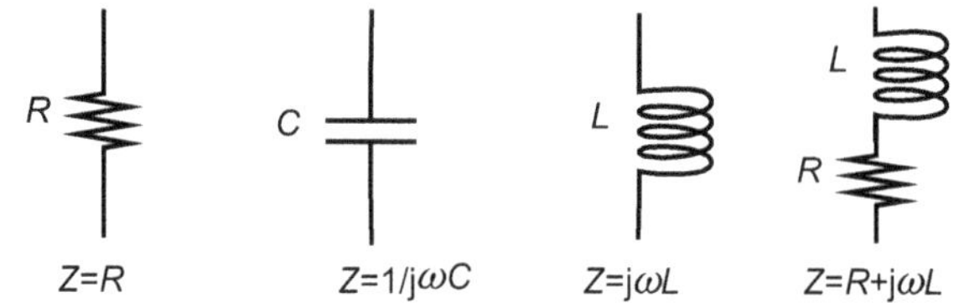

Fig. 3.1 Impedance of various components.

number is of the form $x+jy$ where $j=\sqrt{-1}$ (i.e. $j^2=-1$), x is known as the real part of the number and y is known as the imaginary part. On noting that $\exp(j\theta)=cos(\theta)+j\sin(\theta)$, we can write (3.2) as

$$v(t)=\Re\{V\exp(j\omega t)\} \text{ and } i(t)=\Re\{I\exp(j\omega t)\}, \tag{3.3}$$

where $V=V_1\exp(j\phi_v)$ and $I=I_1\exp(j\phi_i)$ are complex amplitudes that have absorbed the phases ϕ_v and ϕ_i. (It should be noted that $\Re\{A\}$ is used to signify the real part of complex number A and $\Im\{A\}$ to signify the imaginary part, i.e. $A=\Re\{A\}+j\Im\{A\}$.)

If we consider v to be the RF voltage drop across a resistor R carrying RF current i, then Ohm's law implies that $v=iR$ from which $\Re\{V\exp(j\omega t)\}=\Re\{I\exp(j\omega t)\}R$ and $V=IR$. Now consider the voltage drop across an inductor L, then $v=L\mathrm{d}i/\mathrm{d}t$ from which $\Re\{V\exp(j\omega t)\}=L\Re\{j\omega I\exp(j\omega t)\}$ and hence $V=j\omega LI$. Finally consider the voltage drop across a capacitor C, then $C\mathrm{d}v/\mathrm{d}t=i$ from which $C\Re\{j\omega V\exp(j\omega t)\}=\Re\{I\exp(j\omega t)\}$ and hence $V=I/j\omega C$. What we conclude is that both inductors and capacitors behave as if they have an imaginary resistance of the form jX. Quantity X is termed the *reactance* and has ohms as its units. An inductance has positive reactance ωL and a capacitor has negative reactance $-1/\omega C$. In reality, an inductor, or capacitor, will exhibit some loss that is represented by a resistance (R_{loss} say) and then it will behave like the complex resistance $Z=R_{\text{loss}}+jX$ where X is the reactance. A complex resistor is known, in general, as an *impedance*. We can lump several impedances together to form one single impedance and, for impedances Z_1 and Z_2 joined in series, the combined impedance will simply be their sum (i.e. Z_1+Z_2). For impedances in parallel, often denoted by $Z_1 \parallel Z_2$, the combined impedance Z will satisfy $1/Z=1/Z_1+1/Z_2$.

3.2 Power Transfer

We have already discussed power in the context of radio waves and for which it was defined as the rate at which energy is transported by the wave. In terms of electrical devices, we will understand *power* to mean the rate of energy transport between these devices. Further, we will understand the term *power loss* to mean the energy that is lost per unit time in such devices. The voltage drop $\mathcal{V}$ across a device is the energy lost when a unit charge travels through it and so the power loss $\mathcal{P}$ will be

$$\mathcal{P}=\mathcal{I}\mathcal{V}, \tag{3.4}$$

where $\mathcal{I}$ is the current flowing through the device. For a time-harmonic analysis, the average power lost over one cycle of the source will be

$$P_{av} = \frac{1}{2\pi}\int_0^{2\pi} \mathcal{I}(t)\mathcal{V}(t)\mathrm{d}t$$
$$= \frac{1}{2}\Re\{I\bar{V}\} = \frac{V_L\bar{V}_L}{2}\Re\left\{\frac{1}{Z_L}\right\}, \tag{3.5}$$

where Z is the impedance of the device (NB if $z = x + \mathrm{j}y$ then $\bar{z} = x - \mathrm{j}y$ is the *conjugate* of z and $z\bar{z} = x^2 + y^2$).

Now consider a sinusoidally oscillating source and its load, as shown in the circuit of Figure 3.2. The source consists of an ideal voltage source (zero internal impedance) V_S and a source impedance Z_S. (A source will always have some internal resistance and this is the reason why the open circuit voltage of a battery is always higher than the voltage when current is being drawn.) We will assume the internal impedance of the source is $Z_S = R_S + \mathrm{j}X_S$ and that the load has impedance is $Z_L = R_L + \mathrm{j}X_L$. We now ask the question as to what sort of load is required for there to be maximum power dissipated in the load resistance R_S. The same current I will flow through all components and so the voltage drop around the circuit is zero. As a consequence, we have

$$V_S - R_S I - \mathrm{j}X_S I - \mathrm{j}X_L I - R_L I = 0 \tag{3.6}$$

and the voltage drop V_S across the load will be given by

$$V_L = \frac{V_S(R_L + \mathrm{j}X_L)}{R_S + \mathrm{j}X_S + \mathrm{j}X_L + R_L} \tag{3.7}$$

where V_L is the voltage drop across the load. The power dissipated in the load impedance Z_L will be

$$P_L = \frac{V_L\bar{V}_L}{2}\Re\left\{\frac{1}{Z_L}\right\} = \frac{V_S\bar{V}_S}{2}\frac{R_L}{(R_S + R_L)^2 + (X_S + X_L)^2} \tag{3.8}$$

and will be maximum when $\partial P_L/\partial R_L = \partial P_L/\partial X_L = 0$. This implies that $X_L = -X_L$ and $R_L = R_S$. When these conditions are met, there is said to be a *conjugate match* since $Z_L = \bar{Z}_S$. What is clear from this is that the maximum power that can be transferred to that load is half that which is dissipated in the total circuit (i.e. the load and source dissipate the same amount of power). When the reactances of the source and load cancel

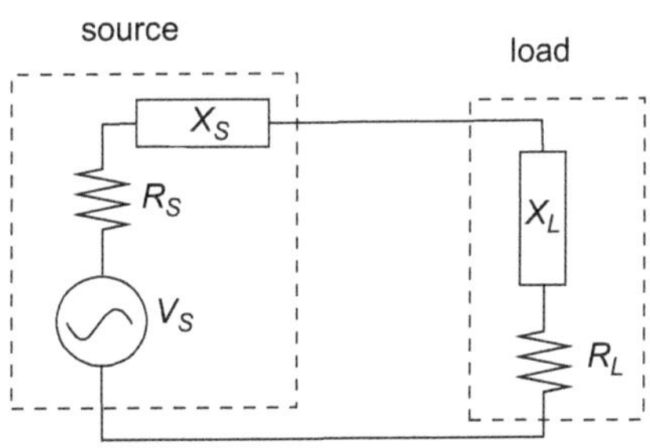

Fig. 3.2 Model of a sinusoidally oscillating source and load.

each other, they are said to resonate and the frequency at which this happens is known as the *resonant frequency*.

3.3 Basic Tuned Circuits

We will now investigate frequency selective circuits, sometimes known as *tuned circuits*. First consider a series capacitor C and inductor L that connect a resistive (totally real impedance) source R_S to a totally resistive load R_L (see Figure 3.3a). At this point we note an important result from circuit theory. If impedances Z_1 and Z_2 have a voltage V across their series combination, then the voltage at their junction will be given $VZ_1/((Z_1+Z_2)$, where Z_1 is the lower impedance (Z_1 and Z_2 are said to act as a *voltage divider*). From this result the voltage across the load will be

$$V_L = \frac{R_L V_S}{R_S + \mathrm{j}\omega L + \frac{1}{\mathrm{j}\omega C} + R_L}. \tag{3.9}$$

We have seen from the above considerations that maximum power transfer to the load resistance will occur when the reactances cancel ($\mathrm{j}\omega L + 1/\mathrm{j}\omega C = 0$) and this will occur at a frequency $\omega_0 = 1/\sqrt{LC}$. The series combination of L and C behaves as a short circuit at frequency ω_0 where it is said to have a *series resonance*. Away from the *resonant frequency*, the power transfer will fall away and so the circuit will act as a *band-pass filter* (i.e. it just passes signals at frequencies around the frequency ω_0). We can rewrite (3.9) as

$$V_L = \frac{R_L V_S}{R_S + R_L} \frac{1}{1 + \mathrm{j}Q\left(\frac{\omega}{\omega_0} - \frac{\omega_0}{\omega}\right)}, \tag{3.10}$$

where $Q = L\omega_0/(R_L + R_S)$. The power that is transferred to the load will then be given by

$$P_L = \frac{V_L \bar{V}_L}{2R_L} = \frac{V_S \bar{V}_S R_L}{2(R_L + R_S)^2} \frac{1}{1 + Q^2\left(\frac{\omega}{\omega_0} - \frac{\omega_0}{\omega}\right)^2}. \tag{3.11}$$

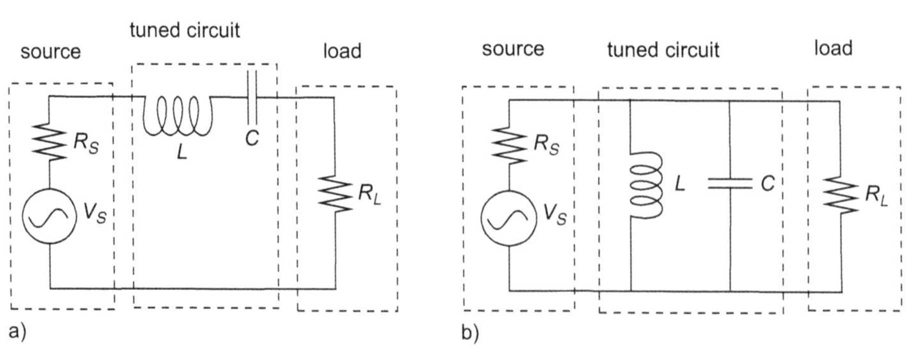

Fig. 3.3 Series resonant circuit and parallel resonant circuit.

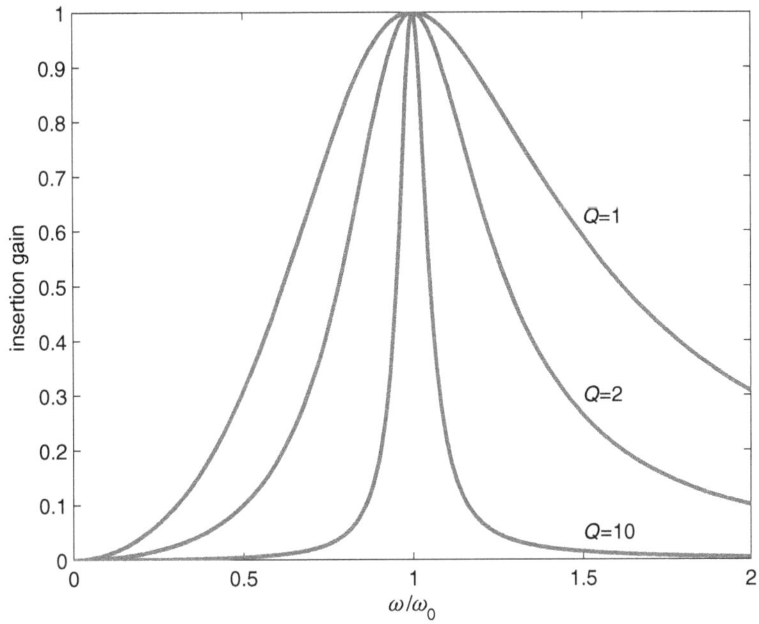

Fig. 3.4 Insertion gain for the series resonant circuit.

The frequency-selective nature of the above tuned circuit can be seen from the plot of Figure 3.4. This Figure shows the *insertion gain* of the circuit (the power delivered to the load through the tuned circuit when normalised by the power delivered without the circuit) for a variety of Q. The *Q factor* is a measure of the *quality* of a tuned circuit and it will be noted that the higher the Q, the sharper the filtering effect. For small deviations from resonance, i.e. $\omega = \omega_0 + \delta\omega$, we can approximate the above expression by

$$P_L \approx \frac{V_S \bar{V}_S R_L}{2(R_L + R_S)^2} \frac{1}{1 + 4Q^2 \frac{\delta\omega^2}{\omega_0^2}} \tag{3.12}$$

on noting that $\omega_0/\omega = 1/(1+\delta\omega/\omega_0) \approx 1 - \delta\omega/\omega_0$ since $1/(1+x) \approx 1-x$ for small x. The frequency selective nature of a filter is often described in terms of its half-power bandwidth B, i.e. the width of the band of frequencies over which the power supplied to the load is more than 50 per cent of the maximum. For the above circuit, the edges of the band will be located at $\omega_0 \pm \omega_0/2Q$ from which $B = \omega_0/Q$. The Q of a circuit is an important quantity as it measures the quality of the circuit when acting as a filter. The *quality factor* Q has a broader meaning and, in general, is defined to be

$$Q = 2\pi \frac{\text{maximum energy stored}}{\text{energy lost per cycle}}. \tag{3.13}$$

It is essentially a measure of the storage efficiency of a circuit. (The definition applies to both mechanical and electrical systems.) We will show that this alternative definition still implies that $Q = L\omega_0/(R_L + R_S)$. Both capacitor and inductor are assumed perfect and so there is no energy loss in either. Any energy supplied to the circuit will oscillate between these devices or be lost in the resistances. The rate at which energy is lost in a resistor R is given by $R\mathcal{I}^2$ and, integrating over a cycle, we get a loss of $\pi RI\bar{I}/\omega$. Consequently, the energy lost in the circuit over a cycle will be $\pi(R_S+R_L)I\bar{I}/\omega$. Furthermore, the maximum energy stored in the inductor will be $LI\bar{I}/2$. Consequently, as expected, we find from (3.13) that $Q=\omega L/(R_S+R_L)$.

An alternative configuration that acts as a band-pass filter is shown in Figure 3.3b. In this case we have a parallel combination of C and L that shunts the load. At frequency $\omega_0 = 1/\sqrt{LC}$, the parallel combination of L and C will have infinite impedance and they are said to have a *parallel resonance* at this frequency. By a similar analysis to that above, we find that

$$V_L = \frac{R_L V_S}{R_S+R_L} \frac{1}{1+\mathrm{j}Q\left(\frac{\omega}{\omega_0}-\frac{\omega_0}{\omega}\right)}, \tag{3.14}$$

where $Q=(R_S \parallel R_L)/\omega_0 L$. This is almost identical to the result for the series LC circuit, except for a difference in the way in which Q is defined.

Up until now, we have assumed our inductors to be ideal, i.e. their impedances are purely imaginary. In reality, the windings of an inductor will add an effective series resistance r_s and the inductor impedance will have a small real part. We often describe this resistance through the *unloaded Q* of the inductor, a quantity which is defined to be $Q_U=\omega L/r_s$. It is sometimes more useful to describe the resistance as an effective parallel resistance r_p and, for large Q_U, this will be given by $r_p=Q_U\omega L$. Consequently, if Q_U is low, the extra resistance could seriously lower the Q of the total circuit and hence lower its filtering capabilities (bandwidth depends on Q). The major aim in the design of inductors is to make their unloaded Q as high as possible, usually by reducing the resistance of the wire as much as possible. This can be done by using wire that is thicker and/or has a high conductivity (silver and gold for example). We can also reduce the resistance by reducing the amount of winding by means of a high-permeability core.

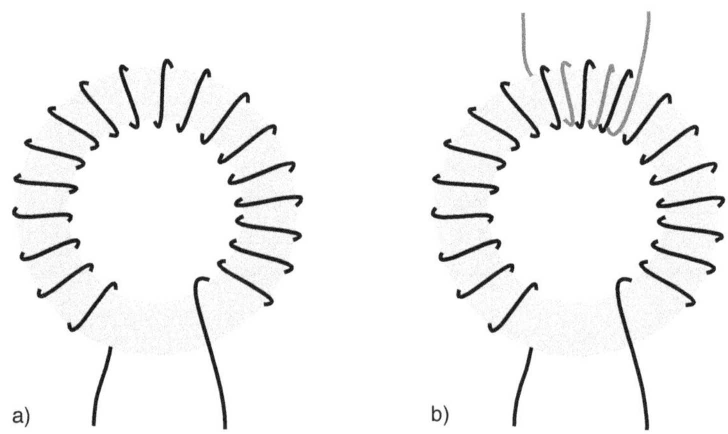

Fig. 3.5 Toroid inductor and toroid transformer.

A ferrite or dust iron core is usually to be preferred (see Figure 3.5a) since this will minimise any additional loss in the core.

3.4 The Inductive Transformer

It is clear that, to transfer maximum power between the source and load, we need to have a conjugate match. We need to somehow convert the impedance of the load into one that is conjugate to the source impedance. We can do this through an inductive transformer (see Figure 3.6a) of the sort we considered in Chapter 1. Let V_1 and V_2 be the complex amplitudes of the voltages applied to the primary and secondary windings of the transformer and I_1 and I_2 be the complex amplitudes of current flowing into these windings. The relations (1.22) and (1.22) of Chapter 1 can then be written as

$$V_2 = \mathrm{j}\omega L_{21} I_1 + \mathrm{j}\omega L_{22} I_2 \tag{3.15}$$

and

$$V_1 = \mathrm{j}\omega L_{11} I_1 + \mathrm{j}\omega L_{12} I_2. \tag{3.16}$$

From Ohm's law we have that $V_2 = -I_2 Z_L$ and so, from (3.15),

$$I_1 = -\frac{\mathrm{j}\omega L_{22} + Z_L}{\mathrm{j}\omega L_{21}} I_2. \tag{3.17}$$

Then, from (3.16) and (3.17),

$$V_1 = (-\mathrm{j}\omega L_{11}\frac{\mathrm{j}\omega L_{22} + Z_L}{\mathrm{j}\omega L_{21}} + \mathrm{j}\omega L_{12}) I_2 = \frac{L_{11}}{L_{21}} V_2 + \frac{1}{Z_L}(\frac{\mathrm{j}\omega L_{11} L_{22}}{L_{21}} - \mathrm{j}\omega L_{12}) V_2. \tag{3.18}$$

It will be noted, from the simple theory of Chapter 1, that $L_{12} = \sqrt{L_{11}L_{22}}$ and $L_{21} = L_{12}$ (a transformer satisfying this condition is said to be *strongly coupled*). Real transformers can have some deviation from strong coupling, but something fairly close to this can be achieved when the coils are wound over each other and, preferably, have a low-loss iron core (a ferrite or iron dust toroid is usually preferred) to increase the field that couples them (see Figure 3.5b). For such transformers, the last term of (3.18) disappears and we

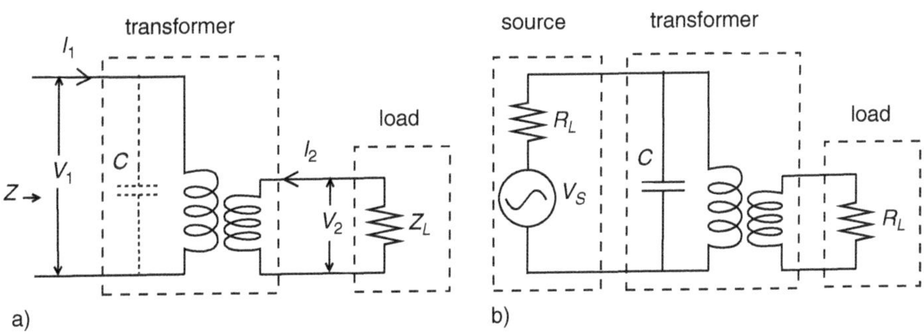

Fig. 3.6 The inductive transformer.

have $V_1 = V_2 L_{11}/L_{21}$ and so $V_1 = (N_1/N_2)V_2$, where N_1 is the number of turns on the primary of the transformer and N_2 is the number of turns on the secondary. Furthermore, if ωL_{22} is much larger than $|Z_L|$, (3.17) reduces to $I_1 = -(N_2/N_1)I_2$ and we have an *ideal transformer*. A significant increase in L_{22} can be attained by winding the transformer on a high-permeability core and is another reason why RF transformers are often wound on ferrite or iron dust toroids.

Assuming an ideal transformer, and dividing (3.18) by (3.17), we obtain that the impedance Z looking into the input of the transformer is given by

$$Z = \frac{j\omega L_{11}}{j\omega L_{22} + Z_L} Z_L. \tag{3.19}$$

Under the assumption that ωL_{22} is much larger than $|Z_L|$, we then obtain that $Z = (L_{11}/L_{22})Z_L$ (or $Z = (N_1^2/N_2^2)Z_L$ in terms of the turns ratio). If we cannot make L_{22} sufficiently large, we can obtain a similar effect using a shunt capacitor C (shown by broken lines in Figure 3.6a). The impedance looking into the transformer will now become

$$Z = \frac{j\omega L_{11} Z_L}{j\omega L_{22} + Z_L - \omega^2 L_{11} C Z_L} \tag{3.20}$$

and will have the desired properties at the resonant frequency ω_0 of C and L_{11} (i.e. when $\omega_0^2 L_{11} C = 1$). We now have a tuned transformer that also behaves as a band-pass filter around the resonant frequency. Consider the tuned transformer with a resistive source of impedance R_S and a resistive load of impedance R_L (see Figure 3.6b). For small deviations $\delta\omega$ about the resonant frequency ω_0, the impedance Z can be approximated by

$$Z \approx \frac{L_{11}}{L_{22}} \frac{R_L}{1 + 2j\delta\omega C R_L \frac{L_{11}}{L_{22}}}. \tag{3.21}$$

The series combination of R_S and Z will act as a voltage divider and, as a consequence, voltage V_1 will be related to voltage V_S through $V_1 = V_S Z/(Z + R_S)$. From this,

$$V_1 = V_S \frac{Z}{Z + R_S} \approx V_S \frac{1}{1 + \frac{R_S L_{22}}{R_L L_{11}} + 2j\delta\omega C R_S} \tag{3.22}$$

and, on noting that $V_1 = V_2 L_{11}/L_{21}$, the voltage across the load will be

$$V_2 \approx V_S \frac{1/n}{1 + \frac{R_S L_{22}}{R_L L_{11}} + 2j\delta\omega C R_S}, \tag{3.23}$$

where $n = L_{11}/L_{21} = N_1/N_2$ is known as the *turns ratio* of the transformer. For maximum power transfer between source and load, we need to choose our transformer such that $Z = R_S$ (i.e. $R_S = (L_{11}/L_{22})R_L$) and so the power dissipated in the load will be

$$P_L = \frac{V_2 \bar{V}_2}{2R_L} \approx \frac{V_S \bar{V}_S}{8R_L} \frac{1/n^2}{1 + (\delta\omega C R_S)^2} \tag{3.24}$$

It is clear from this that the half-power bandwidth B will be $2/CR_S$ or ω_0/Q where $Q = R_S/2\omega_0 L_{11}$.

3.5 The L Network

It is possible to match a source and load by means of the simple L network shown in Figure 3.7. Looking into the network from the source there will be an impedance

$$\begin{aligned} Z &= \mathrm{j}X_S + \mathrm{j}X_L \parallel R_L \\ &= \mathrm{j}X_S + \frac{\mathrm{j}X_L R_L}{R_L + jX_L} \\ &= \mathrm{j}X_S + \frac{X_L^2 R_L + \mathrm{j}X_L R_L^2}{X_L^2 + R_L^2}. \end{aligned} \tag{3.25}$$

Obviously, to match the source we need to choose

$$X_S = -\frac{X_L R_L^2}{X_L^2 + R_L^2} \text{ and } \frac{X_L^2 R_L}{X_L^2 + R_L^2} = R_S. \tag{3.26}$$

On defining $Q = |R_L/X_L|$, we can rewrite these equations as

$$X_S = -\frac{X_L Q^2}{1+Q^2} \text{ and } \frac{R_L}{1+Q^2} = R_S. \tag{3.27}$$

Rearranging the first of (3.27) we obtain that $Q = \sqrt{R_L/R_S - 1}$ (note that we have assumed $R_L > R_S$ otherwise the network would need to be reversed). Consequently, from R_S and R_L we can calculate Q and then we will have that $X_L = \pm R_L/Q$. The choice of sign is arbitrary, but we will see this choice can be useful. Finally, we can calculate X_S using the relation $X_S = -X_L Q^2/(1+Q^2)$. If we choose X_L to be negative then this will be realised by a capacitor C and X_S will be realised by an inductor L. Obviously, we choose the values of C and L to give the correct X_L and X_S at the desired operating frequency ω_0. This implies that $C = Q/\omega_0 R_L$ and $L = (R_L/\omega_0)Q/(1+Q^2)$.

As with other LC networks, the L network will also have filtering properties. To see this, consider the voltage V_L across the load in terms of the source voltage V_S. Once again we can treat the problem as one of a voltage divider, from which

$$\begin{aligned} V_L &= V_S \frac{\mathrm{j}X_L \parallel R_L}{R_S + \mathrm{j}X_S + \mathrm{j}X_L \parallel R_L} \\ &= V_S \frac{\mathrm{j}X_L R_L}{R_S R_L - X_S X_L + \mathrm{j}(X_S R_L + X_L R_S + X_L R_L)} \end{aligned} \tag{3.28}$$

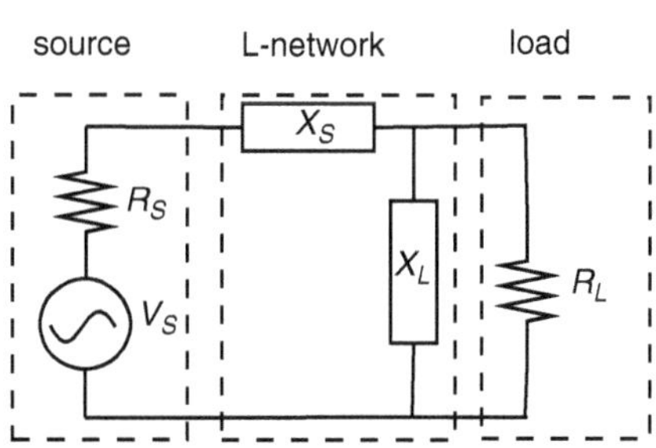

Fig. 3.7 L-network transformer.

and, noting that $X_L = -1/\omega C$ and $X_S = L\omega$,

$$V_L = V_S \frac{R_L}{\mathrm{j}\omega C(R_S R_L + L/C) - CL\omega^2 R_L + R_S + R_L}. \tag{3.29}$$

We further note that $L/C = R_S R_L$ and $LC = Q^2 R_S/\omega_0^2 R_L$, so that

$$V_L = V_S \frac{R_L}{2\mathrm{j}QR_S \frac{\omega}{\omega_0} - Q^2 R_S \frac{\omega^2}{\omega_0^2} + R_S + R_L}. \tag{3.30}$$

The power dissipated in the load will now be given by

$$P_L = \frac{V_S \bar{V}_S}{2R_L} \left(\frac{R_L}{R_L + R_S} \right)^2 \frac{1}{\frac{4Q^2}{(Q^2+2)^2} \frac{\omega^2}{\omega_0^2} + \left(1 - \frac{Q^2}{Q^2+2} \frac{\omega^2}{\omega_0^2} \right)^2} \tag{3.31}$$

on noting that $(R_L + R_S)/R_S = Q^2 + 2$.

Figure 3.8 illustrates the variation of the insertion gain with frequency. For high values of Q, the filter has a band-pass characteristic. For small Q, however, it behaves more like a *low-pass filter* (i.e. it passes all signals at frequencies below the design frequency ω_0 and blocks those above this frequency).

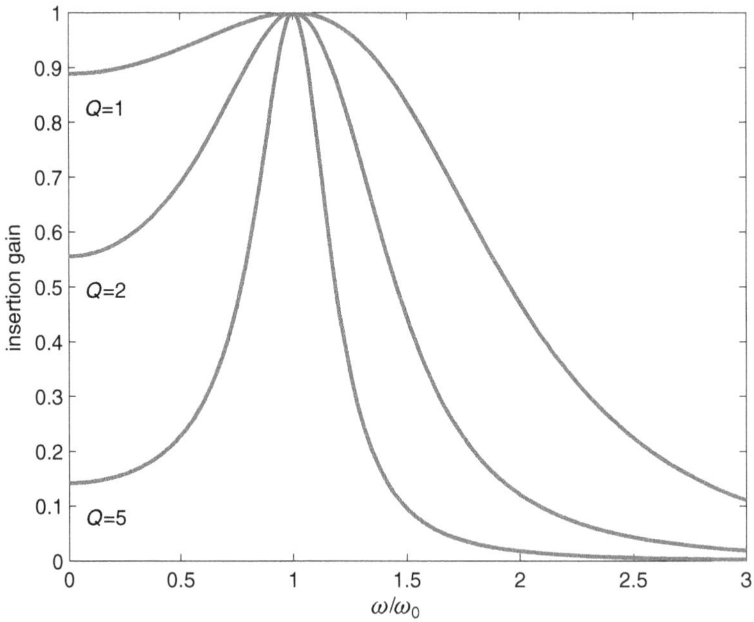

Fig. 3.8 The filtering characteristics of the L network.

3.6 Capacitive Transformer

If a frequency-selective transformer is acceptable, the capacitive transformer of Figure 3.9a is an alternative. This can be regarded as a special case of the general three-element network shown in Figure 3.9b. Looking into the circuit consisting of Z_1,Z_2 and Z_3 we see the impedance

$$Z = Z_1 \parallel (Z_2 + Z_3) = \frac{Z_1(Z_2 + Z_3)}{Z_1 + Z_2 + Z_3}. \tag{3.32}$$

We can analyse the relationship between V_S and V_L by treating the circuit as a pair of voltage dividers. Firstly

$$V_1 = V_S \frac{Z_1 \parallel (Z_2 + Z_3)}{R_S + Z_1 \parallel (Z_2 + Z_3)} = V_S \frac{Z_1(Z_2 + Z_3)}{R_S(Z_1 + Z_2 + Z_3) + Z_1(Z_2 + Z_3)} \tag{3.33}$$

and then

$$V_L = V_1 \frac{Z_3}{Z_2 + Z_3} = V_S \frac{Z_1 Z_3}{R_S(Z_1 + Z_2 + Z_3) + Z_1(Z_2 + Z_3)}. \tag{3.34}$$

Rearranging (3.34), we obtain

$$V_L = V_S \frac{Z_3}{Z_2 + Z_3} \frac{1}{R_S(\frac{1}{Z_2+Z_3} + \frac{1}{Z_1}) + 1}. \tag{3.35}$$

In the case of the capacitive transformer, we will have that $Z_1 = \mathrm{j}\omega L$, $Z_2 = 1/\mathrm{j}\omega C_1$ and $Z_3 = (1/\mathrm{j}\omega C_2)//R_L$. If we make the simplification that $R_L \gg 1/\omega C_2$, we then obtain that

$$V_L \approx V_S \frac{C_1}{C_1 + C_2} \frac{1}{\mathrm{j}\frac{R_S}{\omega L}(\omega^2 LC - 1) + 1}, \tag{3.36}$$

where $C = C_1 C_2/(C_1 + C_2)$. Maximum power transfer will occur around the frequency $\omega_0 = 1/\sqrt{LC}$ and, for small deviations in frequency $\delta\omega$ around this resonance,

$$V_L \approx V_S \frac{1/n}{1 + \mathrm{j}2\frac{R_S}{\omega_0 L}\frac{\delta\omega}{\omega_0}}, \tag{3.37}$$

where $n = (C_1 + C_2)/C_1$, i.e. we effectively have a transformer with turns ratio n.

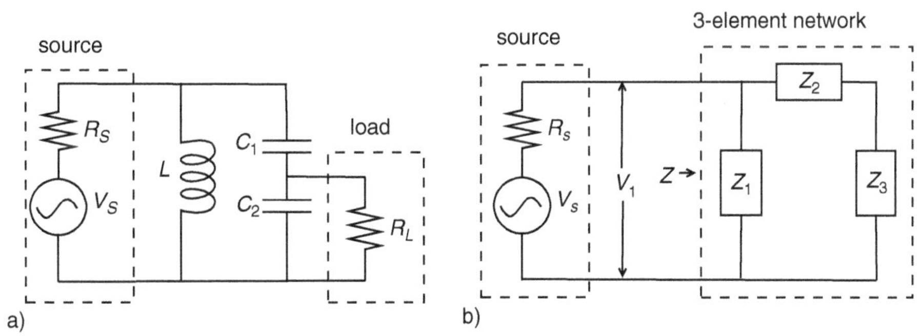

Fig. 3.9 A general three-element network.

3.7 Filters

It is clear that tuned circuits can act as filters, the Q factor measuring how well they perform this function. We now look at how we can improve this filter function by increasing the complexity of the circuit. Consider the circuits shown in Figure 3.10. These can be viewed as a special case of the general three-element network shown in Figure 3.9b. For the filters of Figure 3.10, we will look at the situation where the reactances of C and L are related to the resistance R of the source and load. We will assume that $L = \alpha R/\omega_0$ and $C = \beta/R\omega_0$ where α and β are arbitrary constants and ω_0 is some reference frequency. As a consequence, we can treat the filter as the circuit in Figure 3.9b with $Z_1 = -\mathrm{j}\omega_0 R/\beta\omega$, $Z_2 = \mathrm{j}\alpha R\omega/\omega_0$, $Z_3 = \mathrm{j}\beta\omega_0 R/(\mathrm{j}\beta\omega_0 - \omega\beta^2)$ and $Z_L = Z_S = R$. Then, using (3.34), we can calculate the power dissipated in the load (i.e. $V_L^2/2R_L$) as a function of frequency. Figure 3.11a shows the insertion loss for the case of $\beta = 1$ and $\alpha = 2$, parameter values that define what is known as a *Butterworth filter*. It will be noted that the power falls away very smoothly with a half-power bandwidth ω_0.

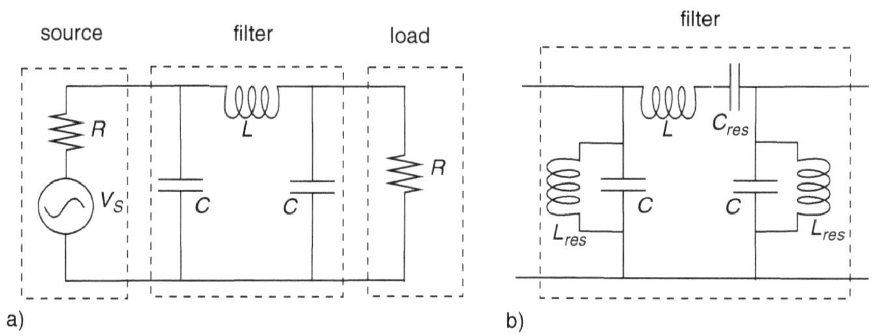

Fig. 3.10 A three-element low-pass filter and its conversion to a band-pass filter.

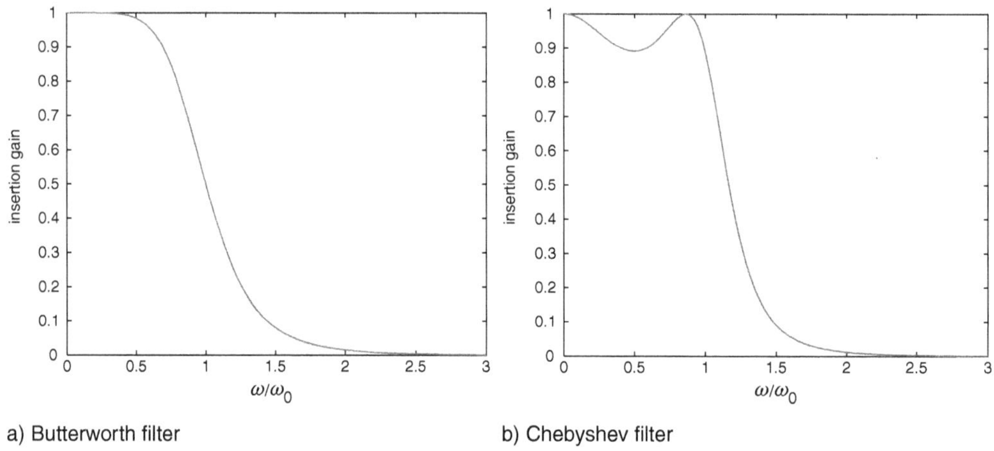

Fig. 3.11 Performance of a three-element low-pass filter.

If we can tolerate some *ripple* in the pass band of the filter, we could use the parameters of a *Chebyshev filter*. Such a filter, with a 0.122 ripple, is obtained when we use the parameters $\beta = 1.5963$ and $\alpha = 1.0967$. Figure 3.11b illustrates the performance of this filter and from which it is now clear that there is a much sharper out-of-band response, providing one can tolerate the ripple.

The above low-pass filter can be turned into a band-pass filter by resonating the capacitors and inductors at the centre frequency ω_C of the desired pass band ($\omega_C = \sqrt{\omega_L \omega_U}$ where ω_L and ω_U and the lower and upper half-power edges of the pass band). The capacitors are parallel-resonated with suitable inductors and the inductors are series-resonated with suitable capacitors. At frequency ω_C, the parallel combinations will act as open circuits and the series combination as a short circuit (i.e. the filter acts as a straight through connector between source and load). Consequently, we will have a band-pass filter with maximum power transferred at frequency ω_C (the half-power pass band will be that of the low-pass filter). Figures 3.12a and 3.12b show the performance of the band-pass filters that are based on the low-pass filters of Figures 3.11a and 3.11b. For this simulation, the centre frequency is twice the bandwidth of the filter ($\omega_C = 2(\omega_U - \omega_L)$) and once again it will be noted that the Chebyshev filter has a much sharper response (at the price of some pass-band ripple).

Better performance could be achieved by cascading several of the above filters, but modern computer techniques allow us to optimise the components in such arrangements and achieve a much better performance than a simple cascade. Figure 3.13a shows the circuit of a five-element low-pass Chebyshev filter with 0.122 pass-band ripple. For this filter $L = 1.2296R/\omega_0$, $C_1 = 1.7058/\omega_0 R$ and $C_2 = 2.5408/\omega_0 R$ (R is the impedance of the filter terminations). The improved performance with the extra elements can be seen in Figure 3.13b and it will be noted that the filter sides are now much sharper, but at the cost of extra ripples. With increasing complexity, however, the losses in realistic components becomes a problem and real filters will exhibit power loss as a result of this

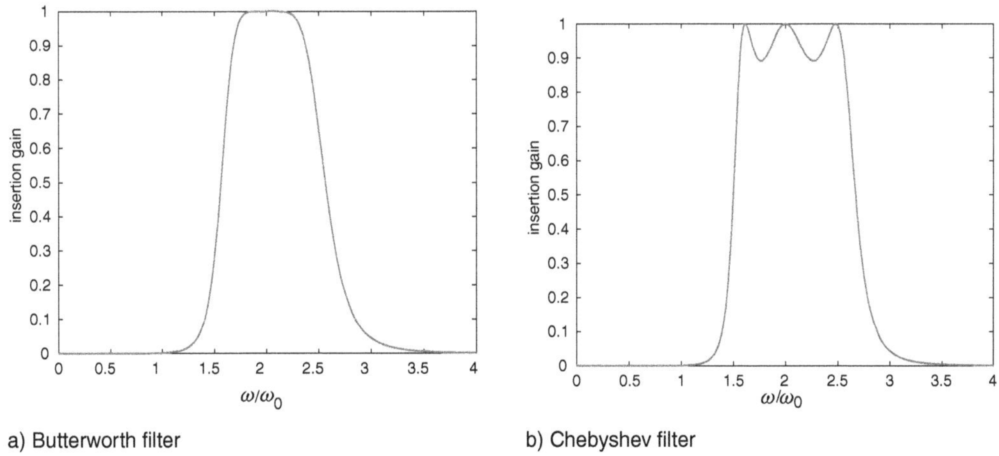

Fig. 3.12 Performance of six-element band-pass filters.

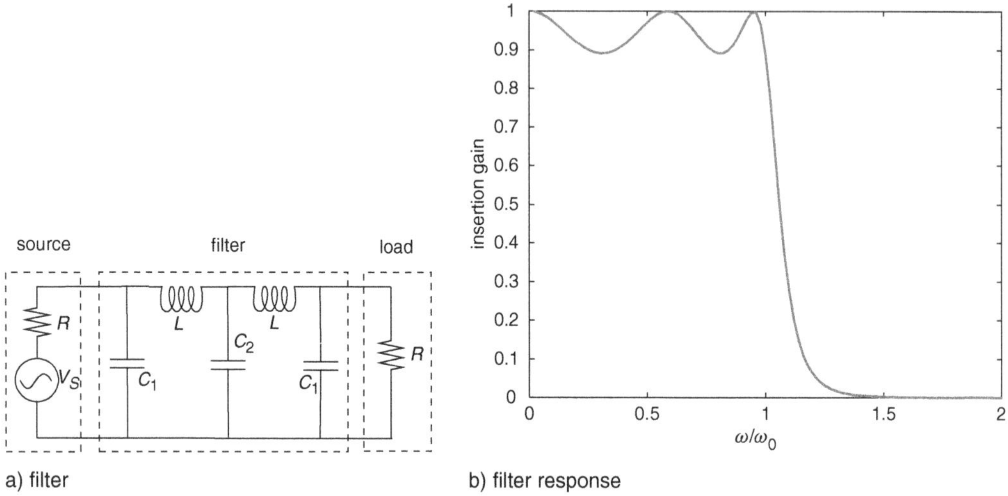

Fig. 3.13 A five-element low-pass Chebyshev filter with 0.122 pass-band ripple.

(i.e. the peak gain is less than 1). The insertion gain G is often expressed in terms of a logarithmic scale in which the value of the gain is given by $10\log_{10}G$ with units of *decibels* (or dB for short). If there is a loss, this will result in a negative value in the dB scales and the magnitude of this number is often termed the *insertion loss*.

3.8 Conclusion

In the current chapter we have developed the techniques that are required for high-fidelity frequency selection, a requirement of modern radio. However, whilst good filtering can help improve the performance of a radio system, the receivers in early radio systems had very low sensitivity and this greatly hindered the development of the technology. The situation changed in 1912 with the development of the first valves that could provide useful amplification and from this point on radio technology greatly accelerated in its progress. In the next chapter we will describe active devices, transistors in particular, and the techniques for their effective use as amplifiers.

4 Amplification

The ability to amplify weak signals was an important milestone in the development of radio technology. This was first achieved around 1912 using a triode thermionic valve and such devices were steadily refined over the next four decades. Around 1947, however, a revolution happened with the development of the transistor and this led to an explosion of semiconductor innovations that eventually resulted in the integrated circuit. This chapter discusses the development of both semiconductor and valve technology. In particular, the development of RF amplification and its impact upon the development of radio. Practical amplifiers suffer from a host of problems that include noise, nonlinearity and parasitic capacitance. The chapter discusses the impact of these factors upon radio performance and, in particular, the ways in which their detrimental effects can be mitigated.

4.1 Thermionic Valves

The first experiments on thermionic valves were essentially carried out by Thomas Edison around 1883 in an attempt to improve the performance of incandescent electric light bulbs. He placed an electrode inside the bulb and found that a current flowed when this was made positive, but no current flowed when it was negative. Edison had created what was essentially a diode. Unfortunately, Edison was not interested in carrying this work forward and it was left to a British physicist, John Ambrose Flemming, to develop Edison's work into something that was much needed in radio, a reliable diode. In the first half of the 1890s, Flemming published several papers on what had become known as the *Edison effect*. We now know that the heating of the filament (the cathode) in the bulb causes electrons to be ejected and these will then be attracted to, or be repelled by, the inserted electrode (the *anode* or plate). In 1904, Flemming patented the first diode valve and the history of radio was changed forever. Figure 4.1a shows the configuration of a typical early diode in conduction. The use of a separate electrode, the *cathode*, to supply the electrons soon became common practice (see Figure 4.1b). The filament now simply heated the cathode to a point where it could release electrons. Such an arrangement meant that the filament supply did not interfere with the electron flow and could be in common with other valves in the circuit. In conduction, the relationship between the current I_P flowing through the valve and the cathode to anode voltage V_P is given by the *Child–Langmuir* law $I_P = \kappa V_P^{3/2}$ where κ is a constant known as the *perveance* of the valve.

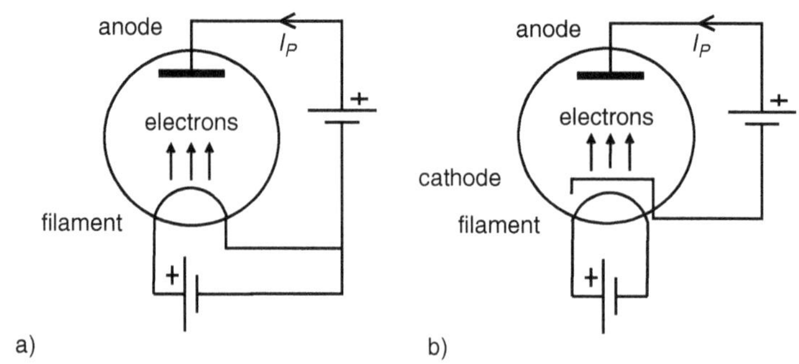

Fig. 4.1 Diode thermionic valve (vacuum tube).

The next major development in the story of thermionic valves (or *vacuum tubes* as they are sometimes known due to the vacuum inside the bulb) was to place an extra electrode (called a *grid*) between the cathode and anode. The grid is a coarse mesh so that it does not impede the flow of electrons from cathode to anode, but is sufficiently close to the cathode for its field to have considerable control over the emission of electrons (a negative grid voltage decreases the emission of electrons, whilst a positive voltage increases the emission). Known as a triode valve, such a device is capable of amplification. A device of this nature was developed by Lee De Forest in 1906 (he called his device an audion), but it took many years before such devices were refined to a point where they could be of much use to radio. By about 1912, however, effective triode valves were available and such valves drove the next stage in the development of radio. Figure 4.2a illustrates the triode (notice we have omitted the filament since it does nothing more than heat the cathode) and 4.2b shows a typical set of characteristic curves. These curves show how the anode current I_P varies with grid voltage V_G and anode voltage V_P. If we insert a load R_L into the anode circuit (see Figure 4.3a), the anode voltage V_P will vary with grid voltage V_G according to the points on the *load line* of Figure 4.3a. For a given grid voltage, the intersection of the corresponding characteristic curve with the load line will provide the anode current I_P and anode voltage V_P. If the grid is negative (the normal mode of operation), the current flow is well approximated by

$$I_P = \kappa(V_P + \mu V_G)^{\frac{3}{2}}, \tag{4.1}$$

where κ is the perveance mentioned above and μ is the *amplification factor*. As a consequence, the voltage drop across the load R_L will be

$$V_L = R_L\kappa(V_P + \mu V_G)^{\frac{3}{2}}. \tag{4.2}$$

The valve will perform voltage amplification, but it will be noted that the amplification can be highly nonlinear. We will return to this issue when we study transistors.

Figure 4.4 is a simple radio that is based on a triode valve and employs what is commonly known as a grid leak detector. The tuned circuit is coupled to the grid through the capacitor C_2 and the grid is biased to the same voltage as the cathode through a very high impedance resistor R. Consequently, the grid will only conduct on positive voltage swings at the output of the tuned circuit. During such periods, the *grid leak capacitor* C_2

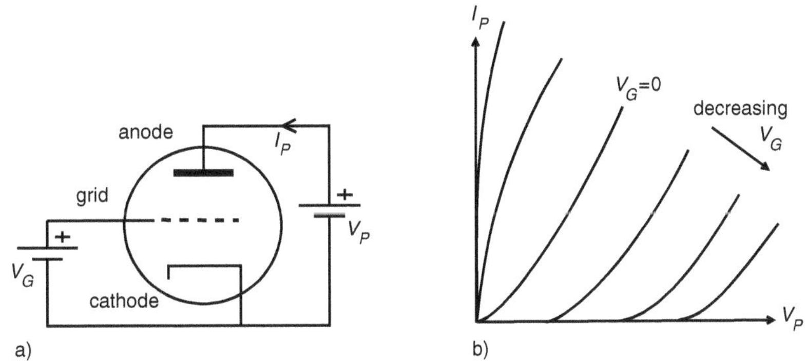

Fig. 4.2 Triode thermionic valve (vacuum tube).

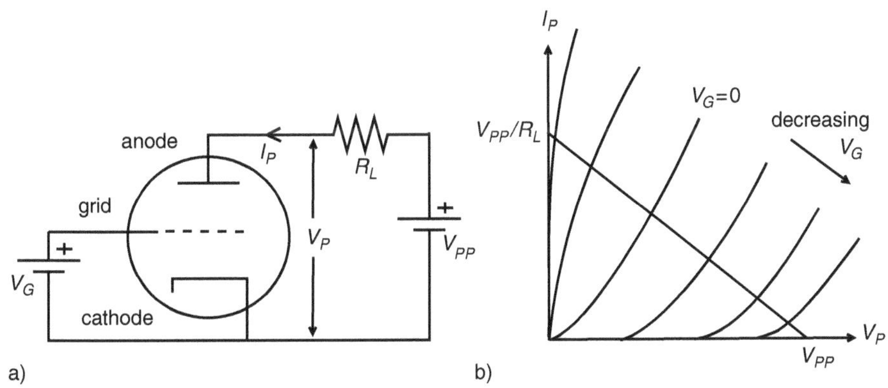

Fig. 4.3 Triode thermionic valve with load.

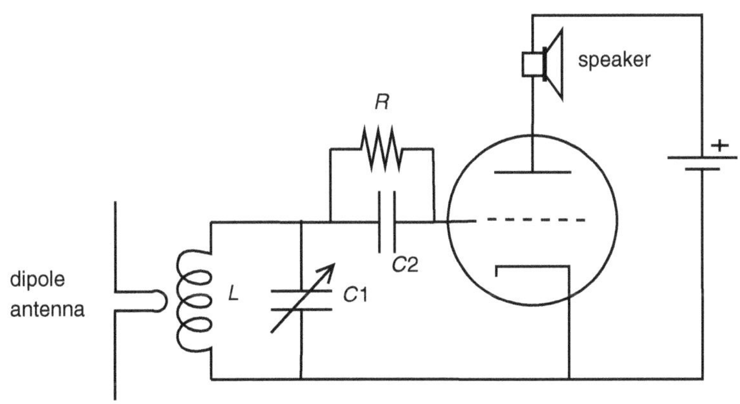

Fig. 4.4 Radio based on a grid leak detector.

will charge up and then maintain a negative bias on the grid during negative swings. The capacitor $C2$ needs to be chosen large enough to smooth out the RF component at the grid, but not so large that it smooths out the modulation. The voltage at the grid will therefore follow the amplitude modulation of the signal and this will be amplified through the valve.

4.2 Semiconductor Diodes

The valve diode was a reliable replacement for the point contact diodes developed by Bose and others, but its heavy power requirement was a severe drawback. However, in 1940, Russell Ohl discovered the semiconductor pn junction, and this led to the development of highly reliable and effective diodes, without the heavy power consumption of valves. To understand the pn junction we first need to understand what is meant by a semiconductor. A crystal form of many materials results when the atoms are bound together by sharing the electrons in their outermost electron shell (this is often known as a covalent bond). Figure 4.5 depicts a crystal solid consisting of atoms with four electrons in their outermost shell. If the solid consisted of carbon atoms, the outermost shell would be an L shell with four electrons (this solid is in fact diamond). Furthermore, once it has shared its electrons with its neighbours, its L shell would be full. As a consequence, there is no option for electrons to move to higher energy levels (they would need to bridge the large energy gap between the L and M shells) and hence gain the kinetic energy of motion, i.e. the material is an insulator. If the crystal consists of silicon atoms, the outermost shell is an M shell containing four electrons. In this case the outer shell is not full and so electrons can more easily enter higher energy levels for which the additional energy is now the kinetic energy of motion. At room temperature, several of the electrons of silicon will have achieved these higher states and will be available for conduction when an electric field is applied. Such a material is known as a *semiconductor*. Another example of a semiconductor material is germanium, an atom for which the outermost shell is an N shell containing four electrons.

The conductivity of semiconductors can be improved by the addition of some impurity into the solid. Consider adding an impurity consisting of phosphorous into a crystal of silicon. The outermost shell of a phosphorous atom is an M shell containing five electrons. When some phosphorous atoms are added to the silicon crystal (i.e. it is *doped*), four of the five phosphorous electrons will be shared with neighbouring silicon atoms (see Figure 4.6). The fifth electron, however, is not bound by the sharing mechanism and can easily move into the available energy levels of the M shell, empty in the case of silicon. Consequently, very little electric field will be required to move an electron through the

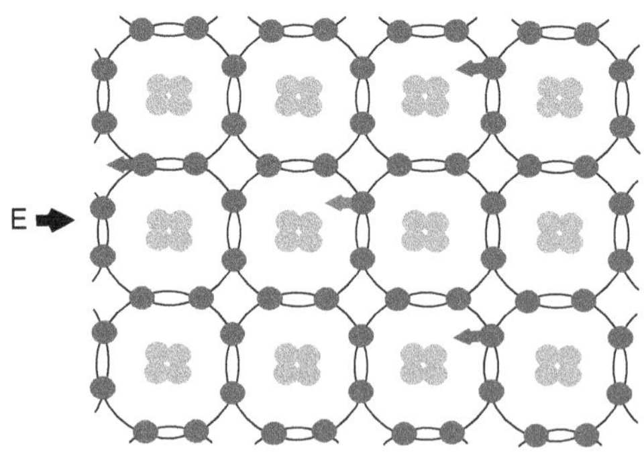

Fig. 4.5 In a semiconductor crystal, some electrons can attain energy levels that allow them to move.

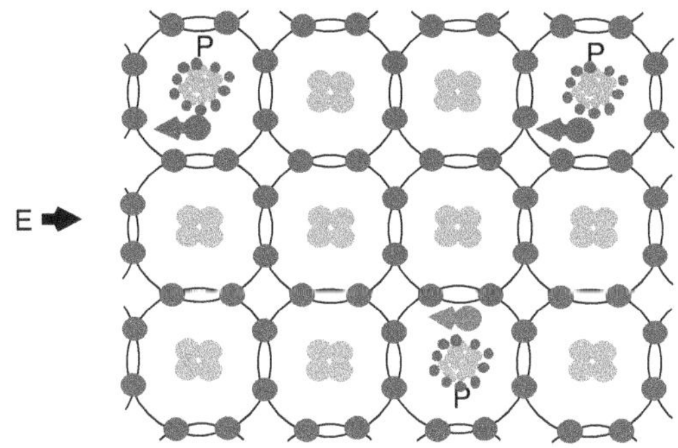

Fig. 4.6 Flow of charge through an n-type semiconductor by movement of electrons.

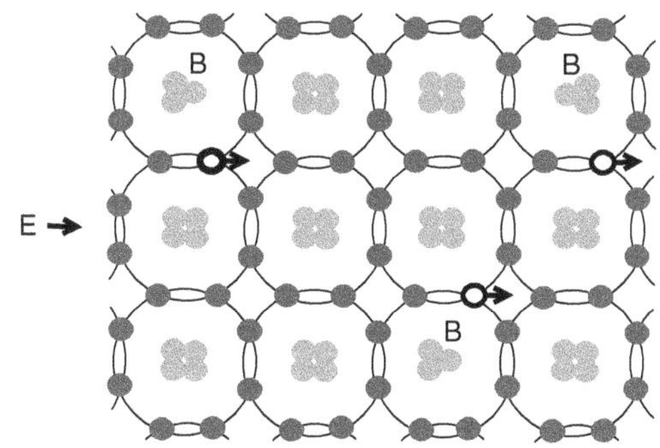

Fig. 4.7 The flow of charge through a p-type semiconductor by movement of holes.

crystal. Now consider adding an impurity consisting of boron to a crystal of silicon. The outermost shell of a boron atom is an L shell containing three electrons. When some boron atoms are added to the silicon crystal, the three electrons will be shared with neighbouring silicon atoms (see Figure 4.7). There will, however, be an electron hole in the L shell of one of the neighbouring silicon atoms. Very little electric field will be required to move an electron of a neighbouring atom into the hole, hence creating a hole in the neighbouring atom. This hole will then be filled by an electron from another neighbouring atom and so on through the crystal. In this way, there will be an effective flow of positive charge through the crystal. In deference to the type of charge carrier, the phosphorous-doped material is known as an n-type semiconductor and the boron-doped material as a p-type semiconductor.

It is possible to grow a semiconductor crystal in which one half is p-type and the other half is n-type. When there is no field across the junction, there is still some charge flow by the process of *diffusion*. Under this process, thermal agitation will cause carriers to flow from a region of strong concentration to a region of low concentration. Electrons will flow from the n-type semiconductor to the p-type where they fill holes and holes will flow from the p-type semiconductor to the n-type. This produces an accumulation of positive charge in the n-type region and negative charge in the p-type region. As a result there is a *depletion region* either side of the junction that is devoid of carriers. The accumulation

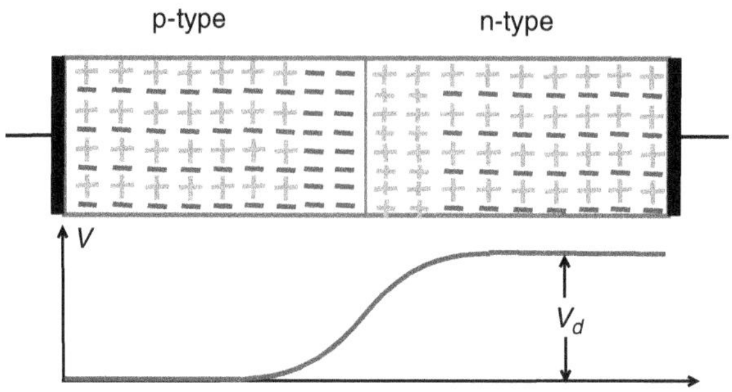

Fig. 4.8 A p-n junction at equilibrium.

of charge on either side of the junction will lead to an electric field that will increasingly oppose the migration of further carriers across the junction until the diffusion is reduced to a trickle. It should be noted, however, that thermal agitation will always produce some minority carriers (holes in the n-type semiconductor and electrons in the p-type) and these will be swept across the junction by this field and form what is known as the *drift current*. Eventually, the drift current and residual diffusion current will balance each other and we will have a state of equilibrium. In this state, the potential difference across the junction will vary as shown in Figure 4.8. The equilibrium potential difference V_d across the junction is known as the diffusion potential and has a value around 0.3 volts for germanium and around 0.8 volts for silicon.

If a voltage is now applied across the diode with the n-type end positive, known as *reverse bias*, it will further impede the diffusion and enhance the drift current. As a consequence, only a small amount of current will flow from n-type to p-type semiconductor (see Figure 4.9b). However, if a voltage is now applied with the p-type end positive, known as *forward bias*, the voltage drop across the depletion layer will be reduced and the diffusion current significantly increased (see Figure 4.9c). The current I through the diode is related to the potential drop across the device through the relation

$$I = I_s\left(\exp\left(\frac{eV}{\eta kT}\right) - 1\right), \tag{4.3}$$

where I_s is the reverse saturation current, e is the charge on an electron, V is the voltage across the device, k is Boltzmann's constant, T is the absolute temperature and η is a constant that can have a value between 1 and 2.

The original semiconductor diodes were of the point-contact variety. In such devices, an n-type semiconductor is fused to a metal contact (often aluminium) to form the cathode and then the anode is formed by contact with a metal wire (usually tungsten) having a very fine point. Some of the metal ions from the point then migrate into the semiconductor and form a small p-type region. As mentioned above, an alternative process is to grow a semiconductor crystal with the desired regions of semiconductor type. The crystal is steadily pulled out of a semiconductor melt to which is added, at appropriate stages, suitable amounts of impurity in order to attain the desired pn structure.

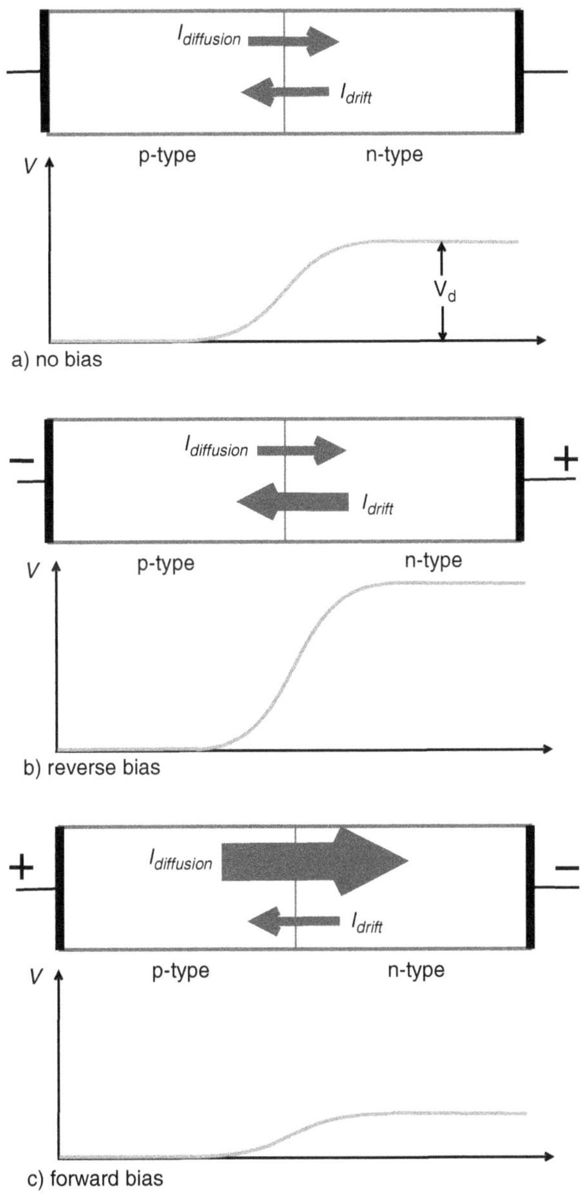

Fig. 4.9 Voltage drop and currents across a pn junction.

4.3 The Bipolar Junction Transistor

The next big step forward was made in 1947 with the invention of the *bipolar junction transistor* (or BJT for short) by the American physicists John Bardeen, Walter Brattain and William Shockley of AT&T's Bell Laboratories. They took the crucial step of adding an additional p-type layer resulting in a germanium device with two junctions. In fact, in the first device invented by Bardeen and Brattain, the p-type regions were created using

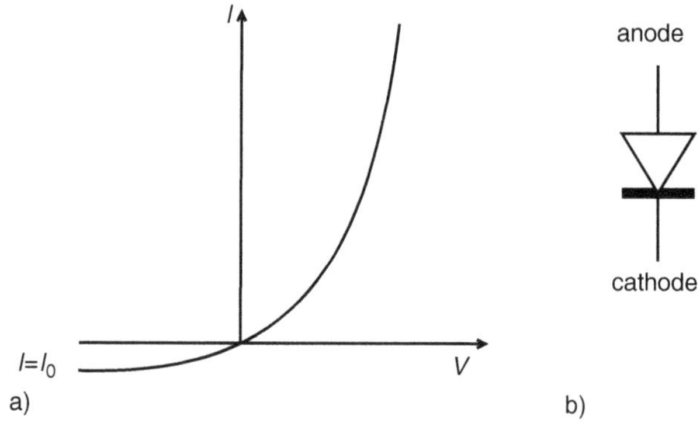

Fig. 4.10 Semiconductor diode characteristic and diode symbol.

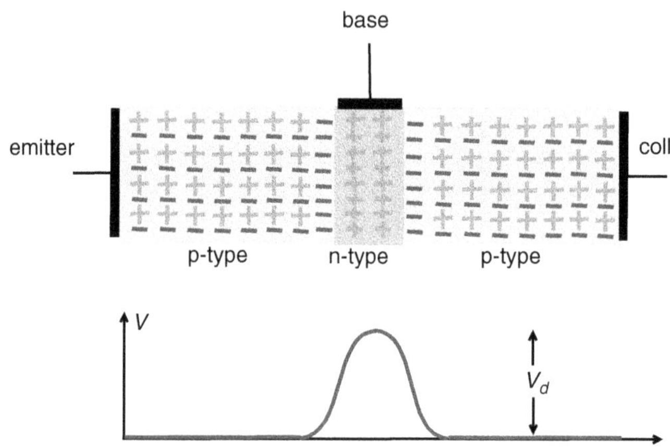

Fig. 4.11 The bipolar junction transistor in an unbiased state.

point contacts with an n-type semiconductor. Shockley, however, was able to improve on this by growing a crystal with an appropriate pnp sandwich. Figure 4.11 shows a pnp transistor geometry in its unbiased state and from which it will be noted that the depletion of carriers from the base area has caused a potential barrier to any further diffusion. The transistor looks like two back-to-back pn junctions, but this analogy is far too simplistic. Crucial to the transistor's operation is the very close proximity of the two p-type regions (not a property of two diodes when connected back-to-back) and a very much lower density of impurity in the n-type region. We now consider what happens when the transistor base and collector are biased by negative voltages with respect to the emitter (see Figure 4.12). It is assumed that the bias voltages are sufficient to make the base forward-biased with respect to the emitter and the collector reverse-biased with respect to the base. Positive carriers will now flood into the base, but will find few electrons to combine with due to the small density of impurity. However, they will feel a strong pull from the collector due to the reverse bias and will readily flow into this region. They will do this in preference to flowing out through the base, due to the smaller distance to be covered. Consequently, a current I_C will flow into the collector which is

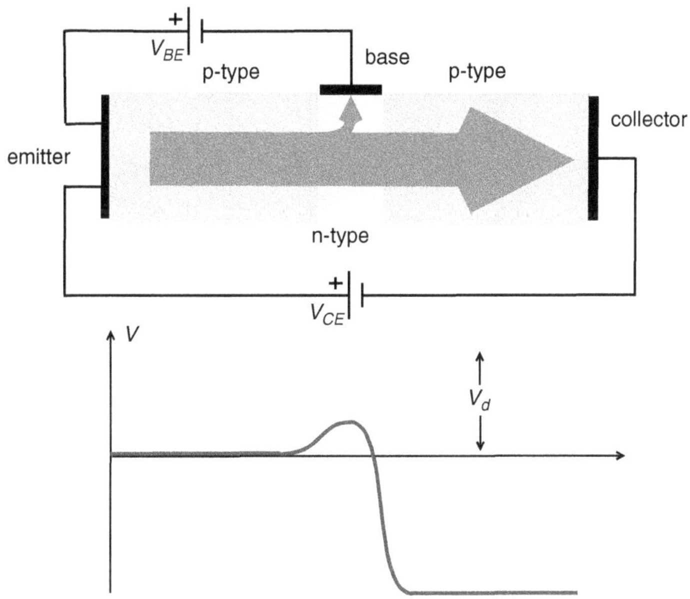

Fig. 4.12 A pnp bipolar junction transistor in a biased state.

β times the current I_B that flows out through the base, i.e. $I_C = \beta I_B$. The quantity β is known as the *current gain* and can have values of the order of 100. Essentially, the BJT is a current amplifying device. It should be noted, however, that the input current is related to the voltage that is imposed between the emitter and the base. The base–emitter junction is essentially a forward biased diode and so the current that flows out of the base is related to the base–emitter voltage V_{BE} through $I_{BE} = I_s\left(\exp\left(\frac{eV_{BE}}{\eta kT}\right) - 1\right)$. The current I_s is very small and so we will have that

$$I_C \approx \beta I_s \exp\left(\frac{V_{BE}}{V_T}\right), \tag{4.4}$$

where $V_T = \frac{kT}{e}$ (approximately 25 mV) and we have taken η to be 1.

It is clear that there needs to be a reasonable bias at the collector in order to maintain a reverse bias on the base–collector junction. Figure 4.13a shows characteristic curves (the relationship between collector current and collector voltage) for a variety of base currents. These show that the characteristics become highly nonlinear for low collector voltages. In reality, the transistor will be used to drive a load and this will affect the voltage at the collector of the transistor. Figure 4.13b illustrates the situation through what is known as the *load line*. For a given base current, the intersection of the corresponding characteristic curve and the load line will provide the collector–emitter voltage and the collector current. As the base voltage varies it is clear that the voltage across the load will vary in sympathy and at a much greater amplitude. This property is put to use in the simple receiver circuit of Figure 4.13c. The input signal is demodulated by the diode consisting of the emitter to base junction and then the demodulated signal is amplified in the transistor (the capacitor $C2$ filters out any residual RF frequencies).

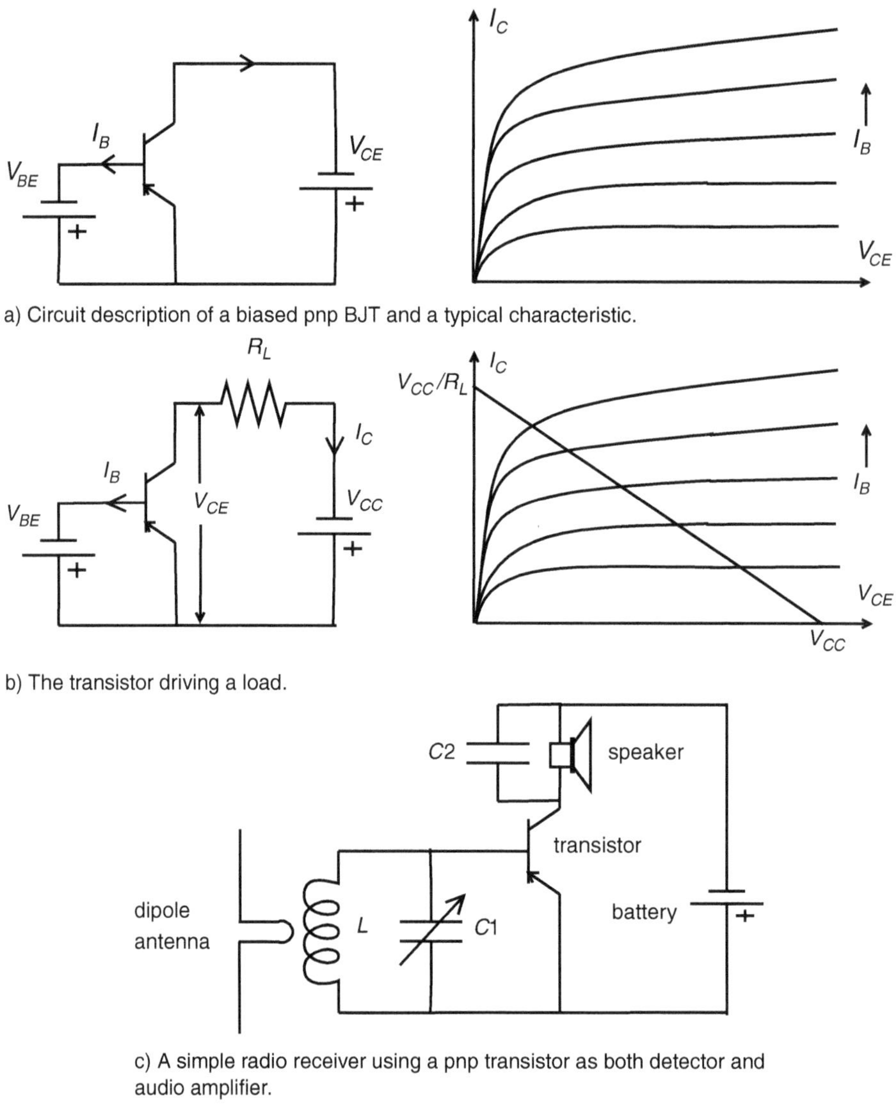

Fig. 4.13 Application of pnp bipolar junction transistor to amplification.

The problem with the amplifier of circuit of Figure 4.13b is that it is highly nonlinear. From (4.4), the relationship between the input voltage V_{BE} and the voltage drop V_L across the load is given by

$$V_L \approx R_L \beta I_s \exp\left(\frac{V_{BE}}{V_T}\right). \tag{4.5}$$

As a consequence, a sinusoidal base voltage V_{BE} would result in a voltage drop V_L with the lower half of the sinusoid missing and the remaining half heavily distorted. In the above receiver, this is put to good use in demodulating the incoming signal. On the whole, however, we would like to amplify without signal distortion. Even without a V_{BE} that goes negative, the output can still become distorted. As the output current rises, the value of V_{CE} can fall and there is the possibility that the lower half of the sinusoid will stray into the highly nonlinear region on the left of the characteristic. In general, an amplifier needs to be designed so that the output voltage does not stray into these strongly nonlinear regions. The solution is to create a DC bias at the transistor base so that the input voltage does not stray into negative territory and to choose the load so that the V_{CE} does not stray into the nonlinear region of small V_{CE}. For appropriate choices of bias and load impedance, it is possible to get variations in voltage drop across the load that are very nearly a constant multiple of quite large variations in the base voltage (see Figure 4.14). Obviously, to choose a suitable bias point, one would need to study the manufacturer's characterisation of the transistor.

The pnp geometry is not the only combination of semiconductors that produces a transistor device and an alternative is the npn combination (see Figure 4.15). The behaviour of the npn device is much the same as a pnp device, except that we must change the polarity of the biases. Figure 4.16a shows an amplifier circuit that uses an npn transistor. It will be noted that we have derived the base bias from the collector supply V_{CC} with a suitable voltage divider (this is a common arrangement). It will be further noted that there is also an emitter resistor that has a bypass capacitor C_{BP} (the value of this capacitor is sufficiently high for it to behave as a short circuit at signal frequencies). This resistor might seem unnecessary, but it provides feedback that stabilises the operation of the transistor. The DC bias voltage V_{BE} is chosen such that the *quiescent current* I_Q through the transistor (the current that flows when there is no signal) places the variations in output voltage v_o well away from the regions of nonlinearity. Once the appropriate quiescent collector current I_Q has been ascertained, the emitter resistor R_E needs to be chosen to give an emitter voltage V_E that is the maximum consistent with the operation of the transistor (it must allow a sufficient swing in the output voltage). The voltage divider (R_1 and R_2) needs to be chosen to give a voltage $V_{BE} + V_E$ at the base and, in

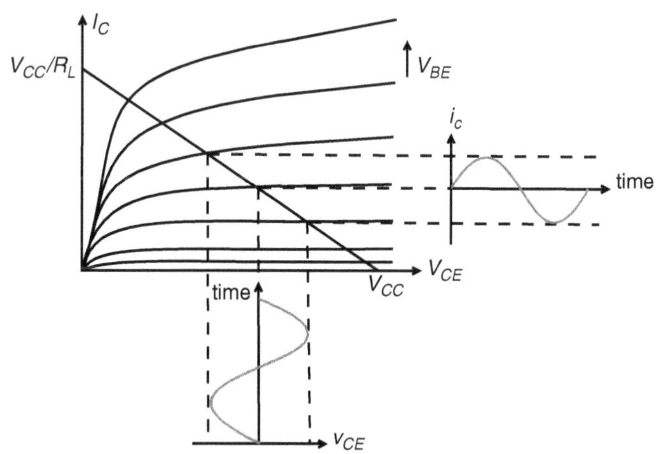

Fig. 4.14 Input/output relationship for a biased amplifier.

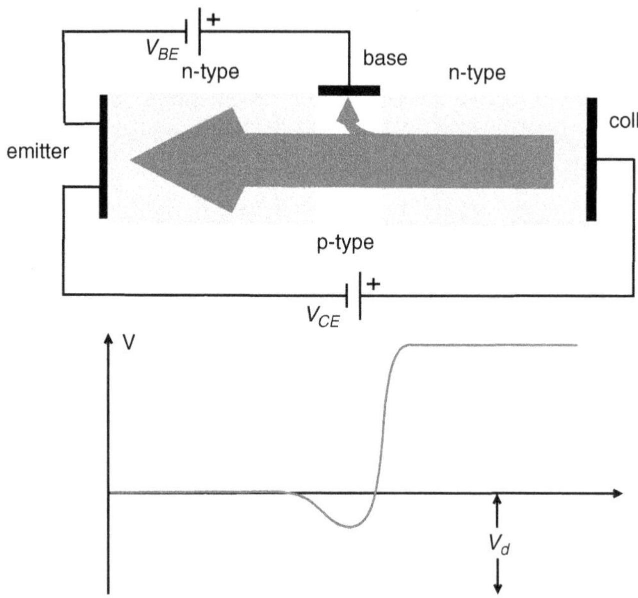

Fig. 4.15 A npn bipolar junction transistor in a biased state.

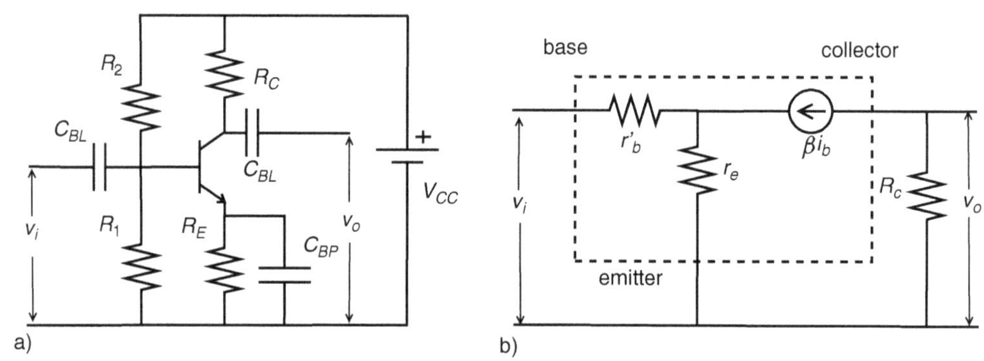

Fig. 4.16 A npn bipolar junction transistor in a biased state and a simple small-signal model.

most circumstance, V_{BE} is sufficiently approximated by the diffusion potential V_d (0.3 volts for germanium transistors and 0.8 volts for silicon). Divider resistances need to be chosen large enough that they do not overload the source (i.e. have a combined resistance well above that of the source), but small enough that they themselves are not overloaded by the transistor.

We now turn to the reason for the emitter resistance R_E. From (4.5) it can be seen that the current gain β, and the temperature T, have a big impact on the quiescent current once V_{BE} has been set. Obviously, temperature will vary considerably, even under normal conditions. Furthermore, there can be a considerable spread of β in manufactured devices. If current rises due to a change in temperature, the voltage across R_E will rise. As a consequence, the voltage V_{BE} will fall and, along with it, the current. Consequently, R_E provides feedback that stabilises a device against fluctuations in temperature. Further, this feedback also helps to stabilise against fluctuations in β. Larger values of R_E provide better feedback, but the value will need to be moderated by the other requirements of the design.

Once we have fixed the DC components of current through the bias circuits, we can ignore this aspect and deal with the RF signal alone, assumed to be a small fluctuation about the quiescent state. For such an analysis, the bypass C_{BP} and blocking capacitors C_{BL} are replaced by short circuits. (The purpose of the blocking capacitors is to prevent the bias of one stage of a more complex circuit affecting that of another stage and their values need to be chosen to be large enough to behave as a short circuit at signal frequencies.) Further, the DC supply is treated as a short circuit (note that we normally place a bypass capacitor across the supply in order to ensure this). The circuit in Figure 4.16a will now reduce to that shown in Figure 4.16b. Note that the transistor has been replaced by a linear model (the circuit within the broken rectangle) and this is valid for *small signals*. In this model, the transistor is effectively a current source with the current proportional to the base current. Resistance r_b' is known as the *base spreading resistance* and r_e as the *emitter resistance*. From the circuit of Figure 4.16b, we find that

$$v_i = i_b r_b' + (\beta + 1) i_b r_e \tag{4.6}$$

and

$$v_o = -\beta i_b R_C, \tag{4.7}$$

where we have used lower case letters to denote RF voltages and currents. Eliminating i_b between (4.6) and (4.7), we obtain that

$$v_o = -R_C \frac{\beta}{r_b' + (\beta + 1) r_e} v_i = -R_C g_m v_i, \tag{4.8}$$

where g_m is known as the *transconductance* of the amplifier. Under normal circumstances, r_b' is very small and can be ignored. On the other hand, β is usually large and so $g_m \approx 1/r_e$.

An alternative form of small-signal model is shown in Figure 4.17. To apply the model, however, we need values for the transconductance g_m, the input resistance r_π and the output resistance r_0. The transconductance is given by $g_m = \mathrm{d}I_C/\mathrm{d}V_{BE}$ when

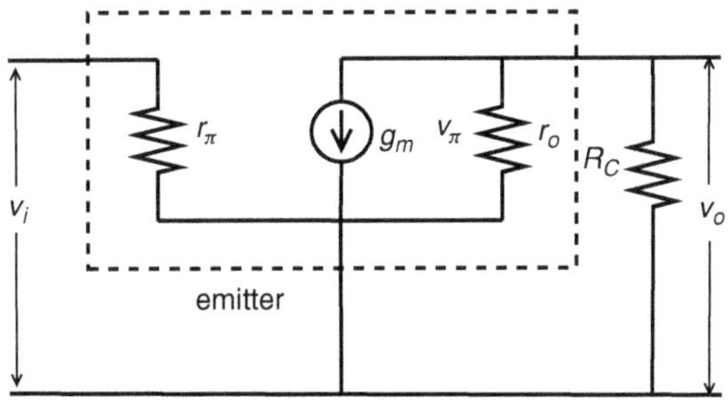

Fig. 4.17 An improved small-signal model.

evaluated at the quiescent current I_Q (i.e. $g_m = I_Q/V_T$) and the input resistance by $r_\pi = r_b' + (\beta + 1)/g_m$ from (4.6). Output resistance r_o is due to the small slope of the characteristic curves at large V_{CE} and can normally be ignored (it has a value of the order of 100 Kiloohms). Referring to Figure 4.16, we have that $v_o = -g_m(R_C \parallel r_0)v_i$, i.e. the voltage gain of the amplifier is $-g_m(R_C \parallel r_0)$. Further, to a load, the amplifier looks like a voltage source with impedance $R_C \parallel r_o$.

4.4 The Field-Effect Transistor

The *field-effect transistor* (FET) is a transistor device with very different physics from that of a BJT. In a BJT the conduction is interrupted by a layer of different-type semiconductor that controls the current through the device by means of a current through a contact that is connected to this intermediate layer. In the FET, however, the flow through a continuous semiconductor is controlled by an electric field that is applied through a contact known as a gate. The gate current is negligible and so the device is voltage-controlled rather than current-controlled. Such transistors can be traced back to 1925 when Julius Lilienfeld took out a patent on such a device. Efforts to create a working transistor were frustrated by the available technology and it was only with developments that came through the BJT that a successful device could be constructed.

There are two major varieties of the device: the junction FET and the insulated gate FET. A junction FET (or JFET) consists of semiconductor channel (the ends of which are known as the source and drain), but with a diode junction on the side of the channel (known as the gate). The diode is reverse-biased and so there is a depletion region whose size is then controlled by the voltage between the source and gate. In order to enhance the size of the depletion region within the channel, the channel is much more lightly doped than the gate region. The size of this depletion region will then control the current that flows through the device (see Figure 4.18a). A typical characteristic behaviour of

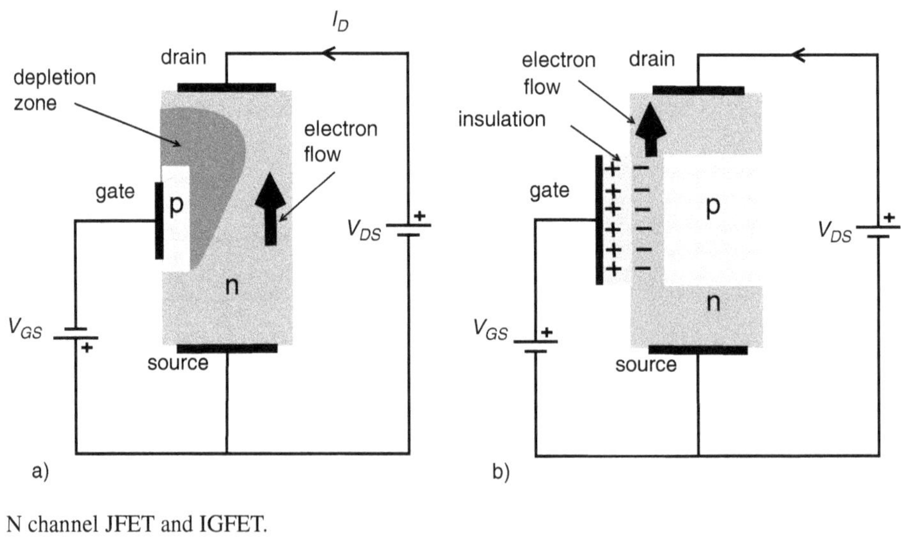

Fig. 4.18 N channel JFET and IGFET.

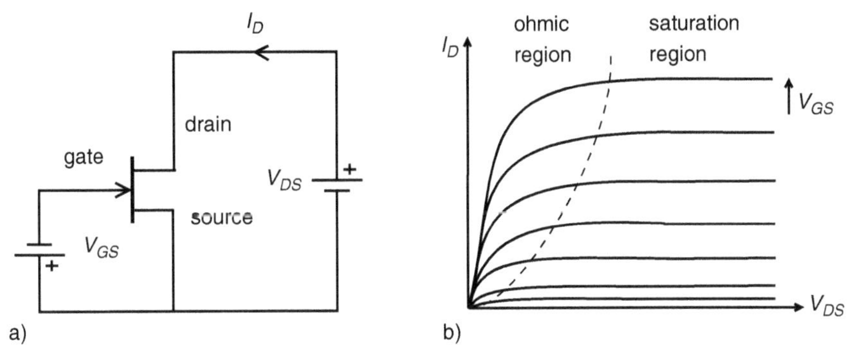

Fig. 4.19 JFET circuit and characteristic.

an n-channel JFET is shown in Figure 4.19, along with a circuit description. Consider a fixed V_{GS} (source to gate voltage), for small V_{DS} (source to drain voltage) the channel will initially behave as a simple resistor with V_{GS} controlling the channel resistance. As the drain voltage V_{DS} increases, the current I_D will steadily increase. However, as V_{DS} gets larger, the depletion region will also grow towards the drain and start to constrict the flow. Eventually, a point is reached where an increase in flow through the channel resistor is countered by an increase in channel resistance due to the channel constriction. At this point, we have reached *saturation* and the current flow remains constant for any increase in drain voltage. The channel is now effectively pinched-off and the drain voltage at which this happens is known as the *pinch-off voltage*. For its operation, the device requires the gate channel junction to be reverse-biased. There will, however, be a threshold V_T below which there will be no flow through the device due to the depletion region covering the total width of the device. For operation in the saturated region, the drain current is related to the gate voltage through the relation

$$I_D = K(V_{GS} - V_T)^2, \tag{4.9}$$

where K is a constant that depends on the device construction.

As with the BJT, we will need to keep the operating range well away from the nonlinear regions of the transistor characteristic if we are going to achieve linear amplification. We can do this by choosing a suitable quiescent drain current I_{DQ} through a biasing network of the sort shown in Figure 4.20a. It will be noted that, since the input impedance of a FET is very high (of the order of megaohms), the resistors of the voltage divider (R_1 and R_2) can also be very large. As with the BJT, manufactured devices can exhibit a great range of material parameters, in particular V_T, and the source resistor R_S provides the feedback that guards against these variations. If ΔV_T is the variation expected from the devices, and ΔI_D is the allowable variation in drain current, we will need to choose R_S greater than $\Delta V_T/\Delta I_D$.

As with the BJT, once we have fixed the quiescent component of current through the bias circuits, we can ignore this aspect and deal with the signal fluctuations alone. To do this, we will need a transistor model of the sort shown Figure 4.20b, valid for small signals. Like the BJT, the FET behaves like a current source. However, the input

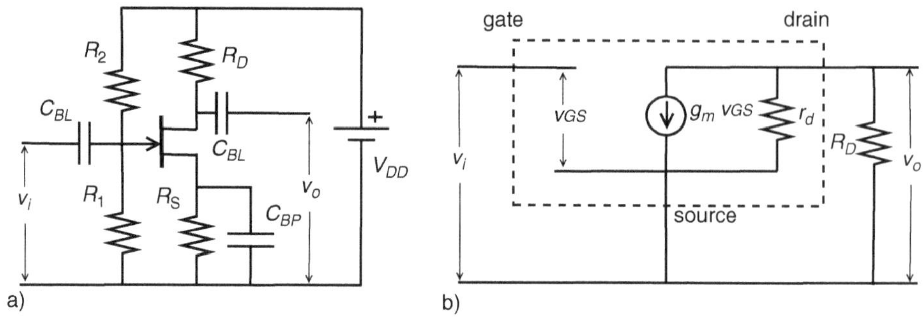

Fig. 4.20 A JFET common-source amplifier and JFET small-signal model.

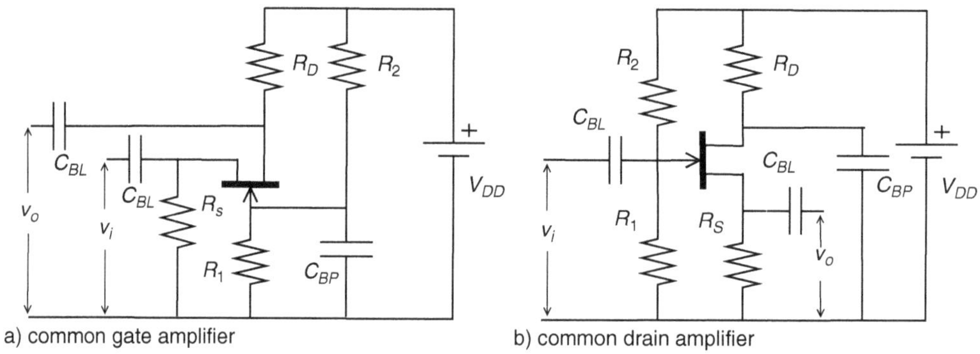

Fig. 4.21 Other varieties of FET amplifier.

resistance is now so high that we can ignore it. As with the output resistance of the BJT, the value of r_d is fairly large and can be ignored to a first approximation. The transconductance is given by $g_m = \mathrm{d}I_D/\mathrm{d}V_{GS}$ and is evaluated at the quiescent current I_{DQ}. Consequently, from 4.9, we obtain that $g_m = 2\sqrt{KI_{DQ}}$. Referring to Figure 4.20, we have that $v_o = -g_m(R_D \parallel r_d)v_i$, i.e. the voltage gain of the amplifier is $-g_m(R_D \parallel r_d)$. Further, to a load, the amplifier looks like a voltage source with impedance $R_D \parallel r_d$.

Up to now we have concentrated on what is commonly known as the common-source amplifier, but for small-signal amplifiers there are other options. Figure 4.21 shows two other varieties of amplifier (common-gate and common-drain amplifiers), both with very different properties from the common-source amplifier. The output of the common-gate amplifier behaves very much like the output of the common-source amplifier and is a voltage source $v_o = -g_m R_D v_i$ with impedance $R_D \parallel r_d$. Looking into the input, things are a little different and the relatively high-input impedance of the common-source amplifier gives way to a relatively low input impedance of $R_S \parallel g_m^{-1}$. For the common-drain amplifier, the input looks like that of the common-source amplifier, but everything changes at the output. The amplifier looks like a voltage source v_i (i.e. a voltage gain of 1) with an impedance of $R_S \parallel g_m^{-1}$. The common-gate and common-drain amplifiers have their counterparts in the BJT world, i.e. the common-base and common-collector

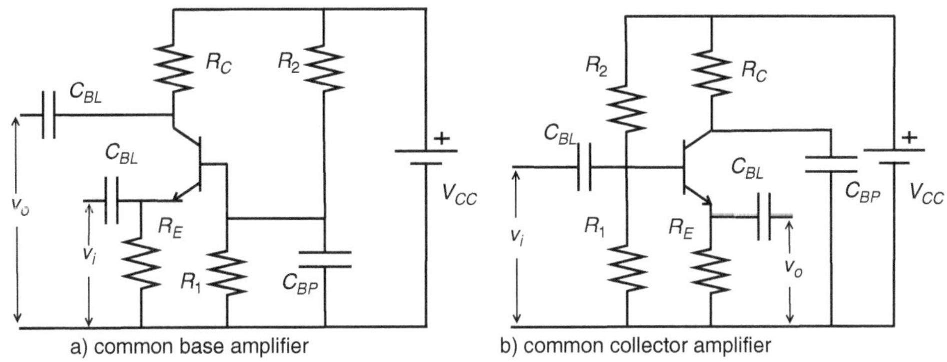

Fig. 4.22 Other varieties of BJT amplifier.

amplifiers (see Figure 4.22). Further, given the similarity of the BJT small-signal model to that of the FET, they will possess similar properties.

An insulated-gate FET (or IGFET) is a semiconductor bar (assumed to be p-type for the present) in which two junctions have been formed (the source and the drain) by infusing n-type semiconductor. The junctions are joined by an n-type layer, so thin that almost no current flows between the junctions (see Figure 4.18b). Placed over this thin region is a metal electrode that is separated from it by an insulating layer. If a positive voltage is now applied to this electrode, it behaves like a capacitor and positive charge accumulates. This charge is then counterbalanced by electrons that are drawn into the thin layer. This enhances the current-carrying capacity of the thin layer by effectively widening it.

The characteristics for the IGFET are similar to those of the JFET and can be explained as follows. We consider the capacitor that is formed between the gate electrode and the thin n-type layer. The average voltage on the channel side will be $V_{DS}/2$ and so we have $Q = C(V_{GS} - V_{DS}/2)$ negative charge formed on the channel side (C is the capacity of the gate channel capacitor). However, current does not start flowing until all the remaining holes in the thin channel are filled up, i.e. CV_T of the above negative charge is not available as carriers where V_T is the gate voltage at which charge starts to flow. We have charge $Q_c = C(V_{GS} - V_T - V_{DS}/2)$ that is available as carriers. Most of the voltage drop across the device will occur in the thin region and so the electric field to which the charge is subjected will be $E_c = V_{DS}/L$ where L is the length of the thin channel. The electron field will cause the available charge to accelerate, but this will be moderated by collisions and the net effect will be a drift velocity $v_d = \mu_n E_c$, where μ_n is a quantity known as the electron mobility. As a consequence there will be a current $\mu_n Q_c E_C/L$, i.e

$$I_D = 2K(V_{GS} - V_T - V_{DS}/2)V_{DS}, \tag{4.10}$$

where K is a constant that depends on the device construction. In reality, the voltage in the gate will not be uniform across the gate, but increase linearly from source to drain. This will mean that the channel width will decrease as we move from the source to drain. Further, as V_{DS} increases, there will eventually be a point where the channel is *pinched-off* at the drain end. At this point there will be no further increase in current with V_{DS}. At the point where we have pinch-off we will have $V_{DS} = V_{GS} - V_T$. Consequently,

$$I_D = K(V_{GS} - V_T)^2 \tag{4.11}$$

and this remains the same throughout the saturation region. Due to the ever-decreasing size of devices, mainly driven by the ever-increasing complexity of integrated circuits (ICs), the channel length L has continued to be reduced and this has resulted in ever-larger values of channel field E_c. However, at a certain level of the field, the carriers will attain enough energy to set the semiconductor lattice into a quantum mechanical mode of vibration and hence lose energy to this mechanism. In other words, the velocity of the carriers will saturate (at a level of about 10^5 m/s for silicon). The saturation velocity v_{sat} will correspond to a field level $E_{\text{sat}} = v_{\text{sat}}/\mu_n$. If this saturation occurs before pinch-off, it will result in a modification to (4.11) of the form

$$I_D = \frac{K(V_{GS} - V_T)^2}{1 + \frac{V_{GS} - V_T}{LE_{\text{sat}}}}. \tag{4.12}$$

In Section 4.7, we will find that such nonlinearity in transistor behaviour can have far-reaching consequences for radio receiving systems.

The gate of an IGFET is frequently made by using a thin layer of oxide to insulate the gate electrode from the semiconductor and, for this reason, these devices are often known as MOS (metal oxide silicon) transistors. The above n-type channel FET is then known as an NMOS transistor. A PMOS transistor is an alternative that has a p-type channel and an n-type substrate. Typical common-source amplifiers for these two transistors are shown in Figure 4.23 and from which it will be noted that the PMOS drain needs to have a negative bias (in addition to a negative bias on the gate). The different bias requirements of the various types of FETs are summarised in Figure 4.24 which shows the characteristics of these devices in saturation. MOS FETs were invented by Dawon Kahng and Martin Atalla in 1959 and have now largely superseded the JFET due to their ease of construction.

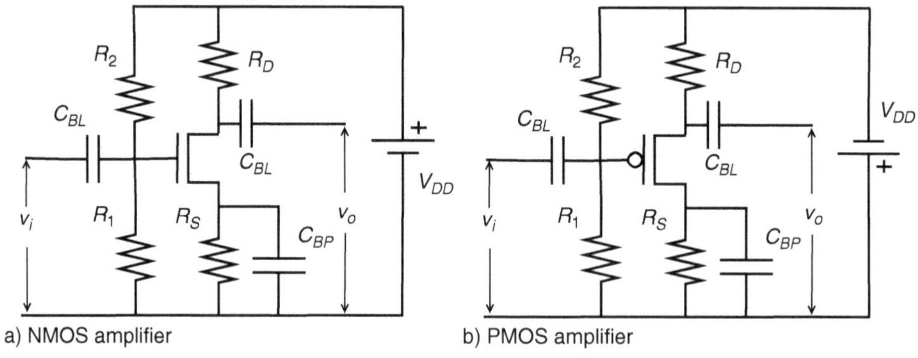

Fig. 4.23 NMOS and PMOS amplifiers.

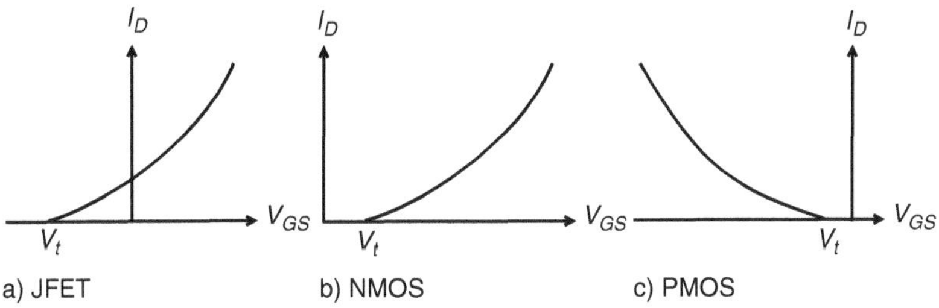

Fig. 4.24 Characteristic curves for various FETs in saturation.

4.5 Radio Frequency Amplifiers

At high frequencies, a major problem arises due to the internal capacitances of a transistor. There is obviously capacitance across the junctions of a BJT and a JFET. Furthermore, CMOS depends on the capacitance between gate and channel. As a consequence, at RF frequencies, we need a model of transistor behaviour of the sort shown in Figure 4.25a (this is appropriate for a FET, but a similar model will also apply to the BJT). The problem with this model is that the analysis of such circuits is difficult due to the capacitive feedback that is inherent in the transistor model. It turns out that the capacitances C_{GS} and C_{DS} are the least problematic and so we ignore them for the moment. Fortunately, we can further simplify feedback using a result that is known as *Miller's theorem* (Miller, 1920). We first note that the source in the FET model can be changed from a current source to an equivalent voltage source as shown in Figure 4.25b. Consequently, we consider the amplifier shown in the first circuit of Figure 4.26. It turns out that the feedback impedance Z can be replaced by impedances Z_1 and Z_2 at the input and output of the transistor (see Figure 4.26), providing they are suitably chosen. From the Kirchhoff voltage law we have,

$$V_o = AV_i + I_f r_o \text{ and } V_i = V_o + I_f Z \tag{4.13}$$

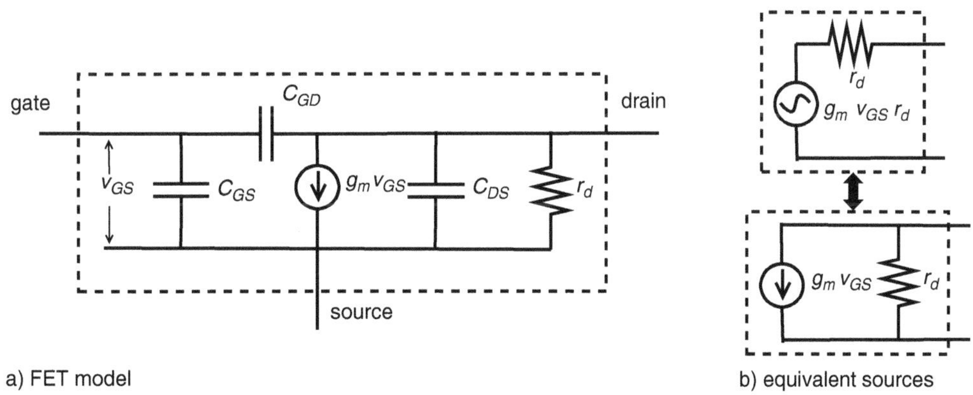

Fig. 4.25 High-frequency model of a FET.

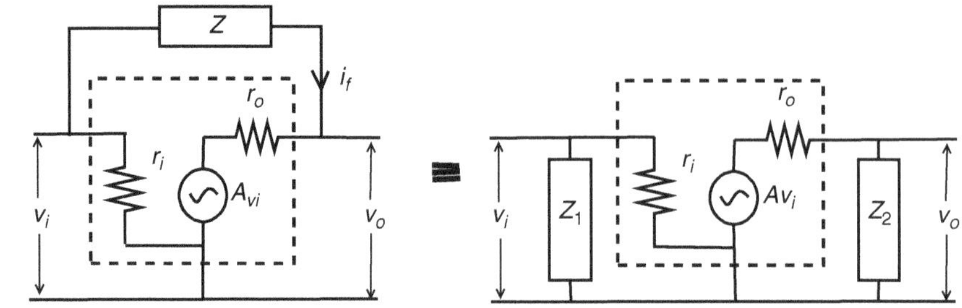

Fig. 4.26 Miller's result.

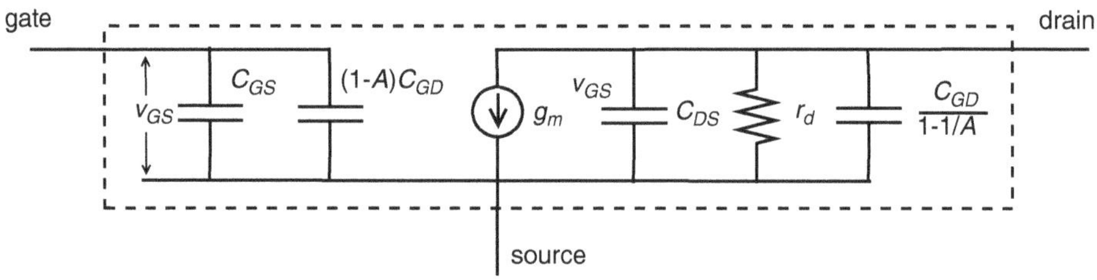

Fig. 4.27 Alternative high-frequency model of a FET based on Miller's theorem.

where $v_o = \Re\{V_o \exp(j\omega t)\}$, $v_i = \Re\{V_i \exp(j\omega t)\}$ and $i_f = \Re\{I_f \exp(j\omega t)\}$ (i.e. we assume time-harmonic signals). From Figure 4.25b, we need Z_1 to draw the same current as the feedback circuit, i.e. $I_f = V_i/Z_1$. Further, from (4.13), we have that $I_f/V_i = (1-A)/(r_o + Z)$ and so

$$Z_1 = \frac{r_o + Z}{1-A}. \tag{4.14}$$

Similar arguments at the output also imply that

$$Z_2 = \frac{AZ + r_o}{A-1}. \tag{4.15}$$

For an ideal amplifier ($r_o = 0$ and $r_i = \infty$), we will have $Z_1 = Z/(1-A)$ and $Z_2 = Z/(1-1/A)$. For the model of Figure 4.25, the feedback is the capacitance C_{GD} and this, and the other capacitances, are usually only a few picofarad. As a consequence, the additional impedance at the output is of the same order as existing impedance. At the input, however, there is a different story. Assuming we can treat the amplifier as ideal, the additional impedance will be capacitative (we assume that we have a common-source or common-emitter amplifier) with the value of the capacitor $(1-A)C_{GD}$ (see Figure 4.27). For an amplifier with reasonable gain we have a large shunt capacitance at the input, hence negating the gain of the amplifier by grounding the input at high frequencies. This is known as the *Miller effect* and we need to find strategies for countering it.

One possible approach is to cancel out the input capacitance with a parallel tuned circuit. Figure 4.28 shows the circuit of a simple receiver in which this approach has been used. In this receiver there is a second amplifier which also acts as a detector. The operating point of the second amplifier is set to be just above that at which the

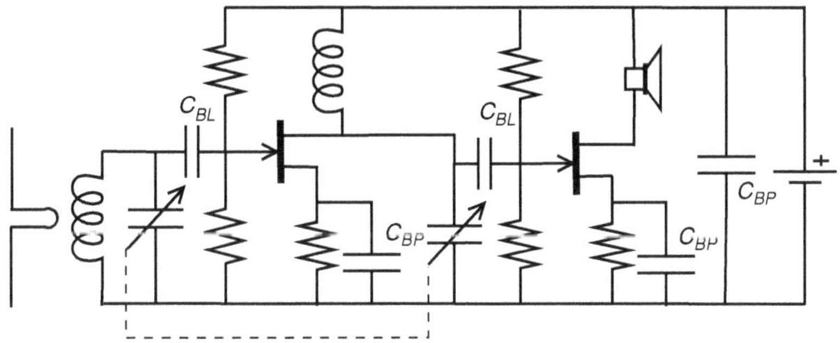

Fig. 4.28 Tuned radio frequency (TRF) receiver.

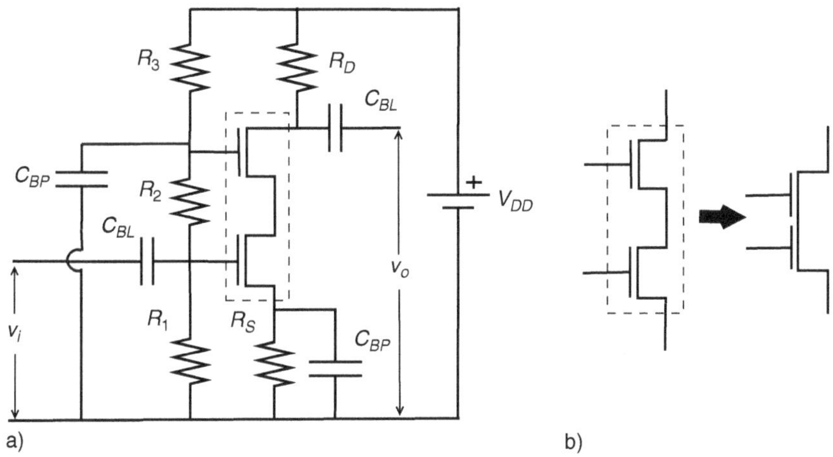

Fig. 4.29 Cascode amplifier and dual gate FET.

transistor starts to conduct and so the transistor will only conduct on the positive part of the RF cycle. Consequently, the average current at the drain will consist of the amplitude modulation of the signal. The important thing to be noted is that, besides overcoming the Miller effect, the additional tuned circuits will increase the selectivity of the receiver, i.e. they will better reject unwanted signals on nearby frequencies. Later we will see that the ability to reject nearby signals is crucial to the performance of a radio receiver. The receiver shown in Figure 4.28 is known as a tuned radio frequency (TRF) receiver and was popular (in its valve form) during the decade after the first world war. The drawback, however, is that all the tuned circuits need to be adjusted together (the dotted line in the Figure indicates a 'ganged' tuning capacitor) if the frequency of reception is to be changed.

The Miller effect can be overcome without the use of tuned circuits through the cascode amplifier (see Figure 4.29a). This circuit is essentially a common-source amplifier that is followed by a common-gate amplifier (a common-emitter amplifier followed by a common-base amplifier is used in the case of BJTs). The input impedance of the common-gate amplifier will be g_m^{-1} and so this will be the load of the common-source amplifier, i.e. the gain of the common-source amplifier will be 1. As a consequence, the common-source amplifier will have no gain and hence no Miller effect. On the other hand, the common-gate amplifier will have a very low input impedance and so the Miller effect will have to be extreme in order to have an effect. Furthermore, the total gain of the amplifier will be that of the common-gate amplifier. By combining two transistors, we now have an amplifier with the same behaviour as the common-source amplifier, but without the Miller effect. Since the drain of the bottom FET and the source of the upper FET are directly connected, the two transistors can be combined into a single device known as a dual-gate FET. (In the case of valve amplifiers, a similar trick to avoid the Miller effect consists of adding an additional grid to form a *tetrode valve*.)

A further example of amplifier topology is the differential pair, as shown in Figure 4.30a. Consider the RF components of current and voltage. There is a constant-bias current I_B that is set by V_{bias} and so $i_D^a + i_D^b = 0$. Further, $v_o^a = -R_D i_D^a$ and $v_o^b = -R_D i_D^b$ and, from the small-signal model of a FET, $i_D^a = g_m v_{GS}^a$ and $i_D^b = g_m v_{GS}^b$. Bringing these relations together, we find that

$$v_o^b - v_o^a = -g_m R_D (v_i^b - v_i^a), \tag{4.16}$$

i.e. the difference in output voltages is a multiple of the difference in input voltages.

Interestingly, the same topology can be used to produce another amplifier that can overcome the Miller effect. Consider the amplifier shown in Figure 4.30b. In this case the transistor loads are constructed from transistors (the load on the left has resistance g_m^{-1} and the load on the right has resistance r_d). This is a useful approach in integrated circuit technology where transistors are far easier to fabricate than resistors. The input transistor has unity gain and is thus immune from the Miller effect. In the case of the output transistor, the base voltage is fixed and so the Miller effect is irrelevant. As with the cascode amplifier, this alternative amplifier will have a gain of $-g_m R_D$.

The amplifier configuration of Figure 4.30a can also be used to make a two-input, one-output, amplifier that is commonly known as an *operational amplifier*. If a buffer

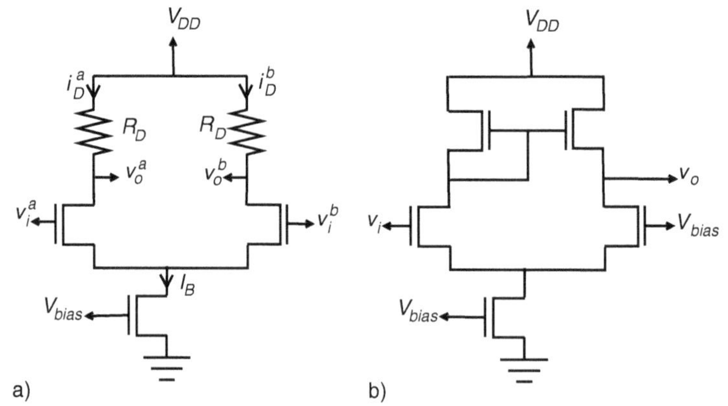

Fig. 4.30 An CMOS differential amplifier and alternate with active loads.

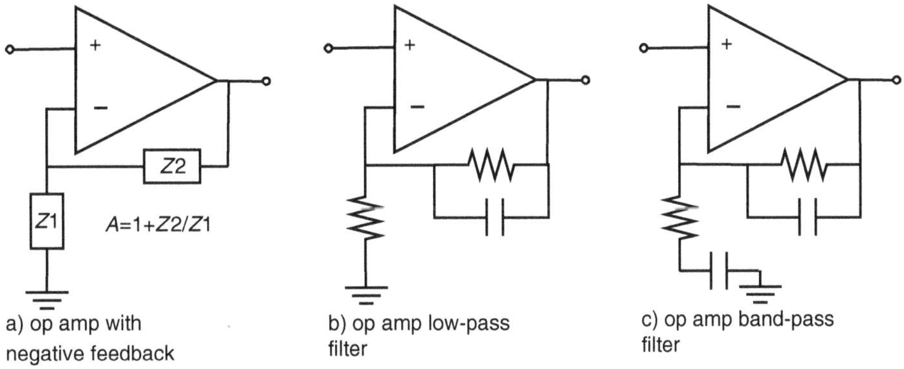

Fig. 4.31 General configuration of an op amp together with low- and band-pass filters.

amplifier is added at the output, we have a high-gain amplifier with high-input impedances and low output impedances. Importantly, for one of the inputs (the one normally labelled −), the output is phase-shifted by 180°. Figure 4.31a shows an operational amplifier with a general negative-feedback configuration (the voltage gain of the amplifier is given by $A = 1 + Z2/Z1$) and Figures 4.31b and 4.31c show the feedback configured so that the amplifier acts as low-pass and band-pass filters respectively.

4.6 Noise

Besides the desired signal, there will always be *noise*, i.e. competing unwanted signals. The electronics of a radio can itself be the source of considerable noise, usually known as *internal noise* in order to distinguish it from that which comes in through the antenna. Even the humble resistor will create noise due to the thermal motion of its electrons (*Johnson noise*). In reality, a resistor should be modelled as an ideal resistance in series with a random noise source (see Figure 4.32a) of mean square voltage

$$\overline{v_n^2} = 4kTBR, \tag{4.17}$$

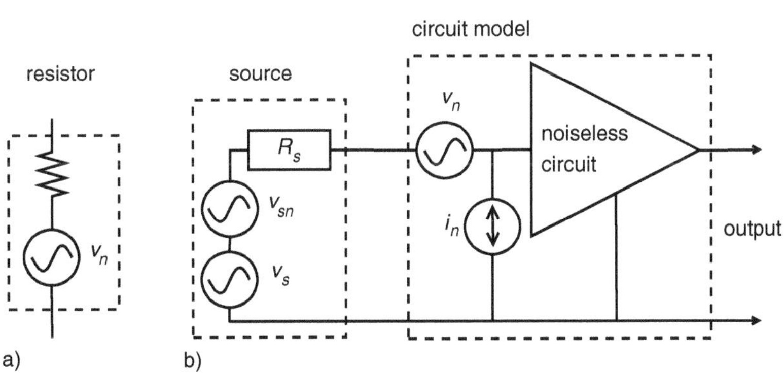

Fig. 4.32 Noise models (a resistor and a general circuit).

where T (in kelvins) is the absolute temperature, B (in hertz) is the bandwidth of the receiver, R (in ohms) is the resistance and k is the Boltzmann constant (1.38×10^{-23} joules per kelvin). (A general impedance Z will behave as the noise source of its resistive part in series with the ideal impedance Z.) Semiconductors are also the source of a considerable variety of noises (popcorn noise and flicker noise to name but two). In particular, *shot noise* occurs at semiconductor junctions due to the discrete nature of the carriers and their independence. (For this noise, the mean-square current is given by $\overline{i_n^2} = 2eI_0B$ where I_0 is the quiescent current.) As a consequence, complex electronic circuits can be quite difficult to analyse from a noise perspective. Fortunately, it can be shown that a complex circuit can be modelled as a noise-free circuit with a series voltage noise source at its input and current noise source across the input (see Figure 4.32b). In the case of a BJT amplifier, the main sources of noise are the shot noises in the junctions and the noise in the base resistance r_b'. At radio frequencies, these noises can be approximated by

$$\overline{v_n^2} = 4kTB\left(r_b' + \frac{r_e}{2}\right) \text{ and } \overline{i_n^2} = 2eI_bB\left(1 + \beta\frac{\omega^2}{\omega_T^2}\right), \tag{4.18}$$

where ω_T is the cut-off frequency of the transistor (the frequency at which the internal capacitances of a BJT have reduced its current gain to unity). In the case of a FET amplifier, the dominant sources of noise are due to the channel resistance and shot noise at the gate in the case of a JFET. At radio frequencies, the noise can be approximated by

$$\overline{v_n^2} = \frac{8kTB}{3g_m} \text{ and } \overline{i_n^2} \approx 2eI_gB + \frac{4}{9}\frac{8kTB}{3g_m}\omega^2C_{GS}^2, \tag{4.19}$$

where I_g is the gate current (zero in the case of CMOS). It will be noted that, for a BJT, the collector current is a key element in deciding the level of noise and so one might ask whether there is an optimum quiescent collector current I_C from the viewpoint of noise. The available equivalent input noise power for a common-source BJT amplifier will be

$$N_i = 4kTB + \frac{4kTB}{R_S}\left(r_b' + \frac{V_T}{2I_C}\right) + 2e\frac{I_C}{\beta}BR_S, \tag{4.20}$$

where R_S is the source resistance (we assume the frequency is well below ω_T). For N_I to be minimum, we require that $\partial N_i/\partial I_C = 0$ and from this we find that $I_C = \sqrt{\beta}V_T/R_S$. This will usually imply a small quiescent current for minimum noise and this can be at odds with other requirements of the amplifier, linearity requirements in particular. (The reader should consult van der Ziel (1986) for a more detailed discussion of noise in semiconductor devices.)

In a realistic radio system, *external noise* (that which comes through the antenna) is just as important as the internal variety. This can arise from man-made sources (computers and ignition interference for example), natural extraterrestrial sources (galactic noise) and natural terrestrial sources (lightning for example). Lightning as a source of noise is complex since, for frequencies below 30 MHz, the noise at a single point can be the accumulation of the effect of lightning strikes across the globe. This arises due to propagation via the ionospheric duct. If an antenna has an effective resistance R_A, the noise coming in through the antenna is often described in terms of *antenna temperature*

T_A, i.e. the temperature that a resistor R_A would need to have in order to produce the same noise as that coming in through the antenna. External noise is the ultimate constraint on a receiver system and it is usual to design a radio receiver to be *externally noise limited* (i.e. the internal noise is below the level of external noise).

The signal-to-noise ratio (SNR) is the crucial parameter in quantifying the effect of noise upon a receiver system and is defined by

$$\text{SNR} = \frac{\text{signal power}}{\text{noise power}}. \tag{4.21}$$

How a stage of a receiver affects the SNR is usually measured in terms of its *noise factor* F (known as the *noise figure* when expressed in dB terms). This is defined by

$$\text{F} = \frac{\text{available output noise power}}{\text{available output noise power due to the source alone}}. \tag{4.22}$$

The noise factor needs to be referenced to a well-defined noise source, usually a resistor at ambient temperature T (defined to be 290 K for convenience). In this case, the noise can be represented as a noise source of power $FkTB$ at the input of a noise-free circuit. If the circuit has a noise source at a temperature T_S other than ambient (an antenna for example), the total noise power at the input to the noise-free circuit will then be

$$F = kT_SB + (F-1)kTB. \tag{4.23}$$

If we apply definition (4.22) to the model of noise shown in Figure 4.32b, we obtain that

$$\text{F} = 1 + \frac{\overline{v_n^2} + \overline{i_n^2}R_s^2}{4kTBR_s}, \tag{4.24}$$

where R_s is the source resistance (note that we have assumed there to be no correlation between the equivalent voltage- and current-noise sources). Obviously, we would like to minimise the noise factor. Since the source impedance is under our control, we could ask the question as to what source impedance would provide the lowest noise. We need to find the R_s for which $\partial F/\partial R_s = 0$ and this turns out to be when $R_s^2 = \overline{v_n^2}/\overline{i_n^2}$. Unfortunately, this is not necessarily the source impedance that would give a power match. Fortuitously, however, it is the value that gives the best SNR. To see this consider the situation shown in Figure 4.33 in which the source is connected to the input of the circuit through a transformer with turns ratio N. Noting that impedance is scaled by N^2 through the transformer and voltage by N, we find that

$$\text{SNR} = \frac{\frac{v_s^2}{R_s}}{\frac{\overline{v_{sn}^2}}{R_s} + \frac{\overline{v_n^2}}{N^2R_s} + \overline{i_n^2}R_sN^2}. \tag{4.25}$$

Minimising SNR with respect to N (i.e. we solve for $\partial\text{SNR}/\partial N = 0$), we find that $R_sN^2 = \sqrt{\overline{v_n^2}/\overline{i_n^2}}$, i.e the transformer needs to change the source impedance to $\sqrt{\overline{v_n^2}/\overline{i_n^2}}$.

As a signal passes through the circuits of the receiver, the SNR will degrade through contributions from various noise sources within the circuits. Further, in many circumstances, the received signal will be very weak and therefore need to pass through several stages of amplification. Consequently, we need to know how the noise factor will

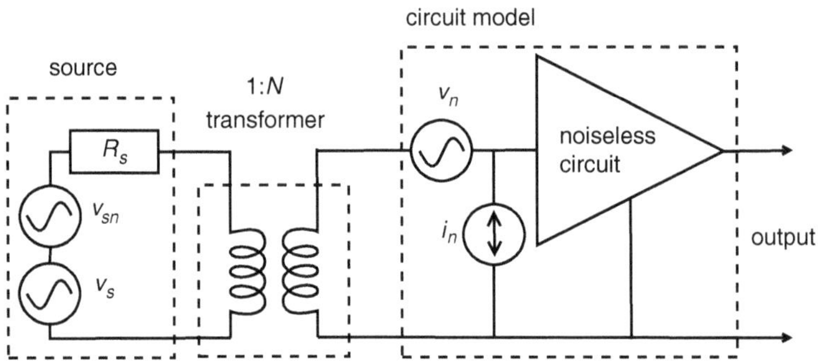

Fig. 4.33 Noise model with source-matching circuit.

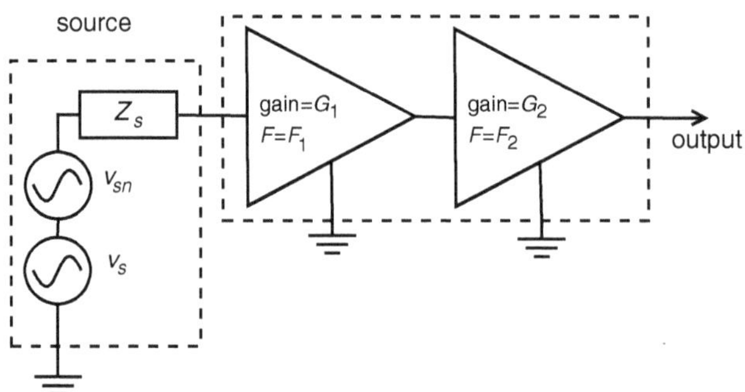

Fig. 4.34 Noise figure of cascaded circuits.

change through a cascade of circuits. Firstly, if a circuit is driven by a source at ambient temperature T, we have from the definition of the noise factor F that the noise power added by the circuit will be an equivalent noise power $(F-1)kTB$ at the circuit input. Now consider the cascaded circuits shown in Figure 4.34. The noise at the output of the first circuit will be $G_1kTB+G_1(F_1-1)kTB$. In the case of the second circuit, the noise power added will be $(F_2-1)kTB$ at the input. Consequently, the total noise at the output of the second circuit will be $G_1G_2kTB+G_1G_2(F_1-1)kTB+G_2(F_2-1)kTB$. The noise due to the source alone will be G_1G_2kTB and so, from the definition of the noise factor,

$$\begin{aligned} \mathrm{F} &= \frac{G_1G_2kTB+G_1G_2(F_1-1)kTB+G_2(F_2-1)kTB}{G_1G_2kTB} \\ &= F_1+\frac{F_2-1}{G_1}, \end{aligned} \tag{4.26}$$

where F_1 and F_2 are the noise factors of the separate amplifiers and G_1 and G_2 are their power gains. An important deduction from this result is that the first amplifier in a radio receiver always needs to be the one with the best noise figure.

An important factor in the characterisation of a radio receiver is its sensitivity. This is dictated by the level of noise with which the incoming signal must compete and is often described in terms of the *noise floor* N_f. For a receiver with noise factor F, and bandwidth

B, the noise floor is total equivalent noise at the input of a noiseless receiver, i.e.

$$N_f = F = kT_AB + (F-1)kTB, \tag{4.27}$$

where T_A is the antenna temperature and T is usually assumed to be the ambient noise temperature, i.e. 290 K. Another measure of sensitivity the *minimum detectable signal* (MDS) which is defined to be the level of signal that is equal to the noise (i.e. it is the same number as the noise floor). The sensitivity of a receiver can sometimes be quoted as a voltage $v_f = \sqrt{N_fR_I}$ where R_I is the input resistance of the receiver.

The relation between SNR and detection was quantified by Claude Shannon of Bell Laboratories in 1948. Shannon showed that C, the maximum capacity of a communication channel in bits per second, was related to the SNR through (Shannon, 1947)

$$C = B\log_2(1+\mathrm{SNR}), \tag{4.28}$$

where B is the channel bandwidth in Hertz and $\log_2$ indicates a logarithm to the base 2 ($\log_2 x = 1.4427\ln x$ in terms of natural logarithms). Although stated in terms of the language of digital communication (bits per second), this is a very general result that sets an ultimate limit on the rate of communications for a given bandwidth and SNR.

4.7 The Effect of Nonlinearity

As we have seen, in the case of both FET and BJT amplifiers, the relationship between input and output voltages is by no means linear. This made it necessary to bias circuits to operate within the most linear part of the circuit characteristic. Unfortunately, it is not possible to remove all nonlinearity and so we need to investigate the consequences of the nonlinearity for circuit performance. We will assume that the output voltage v_o is related to the input voltage v_i through the relationship

$$v_o = k_o + k_1v_i + k_2v_i^2 + k_3v_i^3 + \cdots \tag{4.29}$$

If we assume a sinusoidal input voltage of the form

$$v_i = V\cos\omega t \tag{4.30}$$

the output voltage will take the form

$$v_o = (k_o + \frac{k_2}{2}V^2) + (k_1 + \frac{3k_3}{4}V^2)V\cos\omega t + \frac{k_2}{2}V^2\cos 2\omega t + \frac{k_3}{4}V^3\cos 3\omega t + \cdots \tag{4.31}$$

after the application of some standard trigonometric identities. The output contains harmonics (signals at multiples of the input frequency), a constant offset voltage and an amplification term with voltage gain $k_1 + \frac{3k_3}{4}V^2$. The harmonics can be removed with suitable filtering and so pose no particular problem. The amplification term, however, is problematic due to the dependence of the gain upon the level of input voltage. The value of k_3 is usually negative and this will lead to a reduction in gain with increasing input voltage, i.e. we have *gain compression*. Gain compression is an important consequence of nonlinearity and is normally measured in terms of the 1 dB compression point P_{1dB}, i.e. the input power level at which the amplifier gain is reduced by 1 dB.

Whilst gain compression is a problem for power amplification, even weak signals can suffer from degradation due to nonlinearity. Consider an input signal v_i that is a combination of two sinusoidal signals

$$v_i = V_1 \cos \omega_1 t + V_2 \cos \omega_2 t. \tag{4.32}$$

We will ignore the effects of coefficients higher than k_3 in (4.29), then

$$\begin{aligned} v_o(t) = {} & k_0 + \frac{k_2}{2}(V_1^2 + V_2^2) \\ & + (k_1 V_1 + \frac{3}{4} k_3 V_1^3 + \frac{3}{2} k_3 V_1 V_2^2) \cos(\omega_1 t) \\ & + (k_1 V_2 + \frac{3}{4} k_3 V_2^3 + \frac{3}{2} k_3 V_1^2 V_2) \cos(\omega_2 t) \\ & + \frac{3k_3}{4} V_1 V_2^2 \cos((2\omega_2 - \omega_1)t) + \frac{3k_3}{4} V_1^2 V_2 \cos((2\omega_1 - \omega_2)t) \\ & + \frac{k_2}{2} V_1^2 \cos(2\omega_1 t) + \frac{k_2}{2} V_2^2 \cos(2\omega_2 t) \\ & + \frac{k_3}{4} V_1^3 \cos(3\omega_1 t) + \frac{k_3}{4} V_2^3 \cos(3\omega_2 t) \\ & + k_2 V_1 V_2 \cos((\omega_1 + \omega_2)t) + k_2 V_1 V_2 \cos((\omega_1 - \omega_2)t) \\ & + \frac{3k_3}{4} V_1 V_2^2 \cos((2\omega_2 + \omega_1)t) + \frac{3k_3}{4} V_1^2 V_2 \cos((2\omega_1 + \omega_2)t) \end{aligned} \tag{4.33}$$

after the application of some standard trigonometric identities. We will consider frequencies ω_1 and ω_2 that are within the pass band of the amplifier. The last four lines of the above expression, and the zero frequency terms of the first line, represent products that can be easily removed by filtering. The terms on lines two and three, however, represent the desired amplified frequencies. Besides gain compression, the main thing to note is that a strong undesired signal at frequency ω_2 could cause a reduced response at a desired frequency ω_1 when k_3 is negative. Such *desensitisation* can be quite a problem for weak signals in a strong signal environment. The terms of line four represent another problem that is caused by a non-zero k_3. If the signals at frequencies ω_1 and ω_2 are strong and undesired, the third-order inter-modulation will produce components at frequencies $|2\omega_1 - \omega_2|$ and $|2\omega_2 - \omega_1|$. In a crowded radio environment, it is possible that such components could be coincident with a desired frequency and so be a source of interference.

A measure of this effect is the *inter-modulation distortion* (IMD) which is the ratio of the output power at the frequency $|2\omega_1 - \omega_2|$ to the output power at the fundamental frequency ω_1,

$$\text{IMD} = \left(\frac{3k_3 V_1 V_2}{4k_1} \right)^2. \tag{4.34}$$

A frequently used measure of third-order effects is the *third-order intercept point IIP3* (see Figure 4.35). If the two unwanted signals have equal amplitude ($V_1 = V_2$), this is

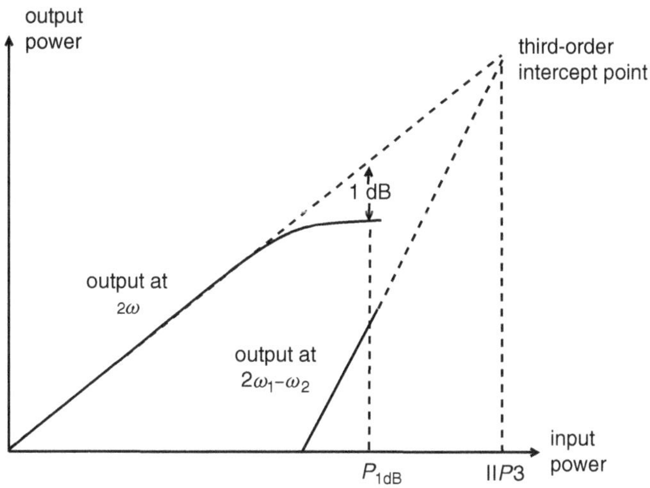

Fig. 4.35 P_{1dB} and third-order intercept point.

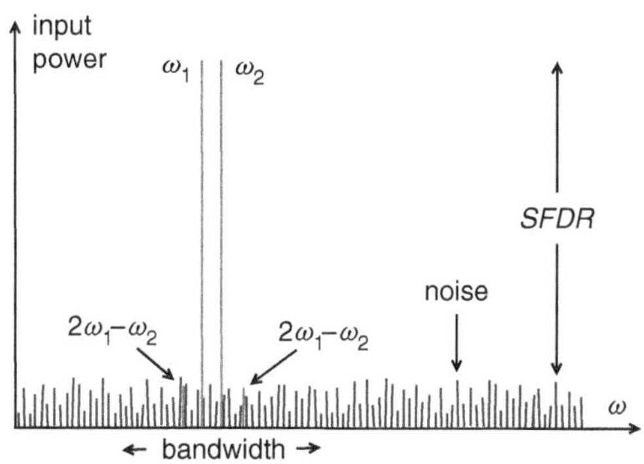

Fig. 4.36 Spurious-free dynamic range (SFDR).

defined to be the input power for which the IMD has value 1 ($IIP3 = |2k_1/3k_3R|$ where R is the input impedance of the receiver). As mentioned in the previous section, we can improve the SNR of a receiver by adding a low-noise preamplifier. We could now ask what impact this will have on the strong-signal behaviour of the receiver, i.e. its IIP3. It turns out that if the receiver consists of two cascaded stages with intercept point $IIP3_1$ for the first stage and $IIP3_2$ for the second stage, the total third-order intercept point $IIP3$ will be given by

$$\frac{1}{IIP3} = \frac{1}{IIP3_1} + \frac{G_1}{IIP3_2}, \tag{4.35}$$

where G_1 is the power gain of the first stage, i.e. the $IIP3$ of the later stages of a receiver has more influence than the earlier stages. As a consequence, whilst the first amplifier of a receiver needs to be designed for minimum noise, later stages should be designed to minimise $IIP3$.

In the case of a receiver, an important measure of nonlinearity is its *spurious-free dynamic range* (SFDR). This is the signal whose third-order distortion just reaches the level of the noise when scaled upon the noise. It can be shown that

$$\mathrm{SFDR} = \left(\frac{IIP3}{N_f}\right)^{2/3}, \tag{4.36}$$

where N_f is the noise floor (measured at the receiver input). The SFDR can be regarded as the range of input powers for which the receiver imperfections will remain hidden by the noise.

4.8 Conclusion

In the present chapter, we have considered the active devices that make the amplification of radio signals possible (i.e. valves and transistors). However, these devices have their limitations which we have discussed in detail, along with techniques for mitigating the worst effects. Whilst amplifiers (both audio and radio frequency) can significantly improve a radio based upon the simple crystal detector, active devices can be used in more imaginative ways to produce high-quality radio receivers and transmitters and this is the subject of the next chapter.

5 Radio Transmitters and Receivers

The advent of reliable amplifiers led to an explosion of new techniques that helped turn radio from a technology for specialised applications into one that has touched the lives of nearly everyone. In particular, it made possible the transmission of high-quality audio that could be received on inexpensive receivers. Radio became a technology for the masses and this changed the nature of the entertainment industry forever. In the current chapter we will look at the technology that made these developments possible. In particular, we will look at the evolution of radio receiver and transmitter designs, culminating in the superheterodyne of Armstrong. We will consider crucial circuit elements such as oscillators, mixers and power amplifiers. Importantly, we will consider various factors that can affect the performance of these elements (factors such as phase noise) and their impact upon the overall performance of receivers and transmitters.

5.1 Feedback and Oscillators

Feedback is the process whereby some of the output of an amplifier is fed back into the input. A general feedback system is illustrated in Figure 5.1, the amplifier having gain A and feedback gain F. Referring to this Figure we have that $v_{out} = A(v_{in} + Fv_{out})$ and so

$$v_{out} = \frac{A}{1 - FA} v_{in}. \tag{5.1}$$

If feedback is positive (i.e. $FA > 0$), and FA is slightly less than 1, we will have a large gain. This fact was recognised by the famous radio engineer Edwin Armstrong around the year 1914. In the early days of radio, valves were very expensive and had low performance. As a consequence, the process of *positive feedback* was able to provide a means of building cheap and sensitive radio receivers with poorly performing valves. The downside, however, was that the feedback had to be adjusted carefully and quite often needed to be readjusted when the radio was tuned to a new station. An amplifier that uses the process of positive feedback has come to be known as a *regenerative amplifier* and a receiver that employs such amplifiers as a *regenerative receiver*. As we shall find out later, there are other advantages to positive feedback besides an increase in gain. Negative feedback (i.e. $FA < 0$) can also be useful. In particular, if the amplifier gain is very large, we will have $v_{out} = -1/F$, i.e. we can control the properties of the amplifier by its feedback.

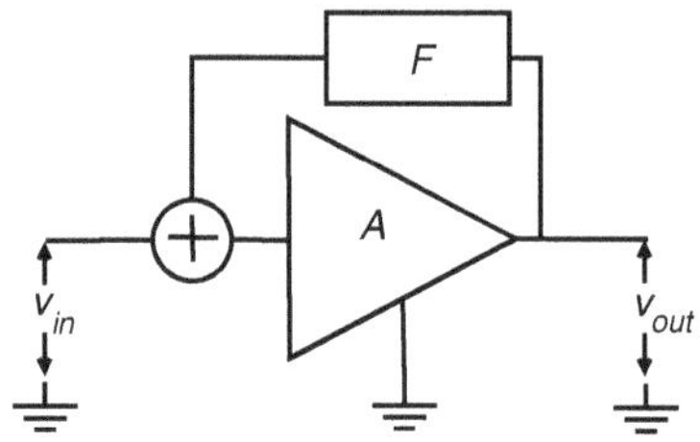

Fig. 5.1 An amplifier with a general feedback network.

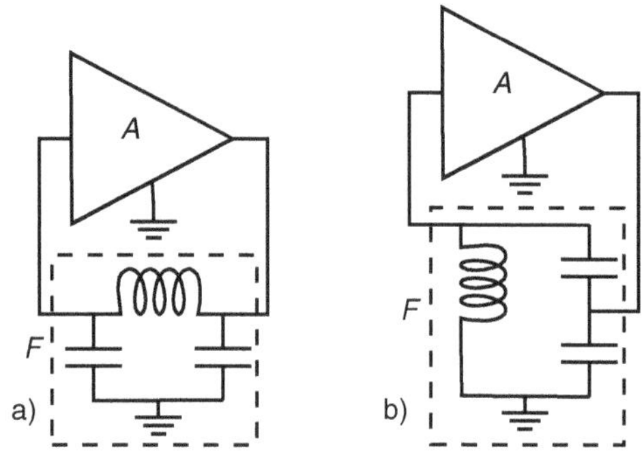

Fig. 5.2 Two forms of Colpitts oscillator.

From Eq. (5.1), we see that, as AF approaches 1, the gain tends towards infinity. Consequently, any small excitation, even circuit noise, will give rise to a substantial output, i.e. the amplifier is unstable. Such instability, however, offers the possibility of signal generation. We consider the situation where the feedback is some kind of frequency-selective circuit. For a harmonic signal ($v_{in} = \Re\{V_{in}\exp(\mathrm{j}\omega t)\}$ and $v_{out} = \Re\{V_{out}\exp(\mathrm{j}\omega t)\}$), Eq. (5.1) can be restated as

$$V_{out} = \frac{A}{1 - FA} V_{in}, \tag{5.2}$$

where F is now a function of frequency and V_{in} and V_{out} are the complex amplitudes of the input and output signals. In order for the amplifier to become unstable, we will need $|FA| \to 1$ and $\arg FA \to 0$ (note that a complex number W can be represented as $|W|\exp(\mathrm{j}\arg W)$ where $\arg W = \arctan(\Im\{W\}/\Re\{W\})$). Let there be a frequency ω_0 at which $\arg(FA) = 0$, then the amplifier will become unstable as $|FA|$ approaches 1, i.e. it will produce oscillations at this frequency. The conditions $|FA| = 1$ and $\arg(FA) = 0$ are known as the *Barkhausen criterion* for oscillation. In a practical amplifier, however, it is difficult to exactly satisfy the condition $|FA| = 1$. Hence, to ensure that the condition can be satisfied, the oscillator designer will choose the gain A such that $|FA|$ is significantly greater than 1. Then, as the amplitude of oscillation rises, gain compression will cause A to fall until the condition is satisfied.

Some examples of feedback are shown in Figure 5.2 and give rise to what is known as a *Colpitts oscillator*. The feedback F in Figure 5.2a is negative and is therefore suitable for use with a common-source (or common-emitter) amplifier (A is negative in such amplifiers). In Figure 5.2b, however, F is positive and so suitable for use with a common-drain (or common-collector) amplifier. We will take a closer look at an oscillator that is based on a common-drain amplifier. A circuit for such an oscillator is shown in Figure 5.3a. It will be noted that we have used a JFET in the design and this allows us to set the quiescent operating point of the transistor by grounding the gate through the inductor L and placing a suitable resistor R_S in the source. In the source circuit there is an RF choke (a large inductance that has effectively infinite impedance at RF frequencies) that isolates the source from the ground at RF frequencies. We will analyse the circuit in terms of the RF model shown in Figure 5.3b (note that the model includes the intrinsic resistance R of the inductor). We assume oscillations at frequency ω and so represent the currents and voltages in terms of their complex amplitudes. From the Kirchhoff current law, we obtain that

$$I_2 = I_1 + V_{GS} g_m. \tag{5.3}$$

Further, from the voltage law we get that

$$RI_1 + j\omega L I_1 + \frac{I_1}{j\omega C_1} + \frac{I_2}{j\omega C_2} = 0 \tag{5.4}$$

and

$$V_{GS} = \frac{I_1}{j\omega C_1}. \tag{5.5}$$

From Eqs. (5.3) and (5.5)

$$I_2 = I_1 \left(1 + \frac{g_m}{j\omega C_1}\right) \tag{5.6}$$

and from Eqs. (5.6) and (5.4)

$$I_1 \left(R + j\omega L + \frac{1}{j\omega C_1} + \frac{1}{j\omega C_2} - \frac{g_m}{\omega^2 C_1 C_2}\right) = 0. \tag{5.7}$$

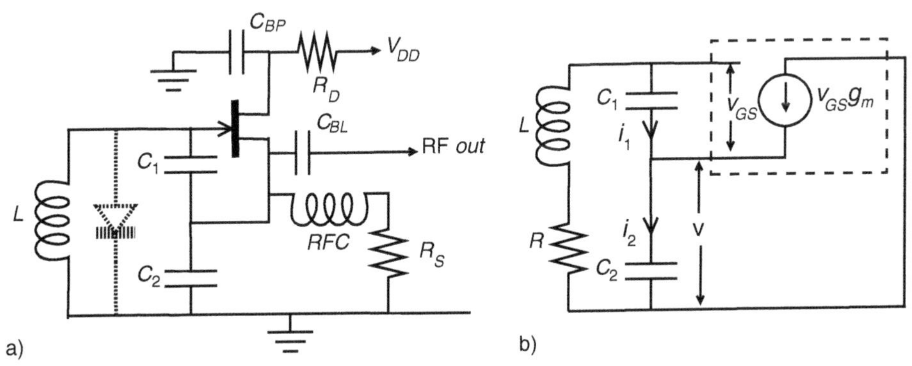

Fig. 5.3 Practical Colpitts oscillator (optional diode shown as dotted lines) and its analysis.

In order for there to be non-zero current flow, it is necessary that

$$R + j\omega L + \frac{1}{j\omega C_1} + \frac{1}{j\omega C_2} - \frac{g_m}{\omega^2 C_1 C_2} = 0. \tag{5.8}$$

From the imaginary part of Eq. (5.7) we have that

$$\omega^2 = \frac{C_1 + C_2}{LC_1C_2} \tag{5.9}$$

and from the real part that

$$g_m = \omega^2 C_1 C_2 R. \tag{5.10}$$

Equations (5.9) and (5.10) are essentially the Barkhausen criterion. As mentioned earlier, it is difficult to satisfy Eq. (5.10), mainly due to the variation in performance between transistors. If $g_m \neq \omega^2 C_1 C_2 R$, we will need ω to be complex in order for Eq. (5.8) to be satisfied. We can rewrite Eq. (5.8) as

$$1 - \frac{g_m}{R\omega^2 C_1 C_2} + j\frac{\omega}{\omega_0} Q \left(1 - \frac{\omega_0^2}{\omega^2}\right) = 0, \tag{5.11}$$

where $Q = \omega_0 L/R$ is the unloaded Q of the inductor and $\omega_0 = \sqrt{(C_1 + C_2)/LC_1C_2}$ is the frequency of oscillation when the Barkhausen criterion is satisfied. If we consider small deviations $\delta\omega$ in frequency about ω_0 (i.e. $\omega = \omega_0 + \delta\omega$), then $\omega/\omega_0 \approx 1 + \delta\omega/\omega_0$ and $\omega_0^2/\omega^2 \approx 1 - 2\delta\omega/\omega_0$. Consequently, Eq. (5.11) can be approximately rewritten as

$$1 - \frac{g_m}{R\omega_0^2 C_1 C_2}\left(1 - 2\frac{\delta\omega}{\omega_0}\right) + 2jQ\frac{\delta\omega}{\omega_0} = 0 \tag{5.12}$$

from which

$$\frac{\delta\omega}{\omega_0} = \frac{\frac{g_m}{R\omega_0^2 C_1 C_2} - 1}{\frac{2g_m}{R\omega_0^2 C_1 C_2} + 2jQ}. \tag{5.13}$$

It will be noted that $g_m/R\omega_0^2 C_1 C_2$ is positive and, from the Barkhausen criterion, has a value around 1 as the oscillator approaches equilibrium. Further, Q is also positive and usually has a fairly large value (10 or more). We let $\delta\omega = \delta\omega_r + j\delta\omega_i$ and then the output voltage of the oscillator, the voltage at the source, will be

$$v = \Re\left\{\frac{i_2}{j\omega C_2}\right\} = \exp(-\delta\omega_i t)\Re\left\{\exp(j(\omega_0 + \delta\omega_r)t)\frac{I_2}{j\omega_0 C_2}\right\}. \tag{5.14}$$

From Eq. (5.13), it will be seen that $\delta\omega_i$ (the imaginary part of $\delta\omega$) will be negative if $g_m/R\omega_0^2 C_1 C_2 > 1$ and so the oscillations will grow with time. Consequently, in order to ensure that oscillations start, we need to choose $g_m/R\omega_0^2 C_1 C_2$ to be suitably greater than 1. Then, as the amplitude builds up, the value of g_m will fall due to gain compression and $g_m/R\omega_0^2 C_1 C_2$ will settle down to a value close to 1. From the above considerations, it is obvious that the gain compression in the amplifier is important to the functioning of the oscillator. In order to help this aspect, a reverse-bias diode is often added at the gate of the oscillator of Figure 5.3a in order to enhance the gain compression of the transistor (the effective g_m will now tend to zero as the voltage becomes large). The optional diode is shown as dotted lines in Figure 5.3a and the gate voltage will limit at around 0.8 V for a silicon diode.

Eq. (5.13) also tells us something about the stability of an oscillator. During startup, the transconductance g_m will steadily fall as the oscillator amplitude settles into its equilibrium state. As can be seen from Eq. (5.13), this will also mean that the oscillation frequency ω will take time to settle into its equilibrium value ω_0, i.e. there will be some initial frequency drift. To this, however, we must add the possibility that there will be additional changes to g_m as a result of the transistor warming up after being switched on. The temperature can take some time to settle down to an equilibrium state and so frequency can drift during this period. What will be noted from Eq. (5.13) is that the larger the value of Q, the smaller the effect of these variations upon frequency. As a consequence it is important to seek the highest possible value of Q in an oscillator circuit. So far we have only considered the inductor resistance to contribute to the Q of the circuit and this is indeed the most problematic resistive component in FET oscillators. However, other resistances can contribute to Q and, in particular, the resistance of any load at the output of the oscillator (the FET source). To reduce the effect of a load, it is prudent to use a high-input impedance buffer amplifier between the oscillator and the load (a common-drain amplifier for example).

Another way of looking at oscillation is through the concept of negative resistance. In Chapter 1, we saw that a tuned circuit will ring at its resonant frequency when excited by an impulse. The oscillations, however, will steadily decay due to the resistance in the circuit. To prevent this decay, we need to introduce a negative resistance that cancels out the inherent resistance in the circuit. How do we create this negative resistance? Consider the circuit shown in Figure 5.4a when analysed through the model in Figure 5.4b. We assume oscillations at frequency ω and so represent the currents and voltages in terms of their complex amplitudes. The voltage at the device input is

$$V = \frac{I}{j\omega C_1} + \left(I + g_m \frac{I}{j\omega C_1}\right) \frac{1}{j\omega C_2} \tag{5.15}$$

so that the impedance looking into the device is given by

$$Z_i = \frac{V}{I} = -\frac{g_m}{C_1 C_2 \omega^2} + \frac{C_1 + C_2}{j\omega C_1 C_2}. \tag{5.16}$$

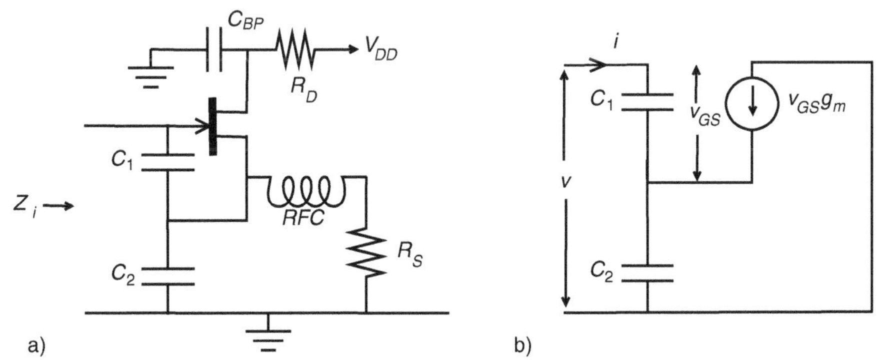

Fig. 5.4 Negative-resistance device and its analysis.

The impedance is the series combination of resistance

$$R_i = -\frac{g_m}{C_1 C_2 \omega^2} \tag{5.17}$$

and the capacitance $C_i = C_1C_2/(C_1 + C_2)$. If we now place an inductor L across this impedance, the circuit can ring at the resonant frequency of C_i and L. In the real world, the inductor will contain some resistance that can be represented by a resistor R in series with the inductor ($R = Q_U \omega L$ where Q_U is the unloaded Q of the inductor) and this will moderate the growth of oscillations. However, the oscillations will still grow provided that, at the resonant frequency of L and C_i, condition $R + R_i < 0$ is satisfied. The oscillations will continue to grow in amplitude until a point is reached where the nonlinearity in the FET has reduced the transconductance to a point where $R + R_i = 0$. Note that, by adding an inductance L across the input to circuit Figure 5.4a, we have turned it into the Colpitts oscillator of Figure 5.3a, i.e. we have a negative resistance explanation of this oscillator.

The phenomenon of negative resistance can also shed further light on the regenerative amplifier that was discussed at the start of this section. We have already seen that positive feedback in an amplifier can generate a large gain. In addition, we have just seen that feedback can generate negative resistance that can cancel the inherent resistance in a tuned circuit at the input to an amplifier, i.e. it can greatly increase the Q of this tuned circuit and hence its selectivity. This combination of beneficial properties made it possible to build the cheap and effective radios that were the mainstay of the consumer radio industry well into the 1930s. Figure 5.5 shows a modern form of regenerative amplifier that is based on FETs. In this configuration, the lower transistor provides the negative feedback (regeneration) and the upper transistor controls the current through this device (i.e. the level of regeneration). The amplifier of Figure 5.5 can be made to perform the additional function of demodulation by setting the bias at the source so that there is strong asymmetry in response to the upswings and downswings of the RF voltage at the gate. To complete the demodulation, however, there will need to be a low-pass filter at the output in order to remove the RF frequencies. As a consequence, it is possible

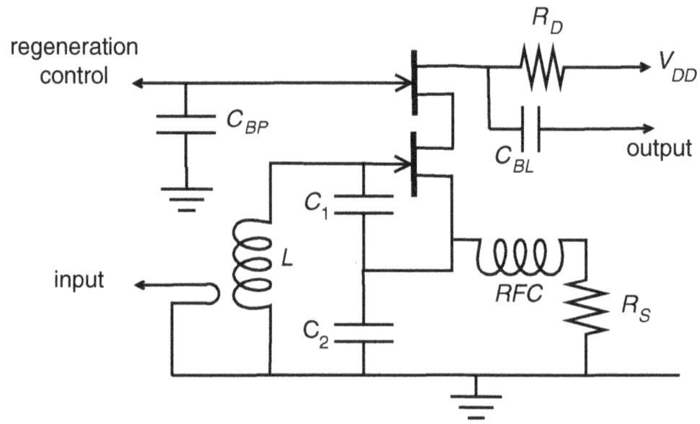

Fig. 5.5 Regenerative amplifier.

to make the amplifier of Figure 5.5 into a simple receiver that is both highly selective and highly sensitive.

In 1922, Armstrong took his concept of regeneration one step further and invented what is known as the super-regenerative amplifier. In the regenerative amplifier of Figure 5.5, we now replace a constant bias at the gate of the upper transistor by an oscillating voltage (i.e. $V_b + V_q \sin(\omega_q t)$). The constant part of the bias voltage V_b is set such that, on the upswing, the voltage at the gate is high enough to set the lower transistor into oscillation and, on the downswing, is low enough to quench the oscillation. A signal that enters the tuned circuit on the upswing will now set the lower transistor into oscillation, the amplitude of which will grow exponentially until quenched by the downswing. It is clear that the level of the signal at the output will depend on the period of oscillation. This should be as long as possible to get the greatest level of output signal, but not so long that it interferes with any modulation of the RF signal (the *quench frequency* is usually set at several times the highest modulating frequency). Phenomenal gains can be attained with a super-regenerative amplifier, but at the cost of an output that is additionally modulated by the quench frequency. The additional modulation, however, is usually not a problem since it can be filtered out after demodulation.

5.2 Mixers

Thus far we have, almost exclusively, considered rectification as the means of extracting the modulating signal from an amplitude-modulated radio signal. In rectification we extinguish the negative swings of the signal and then smooth out (filter) the remaining signal to leave the modulating signal (see Figure 2.16 of Chapter 2). For amplitude modulation (AM), the signal has the form

$$v_{\mathrm{RF}} = V_{\mathrm{RF}}(1 + a(t))\cos(\omega_c t + \phi), \tag{5.18}$$

where ω_c is the carrier frequency, ϕ is a phase offset and $a(t)$ is the modulating signal (we assume $|a(t)| < 1$). If we multiply v_{RF} by the sinusoidal *local oscillator* (LO) signal $v_{\mathrm{LO}} = \cos(\omega_c t)$, we obtain

$$\begin{aligned} v_{\mathrm{RF}} v_{LO} &= V_{\mathrm{RF}}(1 + a(t))\cos(\omega_c t + \phi)\cos(\omega_c t) \\ &= \frac{V_{\mathrm{RF}}}{2}(1 + a(t))(\cos\phi + \cos(2\omega_c t + \phi)), \end{aligned} \tag{5.19}$$

where we have used the trigonometric identity $\cos A \cos B = \frac{1}{2}(\cos(A-B) + \cos(A+B))$. We could then filter out the high-frequency term and be left with a term whose time varying component is proportional to the modulating signal, i.e.

$$v_{out} = \frac{V_{\mathrm{RF}}}{2}(1 + a(t))\cos\phi. \tag{5.20}$$

An AM demodulator that uses a mixer in this fashion is known as a *synchronous demodulator* and the receiver of Figure 5.6 is based upon this concept. Such designs are known as *direct conversion*, or *homodyne*, receivers.

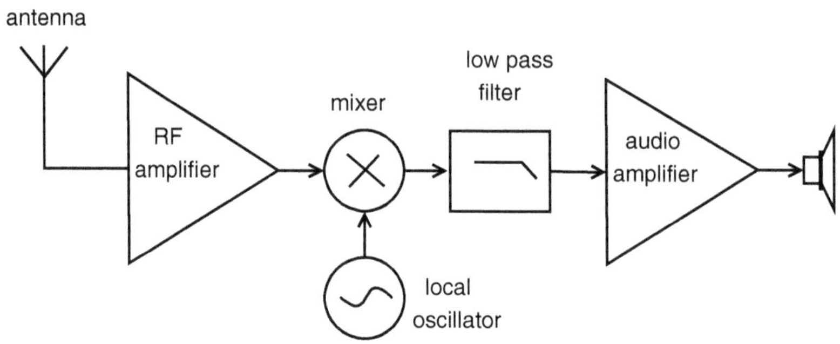

Fig. 5.6 Direct conversion receiver.

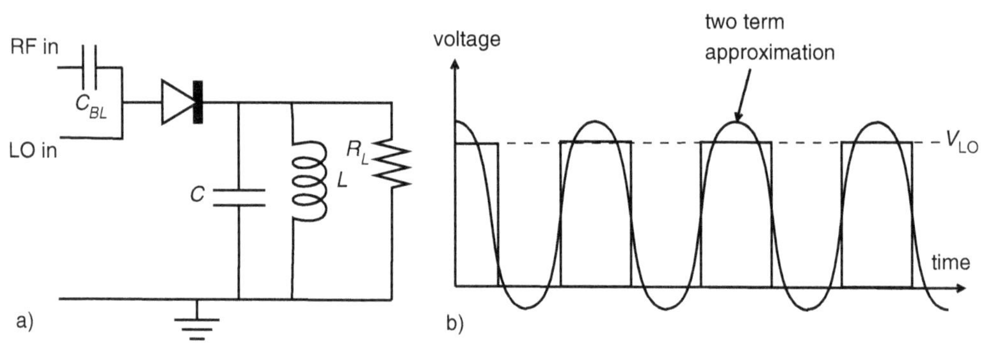

Fig. 5.7 Diode mixer and local oscillator signal.

A device that can achieve the product of two signals is known as a mixer. As we have seen in the previous chapter, a device that contains a square term in its nonlinear behaviour will contain such a product term in its output. Further, all the active devices we have studied thus far have such a term and so could be used as mixers. We will first consider the simplest mixer of all, the single-diode mixer (shown in Figure 5.7a). Diode mixers are usually operated such that the diode is hard on or hard off. This is achieved through a local oscillator voltage that is a sequence of pulses ($v_{LO} = V_{LO}(1+\text{sgn}(\cos(\omega_{LO}t)))/2$) that has an amplitude V_{LO} that is greater than the voltage that switches the diode on (about 0.7 volts for silicon and 0.3 volts for germanium). This will effectively produce a voltage across the load that is a series of pulses multiplied by the incoming RF voltage ($v_{RF} = V_{RF}\cos(\omega_{RF}t)$) when offset by V_{LO}. The sequence of pulses can be represented by a Fourier series, i.e.

$$v_{LO} = V_{LO}\left(1/2+2\cos(\omega_{LO}t)/\pi+2\cos(3\omega_{LO}t)/3\pi+2\cos(5\omega_{LO}t)/5\pi+\cdots\right).$$

As a first approximation we take the first two terms of the Fourier series, i.e. $v_{LO} \approx V_{LO}\left(1/2+2\cos(\omega_{LO}t)/\pi\right)$ (see Figure 5.7b). (In practice, we often make this approximation by using sinusoidal oscillations rather than pulses.) The voltage across the load will now be given by

$$v_{IF} = V_{LO}\left(\frac{1}{2}+\frac{2}{\pi}\cos(\omega_{LO}t)\right)(V_{LO}+V_{RF}\cos(\omega_{RF}t))$$

$$= V_{\text{LO}}^2 \left(\frac{1}{2} + \frac{2}{\pi} \cos(\omega_{\text{LO}} t) \right)$$
$$+ V_{\text{RF}} V_{\text{LO}} \left(\frac{1}{2} \cos(\omega_{\text{RF}} t) + \frac{1}{\pi} \cos((\omega_{\text{RF}} - \omega_{\text{LO}}) t) \right.$$
$$\left. + \frac{1}{\pi} \cos((\omega_{\text{RF}} + \omega_{\text{LO}}) t) \right). \tag{5.21}$$

The major problem with the mixer is that the output contains both the RF signal, LO signal and signals at the frequencies $\omega_{\text{RF}} + \omega_{\text{LO}}$ and $\omega_{\text{RF}} - \omega_{\text{LO}}$. Usually, only one of the product frequencies is required and so all other frequencies are removed by making the parallel tuned circuit resonant at the required frequency. Furthermore, mixer products that were ignored in approximating the LO pulses are also filtered out by this process.

The square-law behaviour of the FET characteristic offers another opportunity for mixing and Figure 5.8 shows an example of a single FET mixer. Let $v_{\text{RF}} = V_{\text{RF}} \cos(\omega_{\text{RF}} t)$ be the incoming RF signal at the FET gate and $v_{\text{LO}} = V_{\text{LO}} \cos(\omega_{\text{LO}} t)$ be the local oscillator signal at the source. If the transistor has a bias current I_D at the drain, then the total current at the drain will be $I_D + i_D$ where i_D is the RF current at the drain. From the characteristic equation of the FET,

$$I_D + i_D = K \left(v_{\text{RF}} - v_{\text{LO}} - I_D R_S - V_T \right)^2$$
$$= K \left(v_{\text{RF}}^2 - 2 v_{\text{RF}} v_{\text{LO}} + v_{\text{LO}}^2 - 2(v_{\text{RF}} - v_{\text{LO}})(I_D R_S + V_T) \right.$$
$$\left. + (I_D R_S + V_T)^2 \right). \tag{5.22}$$

The tuned circuit in the drain of the FET is chosen to be resonant at either $\omega_{\text{RF}} + \omega_{\text{LO}}$ or $\omega_{\text{RF}} - \omega_{\text{LO}}$ and so the $v_{\text{RF}} v_{\text{LO}}$ term in the above expansion is the only term of relevance. Since $v_{\text{RF}} v_{\text{LO}} = V_{\text{RF}} V_{\text{LO}} (\cos((\omega_{\text{RF}} - \omega_{\text{LO}})t)/2 + \cos((\omega_{\text{RF}} + \omega_{\text{LO}})t)/2)$, the current through the load R_L will be $-K V_{\text{RF}} V_{\text{LO}} \cos((\omega_{\text{RF}} - \omega_{\text{LO}})t)$ or $-K V_{\text{RF}} V_{\text{LO}} \cos((\omega_{\text{RF}} + \omega_{\text{LO}})t)$, depending on the resonant frequency of the drain circuit.

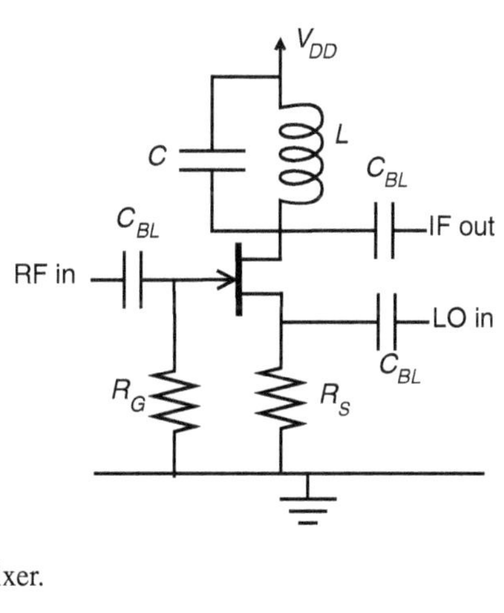

Fig. 5.8 Simple square-law FET mixer.

The major problem with the mixers that we have discussed thus far is that their output contains all the input frequencies, plus the product frequencies. When only a fixed product frequency is required, the unwanted signals can be removed by suitable filtering, as shown in Figures 5.7 and 5.8. When broadband operation is required, we need to consider other means for eliminating the unwanted signals. A solution is found in what is known as a *balanced mixer*, an example of which is shown in Figure 5.9a. Once again, the local oscillator signal is a sequence of pulses that, in this case, swing between V_{LO} and $-V_{\mathrm{LO}}$ (i.e. $v_{\mathrm{LO}} = V_{\mathrm{LO}}\mathrm{sgn}(\cos(\omega_{\mathrm{LO}}t))$. The voltage amplitude V_{LO} is made large enough so that, at any one time, one side of the circuit will be completely on and the other side completely off. As a consequence, the voltage across the output terminals will be given by

$$v_{\mathrm{IF}} = (I_{bias} + v_{\mathrm{RF}}g_m)R_D\mathrm{sgn}(\cos(\omega_{\mathrm{LO}}t)), \tag{5.23}$$

where I_{bias} is the current that flows as a result of the bias that is applied to gate of the lower FET. As in the case of the diode mixer, we approximate the sequence of pulses by a Fourier series, i.e. $\mathrm{sgn}(\cos(\omega_{\mathrm{LO}}t)) = 4\cos(\omega_{\mathrm{LO}}t)/\pi + 4\cos(3\omega_{\mathrm{LO}}t)/3\pi + 4\cos(5\omega_{\mathrm{LO}}t)/5\pi + \cdots$. Then, truncating this series at the first term, and using the identity $\cos A \cos B = \frac{1}{2}(\cos(A-B) + \cos(A+B))$, we obtain

$$\begin{aligned} v_{IF} &\approx (I_{bias} + V_{\mathrm{RF}}g_m\cos(\omega_{\mathrm{RF}}t))R_D\frac{4}{\pi}\cos(\omega_{\mathrm{LO}}t)) \\ &= I_{bias}R_D\frac{4}{\pi}\cos(\omega_{\mathrm{LO}}t) + \frac{2}{\pi}V_{\mathrm{RF}}g_mR_D\left(\cos((\omega_{\mathrm{RF}} - \omega_{\mathrm{LO}})t)\right. \\ &\quad \left.+\cos((\omega_{\mathrm{RF}} + \omega_{\mathrm{LO}})t)\right). \end{aligned} \tag{5.24}$$

The balance between the two sides of the circuit has eliminated RF components to leave the product terms and the LO signal. To eliminate the LO signal, we can use what is known as a *double balanced mixer*. This consists of two single balanced mixers that are connected in such a way that the unwanted LO signal component cancels itself out.

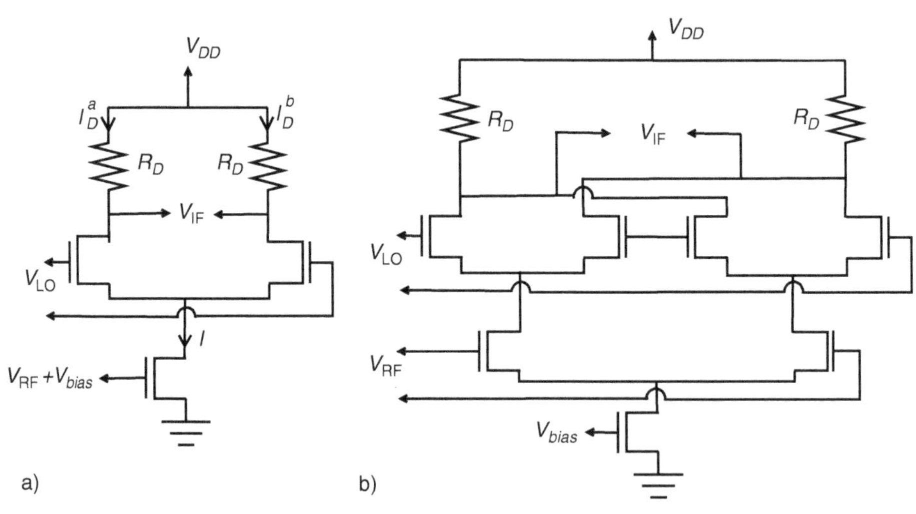

Fig. 5.9 Single and double balanced FET mixers.

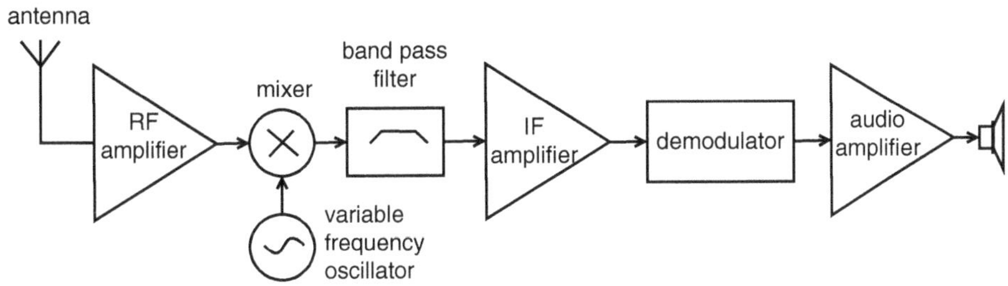

Fig. 5.10 Superheterodyne receiver.

Figure 5.9 shows an example of such a mixer, an architecture that is commonly known as a *Gilbert cell*. The drain resistors can be replaced by transistors, as in the case of the RF amplifier of Figure 4.30b, and so the mixer can be realised completely in terms of transistors. This has important implications. If we take a wafer of silicon, transistors can be constructed on this wafer by infusing impurities into the surface and adding layers of insulation, together with conducting tracks. Complex circuits can then be constructed by connecting these transistors by means of conducting tracks. For example, Gilbert cells and the amplifiers, of Figure 4.30b, can be combined to perform complex radio function on a single silicon wafer. What we have is an integrated circuit (IC) and such circuits have become an essential feature of modern RF electronics.

As we mentioned in Chapter 4, with reference to the TRF receiver, a radio that is designed to receive multiple frequencies needs to be able to simultaneously change several tuned circuits when the frequency is changed. As radio became more popular, the airwaves became more crowded and selectivity became a big issue. Receivers required more and more tuned elements to satisfy the increased selectivity requirements and the cost started to become prohibitive. To solve this problem, Armstrong came up with another of his brilliant ideas, the *superheterodyne receiver*. Consider the receiver architecture shown in Figure 5.10. A signal comes in through the antenna (at frequency $\omega_{\rm RF}$) and is amplified before being fed into a mixer. Here the signal is mixed with an LO signal (at frequency $\omega_{\rm LO}$) from a *variable-frequency oscillator* (VFO). At the output of the mixer, there is a fixed-frequency filter that chooses one of the product signals (either $\omega_{\rm RF} - \omega_{\rm LO}$ or $\omega_{\rm RF} + \omega_{\rm LO}$) and only has a bandwidth that is just wide enough to accommodate the modulation. This *intermediate-frequency* (IF) output is then amplified before being demodulated. Incoming signals are chosen through the correct choice of VFO frequency. The important thing to note is that it is possible to cheaply manufacture highly selective fixed-frequency band-pass filters, unlike the variable variety. The concept of the superheterodyne revolutionised the radio receiver and made it possible to build inexpensive receivers that are highly selective. The superheterodyne receiver, invented by Armstrong in 1918, is still the mainstay of receiver design.

5.3 Modulation and Demodulation

The simplest AM transmitter will have the topology shown in Figure 5.11a. The baseband signal $v_{AF} = 1 + a(t)$, where $a(t)$ is usually an audio signal, is mixed with a carrier

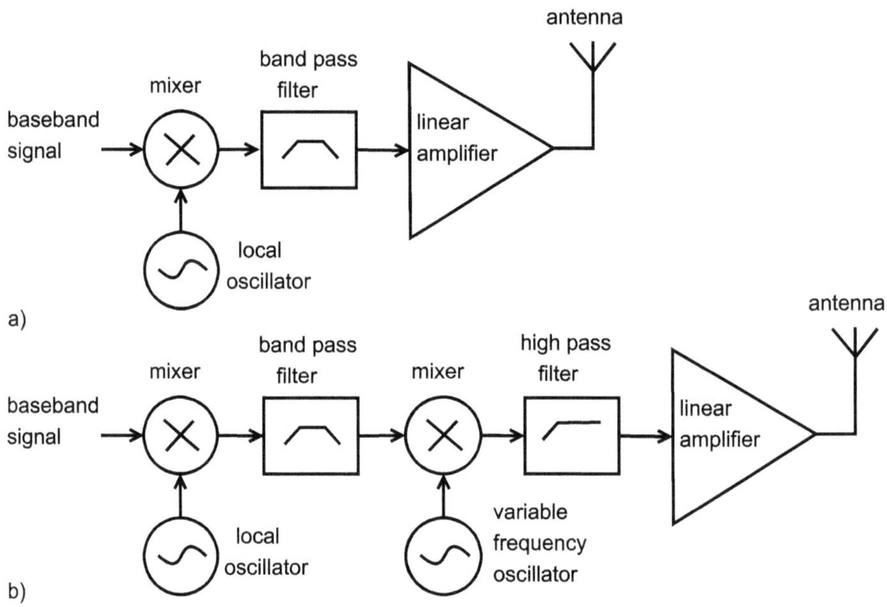

Fig. 5.11 Transmitter topologies.

signal $v_{\mathrm{LO}} = V_{\mathrm{LO}} \cos(\omega_c t)$ at frequency ω_c. The output of the mixer is the signal $v_{\mathrm{RF}} = v_{AF} V_{\mathrm{LO}} \cos(\omega_c t)$ which is then filtered and amplified. A filter is unnecessary for an ideal mixer, but even the best practical mixers will produce undesirable products that will need to be removed. To amplify an AM modulated signal up to the level of power required for transmission, the amplifier will need to be highly linear in order to avoid corrupting the modulation. Consider now the situation where the baseband signal is sinusoidal, i.e. $a(t) = a_0 \cos(\omega_{AF} t)$, then

$$\begin{aligned} v_{\mathrm{RF}} &= V_{\mathrm{LO}}(1 + a_0 \cos(\omega_{AF})) \cos(\omega_c t) \\ &= V_{\mathrm{LO}}\Big(\cos(\omega_c t) \\ &\qquad + \frac{a_0}{2}\cos((\omega_c + \omega_{AF})t) + \frac{a_0}{2}\cos((\omega_c - \omega_{AF})t)\Big). \end{aligned} \tag{5.25}$$

It will be noted that there is a signal at frequency ω_c that contains no information about the modulation (the *carrier signal*), a signal at frequency $\omega_c + \omega_{AF}$ and a signal at the mirror frequency $\omega_c - \omega_{AF}$. In general, the modulating signal will be far more complex (i.e. it will be the sum of sinusoidal signals at a variety of frequencies), but the structure of the modulated RF will be much the same. There will be a carrier at frequency ω_c and two sidebands. The first sideband will consist of a signal with spectrum that is the spectrum of $a(t)$ when translated upwards in frequency by ω_c. The second sideband will consist of a signal with spectrum that is the mirror spectrum of $a(t)$ when translated upwards in frequency by ω_c (see Figure 5.12). The signal below the carrier frequency is known as the *lower sideband* and that above as the *upper sideband*. Importantly, it will be seen that the carrier carries no information about the modulation and the sidebands carry exactly the same information. From an information viewpoint, it is senseless to

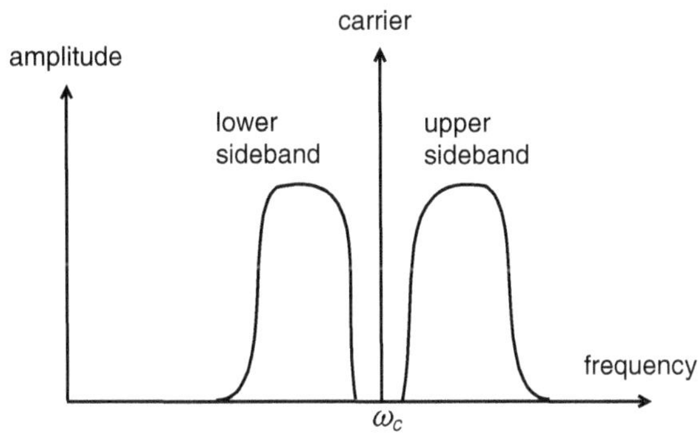

Fig. 5.12 Sideband structure in an AM signal.

transmit anything more than one of the sidebands. This brings us to the concept of single-sideband (*SSB*) modulation in which we merely transmit one of the sidebands (either upper or lower). This mode of operation saves both bandwidth (it is half that of AM) and power. The simplest way to generate SSB is with the transmitter of Figure 5.11a. The band-pass filter is now chosen such that it removes the carrier and the unwanted sideband. Obviously, such a transmitter will require a band-pass filter with very high selectivity and so the transmitter of Figure 5.11a will only be practical for a single frequency. Figure 5.11b, however, shows a transmitter that is suitable for operation over a range of frequencies. This is the transmitter equivalent of the superheterodyne receiver. It consists of a low power SSB transmitter that produces an SSB modulated output at a frequency ω_{IF}. This is then mixed with the output of a VFO at frequency ω_{VFO} to form two signals, one at frequency $\omega_{\text{IF}} + \omega_{\text{VFO}}$ and one at frequency $\omega_{\text{IF}} - \omega_{\text{VFO}}$. The mixer is then followed by a filter that chooses the correct product and then a linear power amplifier brings the signal up to a suitable power level.

Single sideband can be demodulated using synchronous demodulation but, unfortunately, it will respond to both sidebands. Consequently, if the receiver is tuned into a desired sideband, there could be an unwanted signal within the other sideband and this will also be demodulated, hence interfering with the desired signal. A solution is to use the superheterodyne receiver of Figure 5.10 with a demodulator of the synchronous variety. The band-pass filter, however, is now chosen narrow enough to only allow through a single sideband and so there will be no competing signal at the synchronous detector.

The above approach to the generation and demodulation of SSB relies upon the use of highly selective filters, but there is another approach. Figure 5.13a shows what is known as a *phasing SSB generator*. In essence, this circuit consists of two mixer-based AM generators, but with one having both its inputs advanced by 90°. If we consider the modulating signal to be sinusoidal, i.e. $v_B(t) = V_B \cos(\omega_B t)$, the sum of the generator outputs will be

$$\begin{aligned} v_{SSB} = {} & V_{\text{LO}} V_B \cos(\omega_c t) \cos(\omega_B t) \\ & + V_{\text{LO}} V_B \cos\left(\omega_c t + \frac{\pi}{2}\right) \cos\left(\omega_B t + \frac{\pi}{2}\right) \end{aligned}$$

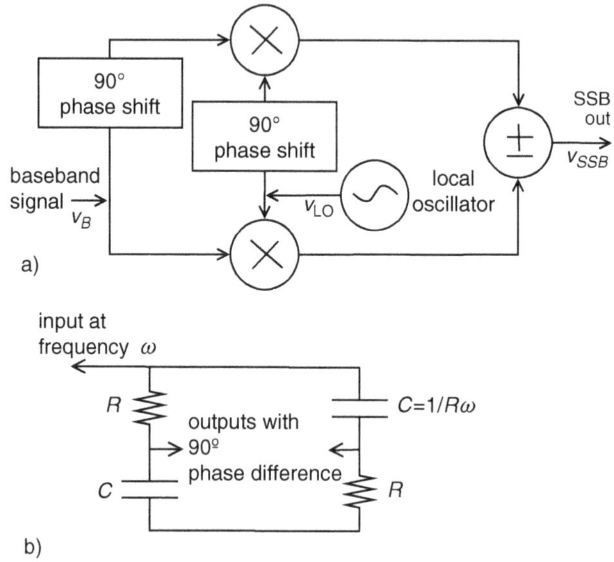

Fig. 5.13 The phasing method for generating SSB.

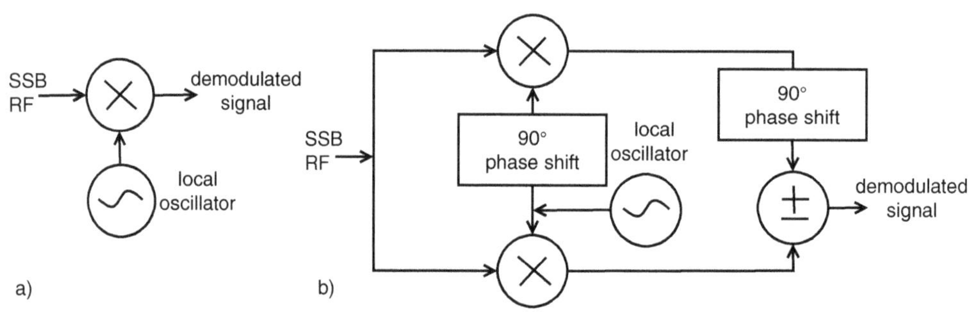

Fig. 5.14 Methods for demodulating SSB.

$$
\begin{aligned}
&= \frac{V_{\mathrm{LO}} V_B}{2} (\cos((\omega_c - \omega_B)t) + \cos((\omega_c + \omega_B)t)) \\
&\quad + \frac{V_{\mathrm{LO}} V_B}{2} (\cos((\omega_c - \omega_B)t) - \cos((\omega_c + \omega_B)t)) \\
&= V_{\mathrm{LO}} V_B \cos((\omega_c - \omega_B)t). \qquad (5.26)
\end{aligned}
$$

Consequently, we have produced the lower sideband alone without the need for filtering (we can produce the upper sideband by merely taking a difference of modulator outputs).

The phasing approach can also be used for the demodulation of SSB by means of the process shown in Figure 5.14b. The demodulator consists of two synchronous detectors, one having its oscillator and output advanced in phase by 90°. If we sum the outputs, the upper sideband will be demodulated and the lower sideband eliminated (if take the difference between the outputs the lower sideband will be demodulated). The major practical problem with the phasing approach comes from the production of the 90° phase change. For the local oscillator, this can be achieved through the circuit of Figure 5.13b. In the case of the baseband signal, however, its bandwidth is often too wide for such a circuit and a polyphase network must be used. In modern communications, the signal is often modulated, and/or demodulated, within a computer. Within the digital domain, it is easy to produce the 90° phase shift over a broad range of frequencies and so, increasingly, the *in-phase* and *quadrature* (90° out-of-phase) baseband signals are produced and/or processed within the digital domain.

Thus far, we have concentrated on AM and its derivative modulation SSB, but we now turn our attention to *angle modulation* (FM and PM). Phase modulation can be generated through the circuit shown in Figure 5.15a. This circuit includes a reverse-biased diode which acts as a voltage-controlled reactance. The reverse-biased PN junction of a diode acts as a capacitor across its depletion region and its capacitance varies with bias due to the fact the width of the depletion region varies with this bias voltage. As a consequence, the capacitance in the circuit will vary with the incoming baseband voltage ($C \approx C_0 + v_B C_1$). Let the RF input to the modulator have the form $v_{LO} = \Re\{V_{LO}\exp(j\omega_{LO}t)\}$ and the output have the form $v_{PM} = \Re\{V_{PM}\exp(j\omega_{LO}t)\}$. The circuit is essentially a baseband-controlled

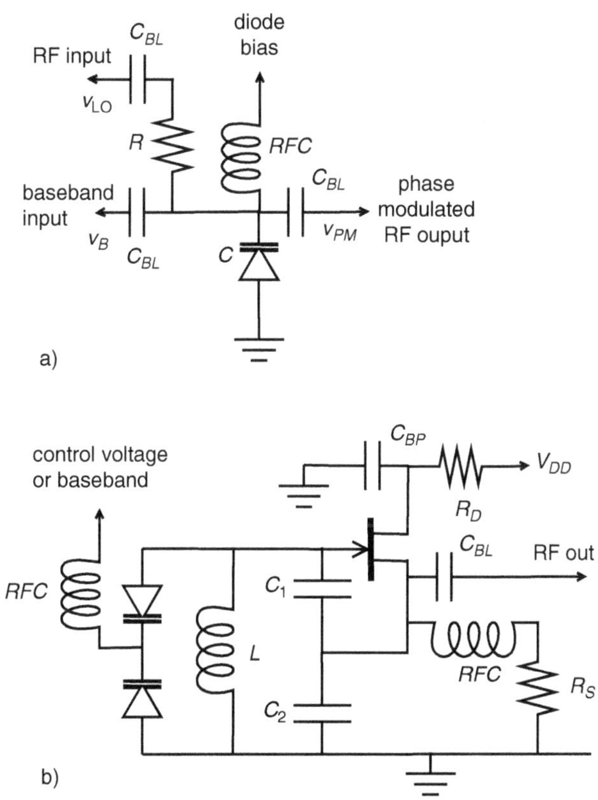

Fig. 5.15 A phase modulator and a voltage controlled oscillator for generating FM.

voltage divider for which the complex amplitude of the output signal is given by

$$V_{PM} = V_{\mathrm{LO}} \frac{\frac{1}{\mathrm{j}\omega_{\mathrm{LO}}C}}{R + \frac{1}{\mathrm{j}\omega_{\mathrm{LO}}C}} = \frac{V_{\mathrm{LO}}}{\mathrm{j}\omega_{\mathrm{LO}}RC + 1}. \tag{5.27}$$

If $\omega_{\mathrm{LO}}RC \ll 1$, then $V_{PM} \approx V_{\mathrm{LO}}(1 - \mathrm{j}v_B\omega_{\mathrm{LO}}RC)$ and, on noting that $\exp x \approx 1 + x$ for small x, we find that

$$\begin{aligned} v_{PM} &= \Re\{V_{PM}\exp(\mathrm{j}\omega_{\mathrm{LO}}t)\} \\ &= \Re\{V_{\mathrm{LO}}\exp(\mathrm{j}(\omega_{\mathrm{LO}}t - v_B\omega_{\mathrm{LO}}RC_1 - \omega_{\mathrm{LO}}RC))\}, \end{aligned} \tag{5.28}$$

i.e. the baseband signal modulates the phase. A synchronous demodulator can be used to demodulate a PM signal since its output is proportional to the cosine of the phase difference between the input and local oscillator signals (see (5.20)). It will be noted, however, that this phase difference needs to be small in order for the demodulation to be linear.

FM can be generated by using what is known as a *voltage-controlled oscillator* (VCO). Figure 5.15b shows the circuit of the Colpitts VCO in which diodes have been added to allow the frequency to be varied through the variation of their bias, and hence capacitance. To generate FM, we simply use the baseband signal as the biasing voltage and then the frequency of the oscillator will vary in sympathy with the baseband signal. Demodulating FM can be done by means of what is known as slope detection. Consider the circuit in Figure 5.16a and for which the tuned circuit has a very high Q. The frequency response of the tuned circuit will look like that shown in Figure 5.16b. If the peak response of the tuned circuit is now offset from the centre of the FM frequencies, as the frequency of the FM signal varies the amplitude of its voltage will vary. Essentially, we have converted an FM signal into an AM signal. The AM signal will now be demodulated by the diode detector shown in Figure 5.16a.

Since we are dealing with angle modulation, the amplitude of the signal is irrelevant. Consequently, it is normal in an FM receiver to remove any possibility of AM interference by placing an amplitude-limiting amplifier prior to the demodulator. A simple example of such an amplifier is shown in Figure 5.17. In this amplifier, negative feedback is used to control the gain with the diode feedback circuit limiting the gain when a threshold output voltage is reached.

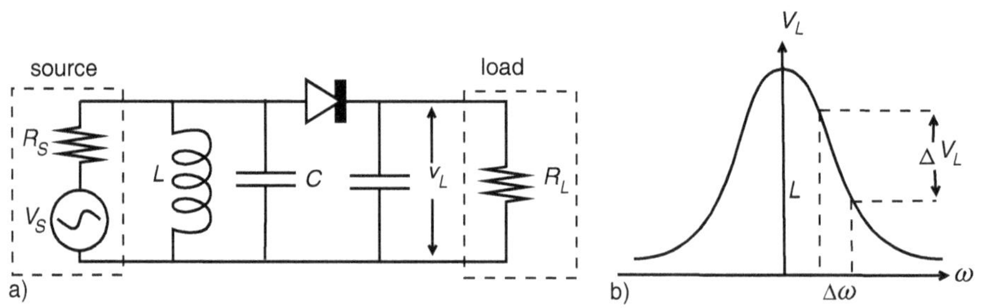

Fig. 5.16 Slope detector for demodulating FM.

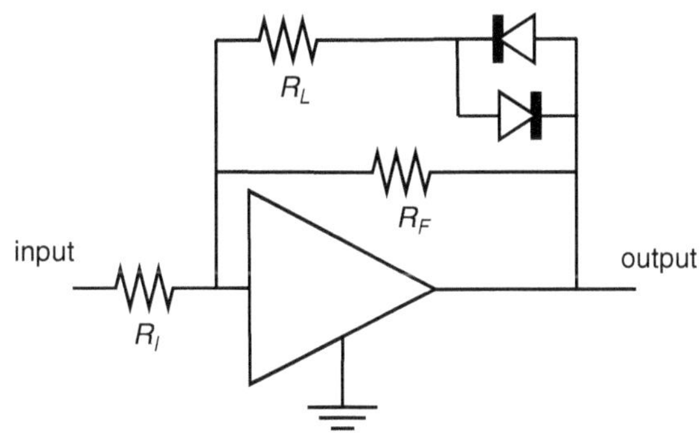

Fig. 5.17 A limiting amplifier.

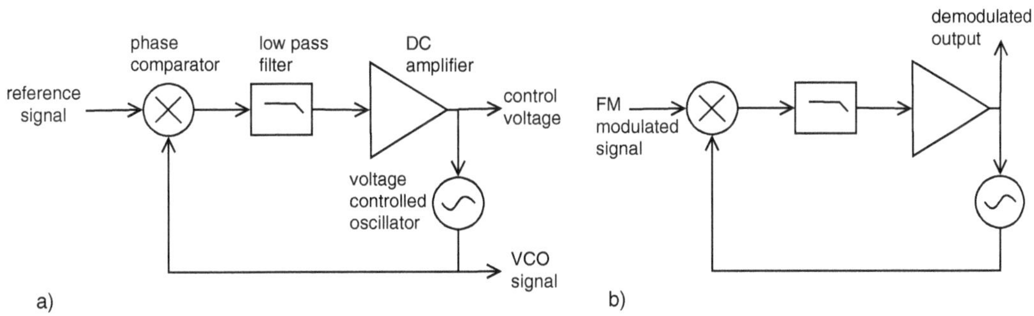

Fig. 5.18 Demodulating FM by means of a phase-locked loop.

FM signals can also be demodulated by means of what is known as a phase-locked loop, an example of which is shown in Figure 5.18a. In such a loop, the phase of the output of a VCO is compared with the phase of a reference signal (this operation is shown as a mixer since this is the simplest example of a phase-comparison device). After filtering and amplification, the phase difference is then passed to the VCO as a control voltage. Any small change in VCO phase is thus fed back as a control voltage and, by this means, the VCO is locked onto the reference signal. If the incoming reference signal is FM-modulated (see Figure 5.18b), the VCO will be locked onto this signal and the control voltage will be the demodulated FM.

FM radio was invented by Edwin Armstrong in 1933 in an effort to overcome some of the drawbacks of AM, and *static noise* in particular. This noise (originating from lightning and man-made sources such as electric switching) detracted from the quality of broadcasts and, in particular, that of music. Armstrong was able to demonstrate that, by increasing the bandwidth of the FM modulation, it was possible to obtain interference-free reception for a relatively low carrier-to-noise ratio. The signal-to-noise ratio after demodulation is given by

$$\mathrm{SNR} \approx 3\left(\frac{B_{FM}}{2B}\right)^3 \mathrm{CNR}, \tag{5.29}$$

where B is the bandwidth of the baseband, B_{FM} is the bandwidth of the FM signal and CNR is the carrier-to-noise ratio in the bandwidth B_{FM}. Unfortunately, the above relation breaks down when CNR is less than 13 dB and the demodulated signal is rapidly overtaken by the noise at lower CNRs.

5.4 Oscillator Noise and Reciprocal Mixing

A particularly useful way of viewing an oscillator is as a filter of the random noise that excites the oscillations. We will consider the Colpitts oscillator of Section 5.1 from this viewpoint. The noise will be represented by a voltage source v_n at the gate of the amplifier and we will study the behaviour of the oscillator in terms of the model of Figure 5.19 (note that the mean square noise voltage per unit bandwidth is given by $\overline{v_n^2} = (F-1)kTR$, where F is the noise factor of the amplifier). We consider noise input at frequency ω and so represent the currents and voltages in terms of their complex amplitudes. The analysis proceeds as for the Colpitts oscillator, except that Eq. (5.5) is replaced by

$$V_{GS} = \frac{I_1}{j\omega C_1} + V_n, \tag{5.30}$$

where V_n is the noise amplitude. From (5.3) and (5.30)

$$I_2 = I_1\left(1 + \frac{g_m}{j\omega C_1}\right) + V_n g_m \tag{5.31}$$

and hence Eq. (5.7) will be replaced by

$$I_1\left(R + j\omega L + \frac{1}{j\omega C_1} + \frac{1}{j\omega C_2} - \frac{g_m}{\omega^2 C_1 C_2}\right) + \frac{V_n g_m}{j\omega C_2} = 0. \tag{5.32}$$

From (5.31), the complex amplitude V of the RF voltage at the output of the amplifier (i.e. the transistor source) is given by

$$V = \frac{I_2}{j\omega C_2} = \frac{I_1}{j\omega C_2}\left(1 + \frac{g_m}{j\omega C_1}\right) + \frac{V_n g_m}{j\omega C_2} \tag{5.33}$$

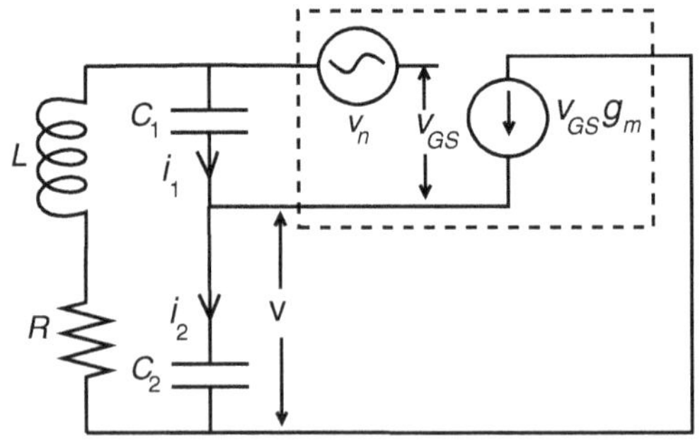

Fig. 5.19 Colpitts oscillator with noise source.

and so, from (5.32) and (5.33), we obtain

$$V = \frac{V_n g_m}{j\omega C_2} \frac{R + j\omega L + \frac{1}{j\omega C_1}}{R + j\omega L + \frac{1}{j\omega C_1} + \frac{1}{j\omega C_2} - \frac{g_m}{\omega^2 C_1 C_2}}. \tag{5.34}$$

This can be rewritten as

$$V = \frac{V_n g_m}{j\omega C_2} \frac{1 + jQ\frac{\omega}{\omega_0}\left(1 - \frac{\omega_0^2}{\omega^2}\right) - \frac{1}{jR\omega C_2}}{1 - \frac{g_m}{R\omega^2 C_1 C_2} + jQ\frac{\omega}{\omega_0}\left(1 - \frac{\omega_0^2}{\omega^2}\right)}, \tag{5.35}$$

where $Q = \omega_0 L/R$ and $\omega_0 = \sqrt{(C_1 + C_2)/LC_1C_2}$. After the oscillator has settled into equilibrium, the parameter $\Delta = 1 - g_m/R\omega_0^2 C_1 C_2$ will be extremely small. However, Δ cannot be exactly zero since this would make V infinite. If we now consider frequencies close to ω_0, the denominator in (5.35) will dominate and so we can set the numerator to its value at $\omega = \omega_0$. Noting that $1 - \omega_0^2/\omega^2 \approx 2\delta\omega/\omega_0$, where $\delta\omega = \omega - \omega_0$, we then find that

$$V \approx \frac{V_n g_m}{j\omega_0 C_2 \Delta} \frac{1 - \frac{1}{jR\omega_0 C_2}}{1 + 2jQ\frac{\delta\omega}{\Delta\omega_0}} \tag{5.36}$$

It will be seen that the amplitude will fall off as $1/\delta\omega$ as we move away from resonance. Since the noise source will produce output on all frequencies, it is clear from (5.36) that the oscillator will produce a spread of frequencies and not a single frequency. Further, the spread in frequency will be of the order of $\Delta\omega_0/Q$.

The spread in frequency at the oscillator output can be regarded as a modulation of the desired signal at frequency ω_0 and, due to the random nature of the noise source, the modulation itself will be random. However, due to gain compression, the amplitude will remain fairly constant and the modulation will essentially consist of phase fluctuations (note that frequency is the rate of change of phase). For this reason, oscillator noise is often termed *phase noise*. From (5.36) it will be noted that the higher the value of Q, the narrower the spread in frequency. As a consequence, it can be seen that a high Q is desirable for low phase noise.

The spread of frequencies in a practical oscillator is obviously a major problem as it prevents us from transmitting a totally clean signal by adding some random modulation onto the signal. Further, at the receive end, the local oscillator can also add additional random modulation to the incoming signal. There is, however, an additional problem at the receiver. Consider a superheterodyne receiver with IF frequency f_{IF} that is tuned to a desired signal S1 on frequency f_1. If there is also a strong undesired signal S2 on another frequency f_2, then this signal will also be present at the IF if frequency $f_{LO} + f_2 - f_1$ is present within the spectrum of the LO. This phenomenon, illustrated in Figure 5.20, is known as *reciprocal mixing*.

From the above considerations, and those of Section 5.1, it is clear that a high-Q resonant circuit is essential for an oscillator to produce a clean and stable output. In a resonant circuit that is formed with capacitors and inductors, a major problem arises due to the fact that it is very difficult to produce inductors with self Q greater than

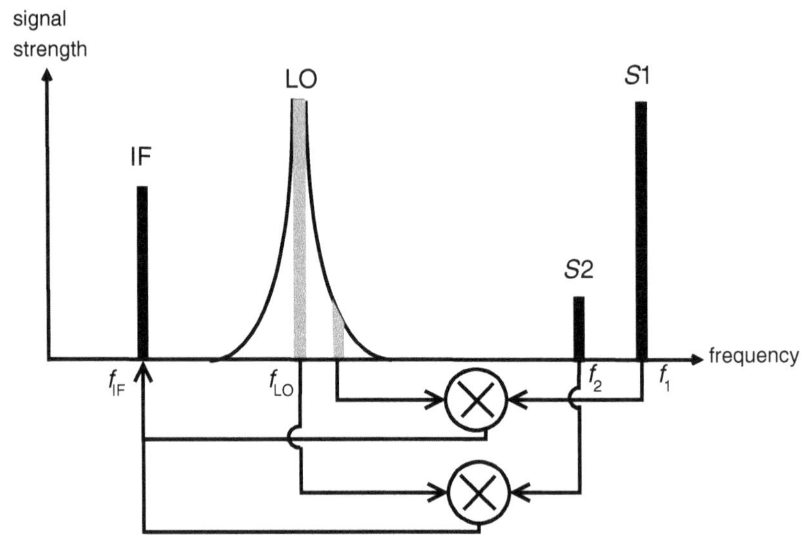

Fig. 5.20 Reciprocal mixing.

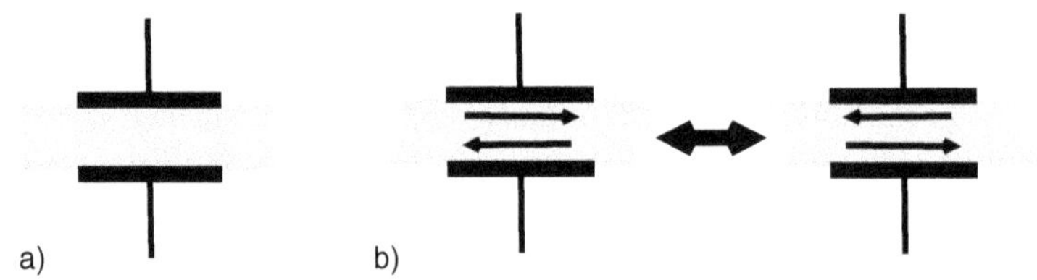

Fig. 5.21 Quartz crystal resonator and shear vibration mode.

about 200. As a consequence, such resonant circuits are often inadequate if we require a pure and stable sinusoid from the oscillator. A solution to this problem can be found in what is known as a *crystal resonator*. The mechanical vibrations of a crystalline solid can often exhibit a very high Q and this can be exploited to produce a very high-Q electrical resonator. This is possible through a phenomenon (discovered by Pierre Curie in 1880) known as the piezoelectric effect. Under this effect, deformations of some crystalline solids will induce an electric field (the deformations cause a realignment of the electric dipoles within the solid and this causes a field). The reverse is also true and an electric field across the same solid will induce a deformation (the fields cause a realignment of the electric dipoles within the solid). Consider a crystalline material that exhibits the piezoelectric effect (quartz is an example of such a material) when placed between two electrodes (see Figure 5.21). If a voltage is placed across the electrodes, this will cause a transverse deformation of the crystal. When the voltage is switched off, the crystal will vibrate at its mechanical resonant frequency ω_0 (this frequency depends on the dimensions and the elastic properties of the crystal). Shear waves will travel backwards and forwards between the upper and lower electrodes, the deformations being transverse to the propagation direction. (The distance between the electrodes will be an odd number of wavelengths, i.e. the crystal is capable of many vibration modes.) Through the piezoelectric effect, this will now mean that there is now an oscillating voltage across the electrodes, i.e. the quartz crystal will behave like a tuned circuit with resonant frequency ω_0. Importantly, this tuned circuit will have an extremely high Q (of the order of 10^7 for a quartz crystal). Figure 5.22a shows a model of the equivalent electrical resonator and Figure 5.22b the behaviour of the reactance of the circuit. It will be noted that the resonator has both series and parallel resonances. Quartz is the material often used for crystal resonators since it exhibits very little change in resonance properties with temperature.

A crystal oscillator circuit can be derived from one which uses inductors and capacitors by replacing the inductor with a crystal. Figure 5.23a shows a crystal oscillator that was derived from the Colpitts oscillator of Figure 5.3a. The fundamental mode of crystal vibration is usually only used for frequencies below 32 MHz, and above this the higher modes (overtones) tend to be used due to difficulty of making fundamental mode crystals for these frequencies. To make sure the circuit oscillates at the desired overtone, we modify the oscillator of Figure 5.23a to that of Figure 5.23b. By choosing capacitor C and inductor L to have a series resonance at the desired overtone, we then encourage oscillation at the desired overtone and block off oscillations at the fundamental and undesired overtones.

The problem with crystal oscillators is that it is difficult to vary the frequency. Some variation can be achieved in the above Colpitts oscillator by adding a variable capacitance in parallel with the crystal, but the achievable variation is small. Phase-lock loops, however, offer some other possibilities for producing stable oscillators. We first note that it is much easier to produce stable variable frequency oscillators (VFOs) at low frequencies. This can then be used to produce a stable oscillation at higher frequencies through the circuit shown in Figure 5.24a. The final signal (frequency f) is produced by a voltage-controlled oscillator (VCO), but a sample of this signal is converted down to a much lower frequency ($f - f_{XTAL}$) using a mixer and a stable crystal oscillator (frequency f_{XTAL}). The phase of this signal will then be compared with that from a stable VFO (frequency f_{VFO}) using a mixer followed by a low-pass filter. This will produce an

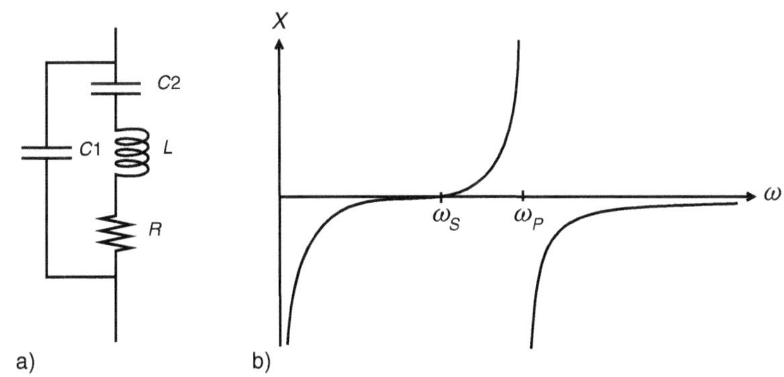

Fig. 5.22 Model of Quartz crystal resonator.

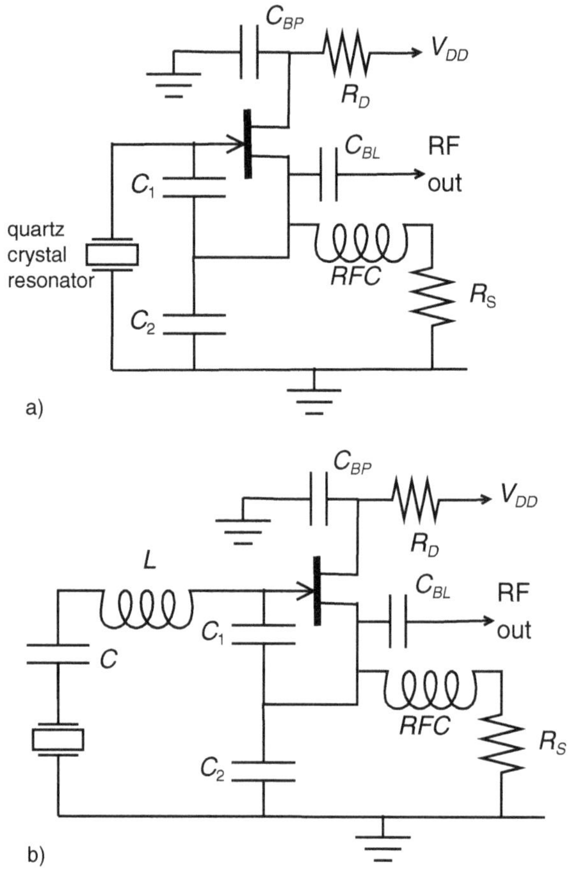

Fig. 5.23 Crystal controlled Colpitts oscillator and an overtone oscillator.

output that is a function of difference in phase, which, after amplification, produces a control voltage for the VCO. At the desired frequency, set by the stable low-frequency VFO, there will be a given phase difference that ensures the VCO runs at the correct frequency of $f_{\mathrm{VFO}} + f_{\mathrm{XTAL}}$. If the VCO drifts, there will be a change in phase difference that causes the VCO frequency to move back towards the desired frequency. In this way the VCO is locked onto the VFO.

Another possibility has come about through developments in digital electronics and, in particular, digital devices that act as programmable frequency dividers. By means of such a device, a phase-locked loop can be turned into a programmable frequency source. Figure 5.24b shows a suitable topology for such a *frequency synthesiser*. As before, the desired signal is generated by a voltage-controlled oscillator (VCO) and this is then divided down to the frequency of a reference oscillator by the programmable divider (the desired frequency needs to be a multiple of the reference frequency). If the VCO starts to drift from the desired frequency, there will be a phase difference between the frequency-divided VCO signal and the reference signal. This will result in a voltage at the

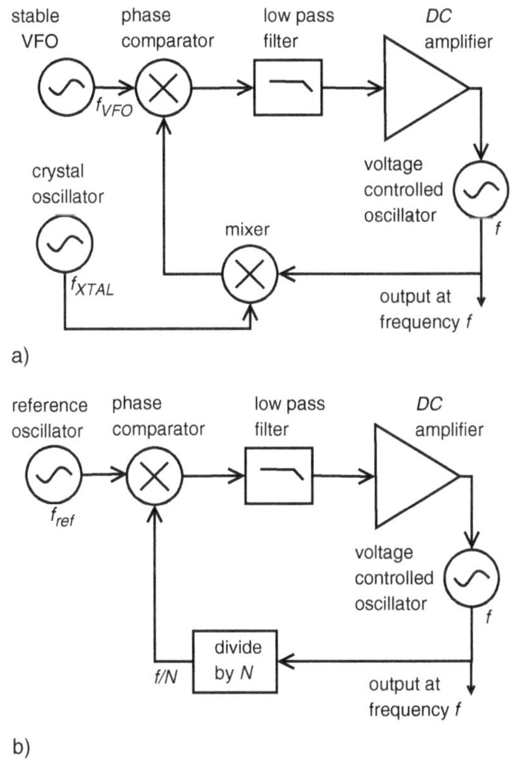

Fig. 5.24 Frequency stabilisation through phase-locked loops.

output of the phase detector and, after amplification, this will provide a control voltage for the VCO. This voltage will then provide the feedback that moves the frequency-divided signal back towards that of the reference frequency. In this way the VCO is locked onto the reference. Consequently, arbitrary frequencies that are a multiple of the reference frequency can be generated by choosing a suitable frequency division. In both circuits of Figure 5.24, a problem arises when the circuit is first switched and the VCO is not locked onto the reference oscillator. A simple solution to this problem is to add some positive feedback to the DC amplifier so that it oscillates at a low frequency until the correct control voltage for the VCO is found. Once lock is achieved, these oscillations will then cease.

5.5 Power Amplifiers

For effective communications, we need to produce RF signals at considerable power. However, the small-signal amplifiers of the previous chapter are no longer suitable. Figure 5.25 shows a typical topology of a single-transistor power-amplifier circuit (an NPN BJT device is shown, but the topology can apply equally well to a FET). To simplify our discussion, we will assume that the transistor amplifies in a linear manner. In general,

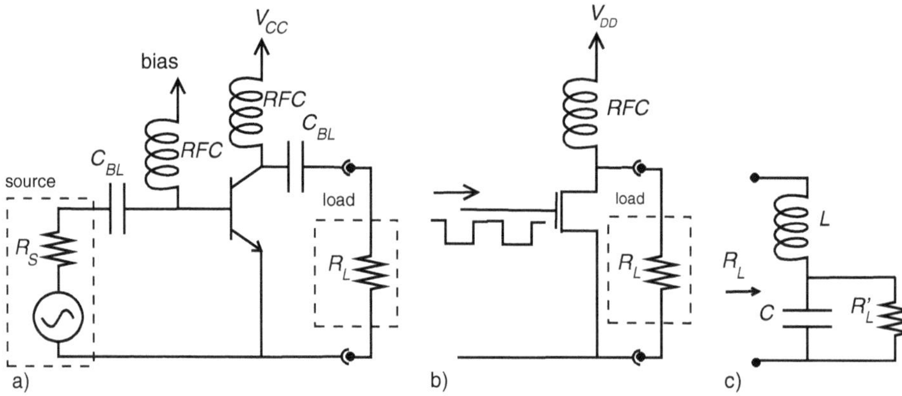

Fig. 5.25 Single-transistor amplifier topology and matching network.

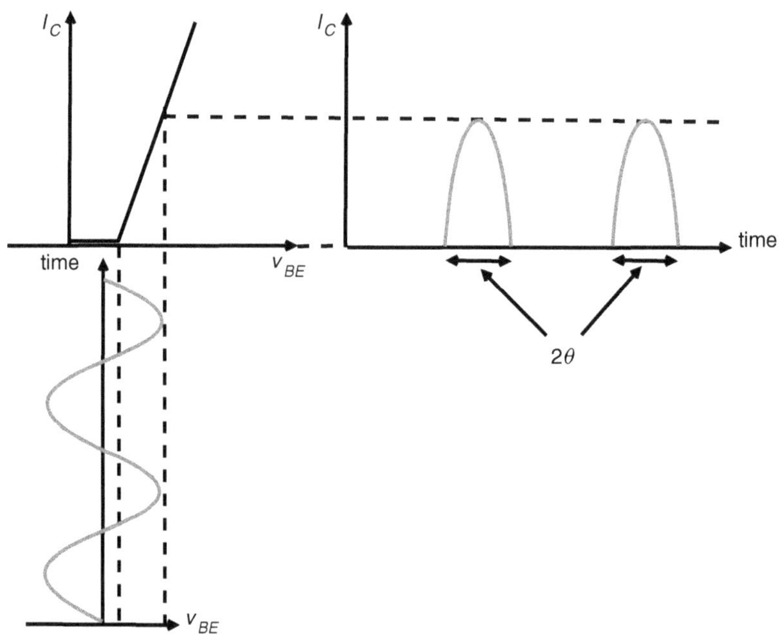

Fig. 5.26 Relation between input and output for a single transistor power amplifier.

the voltage at the base of the transistor is given by

$$\mathcal{V}_{\mathrm{BE}} = V_{\mathrm{bias}} + V_{\mathrm{in}} \sin(\omega t) \tag{5.37}$$

and the current at the collector can be represented by

$$\mathcal{I}_{\mathrm{CE}} = \max(I_{\mathrm{P}}(\sin(\omega t) - cos(\theta)), 0) \tag{5.38}$$

if we assume the amplifier behaves in a linear manner when it conducts. This is illustrated Figure 5.26 where the transistor is shown to conduct over an angle 2θ of each period of the output. The collector emitter current is not necessarily sinusoidal but, since it is still periodic in nature, it can be represented as a Fourier series, i.e.

$$\mathcal{I}_{\text{CE}} = I_0 + I_1 \sin(\omega t) + I_2 \sin(2\omega t) + I_3 \sin(3\omega t) + \cdots \tag{5.39}$$

The harmonic content (terms with coefficients I_2, I_3, etc.) are usually of no interest and are filtered out by the means of a low-pass filter. Consequently, after filtering, the output current will have the form

$$\mathcal{I}_{\text{CE}} = I_{\text{DC}} + I_{\text{RF}} \sin(\omega t), \tag{5.40}$$

where we have, from Fourier theory, that $I_{DC} = I_P(\sin\theta - \theta\cos\theta)/\pi$ and $I_{\text{RF}} = I_P(2\theta - \sin 2\theta)/2\pi$. We will assume that the voltage swing is V_{CC} (the maximum possible) and so $I_{\text{RF}} = V_{CC}/R_L$. One of the major considerations with a power amplifier is its efficiency; this is the ratio of the average output power $P_{\text{RF}} = I_{\text{RF}}^2 R_L/2 = V_{CC} I_{\text{RF}}/2$ to the DC input power $P_{DC} = I_{DC} V_{CC}$, i.e.

$$\text{efficiency} = \frac{P_{\text{RF}}}{P_{DC}} = \frac{2\theta - \sin 2\theta}{4(\sin\theta - \theta\cos\theta)}. \tag{5.41}$$

If we set the transistor bias so that $\theta = \pi$, the transistor will conduct over the whole cycle and the output waveform will be sinusoidal. The amplifier is known as a *class A amplifier* and, from (5.41), we see that it has an efficiency of 50%. If we now set the bias so that $\theta = \pi/2$, the transistor will now only conduct for half a cycle. In this case, the amplifier is known as a *class B amplifier* and, from (5.41), we see that it has an efficiency of 78.5%. The downside is that the output is now highly distorted and needs filtering to remove unwanted harmonics (signals at multiples of the input frequency). We can get even higher efficiencies with amplifier that conducts for less than half a cycle, a *class C amplifier*. In theory, by reducing the conduction period to almost nothing, we can get to almost 100% efficiency. There is a catch, however, in that there will be a very large current during the transistor's on period. The maximum current during the on period will be

$$I_{\max} = \frac{2\pi V_{CC}}{R_L} \frac{1 - \cos\theta}{2\theta - \sin 2\theta} \tag{5.42}$$

and this will become infinite as θ tends to zero. All transistors will have a finite limit of current that they can sustain and so the achievable efficiency of a class C amplifier is limited by the transistor itself. Another problem with the class C amplifier is the high degree of distortion in the output and this precludes it being used in any application that requires linear amplification (SSB for example). For angle modulation (i.e. FM and PM), however, class C can provide extremely efficient power amplification.

The advent of high-speed power-switching transistors has made possible another avenue for highly efficient power amplification: the *class D amplifier* (see Figure 5.25b). The transistor is now driven by a series of pulses (these can be frequency-, phase- or width-modulated). In fact, this is not an amplifier but a high-speed switch. When the transistor is off there is obviously no power dissipated. Further, when on, there is no voltage drop across the transistor and hence no power is dissipated. Obviously, this is an ideal and a real device will dissipate some power. At switch-on the current will take time to rise and at switch-off it will take time to fall. Nevertheless, these amplifiers can

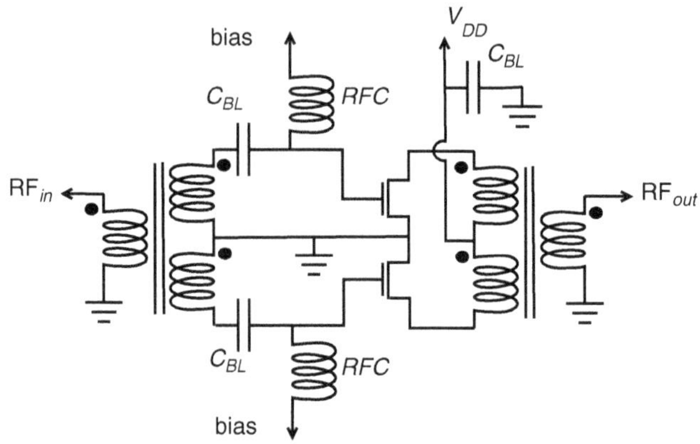

Fig. 5.27 Class B push–pull amplifier.

be extremely efficient and the necessary pulse drive is easily generated with modern digital techniques. As with the class C amplifier, the output will be rich in harmonics and therefore require filtering to remove unwanted harmonics.

The question now arises as to the nature of the load R_L. If the average power output of P_o is specified, we will need to choose R_L to be $V_{CC}^2/2P_o$. This, however, might not be convenient (the impedance of the antenna is often 50 Ω) and so it will be necessary to include a matching network in the amplifier. Figure 5.25c shows a simple L network that might be used. It will be noted that we have chosen the reactances to form a low-pass filter so that the network will also help to remove unwanted harmonics, besides matching the power amplifier to the load R_L'.

Both class A and B amplifiers retain the amplitude information of the original signal, but a class B amplifier will require filtering to remove its unwanted harmonics. This filtering, however, can be a problem if operation is required over a wide range of frequencies. Fortunately, a push–pull amplifier can help solve this problem (see Figure 5.27). Such an amplifier is essentially two class B amplifiers that are 90° out of phase. These generate two half sine waves that combine to form a full sine wave and so, in principle, avoid the problem of harmonics. However, as we have discussed in the previous chapter, a transistor will have a nonlinear characteristic and so the full sinusoid will itself be distorted, especially if we are dealing with large signals. Since a FET has a quadratic characteristic, the distortion will mainly consist of the harmonic at frequency 2ω when the input signal is at frequency ω. Fortunately, a push–pull amplifier has the property that it will cancel out this harmonic between the two sides on the amplifier. Consequently, the push–pull amplifier of Figure 5.27 will produce an output that is relatively free from harmonics.

5.6 Conclusion

In the current chapter we have discussed some of the developments that turned radio into a key technology of the twentieth century. Circuit components such as amplifiers, mixers and oscillators have made possible high-performance radios through radio architectures such as the superheterodyne. However, from the late twentieth century onwards, there have been major advances in computer technology that have now made it possible to build radios in computer software. By digitising a radio signal, it is now possible to carry out within a computer the signal processing that was hitherto carried out in the analogue domain. In the next chapter we will consider the use of digital techniques in radio and the practical realisation of software radio.

6 Digital Techniques and Software-Defined Radio

The advent of digital electronics, and computers, has revolutionised the world of radio. We can now perform nearly all the functions of a radio in the digital domain and functions that are difficult in the analogue domain can become easy in the digital domain. The key to all of this is the ability to transform radio signals back and forth between the digital and analogue domains through analogue-to-digital and digital-to-analogue converters. It is the advances in these converters that have turned radio from a technology of the analogue domain into one of modern computing. Not all analogue radio is redundant, since functions such as filtering, amplification and antennas will always be required. However, modern developments have brought the computer closer and closer to the antenna and have turned a large part of modern radio development into that of software development. In the current chapter we will discuss the elements of what has come to be known as *software-defined radio*.

6.1 Basic Digital Electronics

In the digital domain, a signal will be represented by a sequence of samples at intervals sufficiently small to capture the essential detail of the signal. Our normal way of representing such samples is with numbers in the decimal system with a base of 10. However, such a representation is inconvenient for digital systems and it turns out more convenient to use a binary system with base 2, i.e. numbers are represented in terms of zeros and ones. For example, the numbers 0 to 15 will take the form 0, 1, 10, 11, 100, 101, 110, 111, 1000, 1001, 1010, 1011, 1100, 1101, 1110, 1111. The rightmost digit will represent the number of 1s, the next the number of 2s, the next the number of 4s, the next the number of 8s and so on.

In the electronic world, the number 1 can take the form of an *on state* and 0 the form of an *off state*. On/off switches are clearly a key element in digital electronics and their simplest realisation is the NMOS transistor. In this device, a suitably positive voltage at the gate turns the device on and current flows, but a zero voltage switches the device off. A PMOS device also acts as a switch, but a negative voltage is required at the gate in order to switch the device on. These devices can be used to form *gates* that can be used to combine and manipulate digital signals. The simplest gate is known as a *NAND gate* and the circuit of such a device is shown in Figure 6.1a. Figure 6.1b shows the output Y of the gate for various combinations of its inputs A and B (we have given the

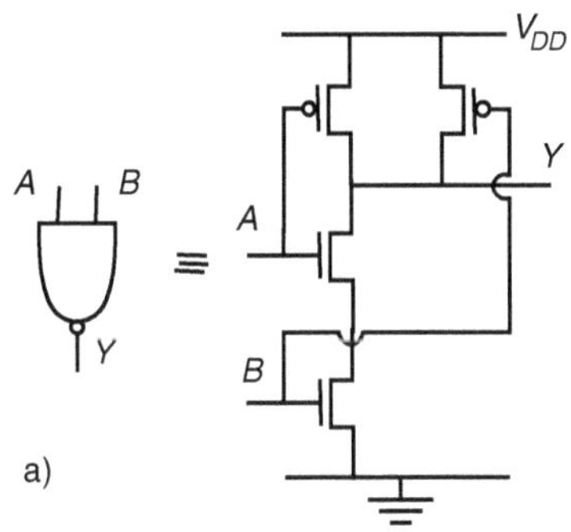

b)

A	B	Y
0	0	1
0	1	1
1	0	1
1	1	0

Fig. 6.1 NAND gate and its CMOS realisation.

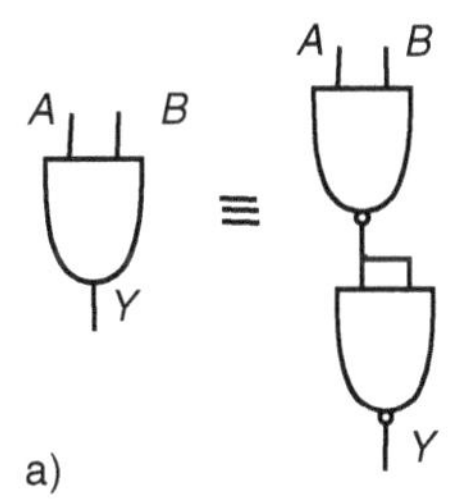

b)

A	B	Y
0	0	0
0	1	0
1	0	0
1	1	1

Fig. 6.2 AND gate in terms of NAND gates.

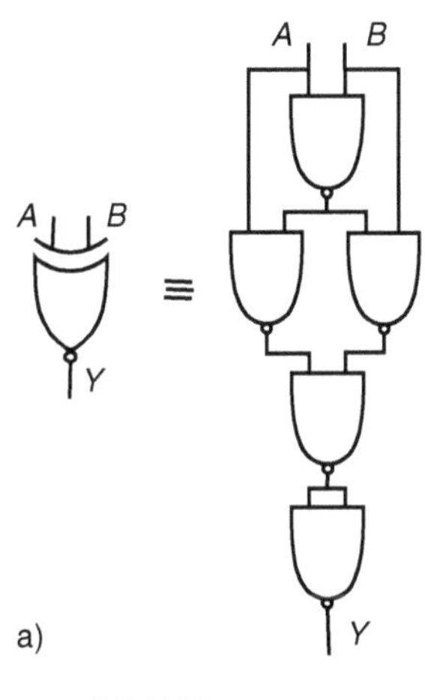

b)

A	B	Y
0	0	1
0	1	0
1	0	0
1	1	1

Fig. 6.3 XNOR gate in terms of NAND gates.

input and output voltages a nominal value of 1). Figures 6.2 to 6.4 show several other gates that can be built from the basic building block of a NAND gate (the additional gates are the *AND*, *XNOR* and *XOR* gates). The gates above are part of what is known as *combinatorial logic* since they are equivalent to a combination of logical operations with the 0 and 1 states regarded as the *false* and *true* states of logic.

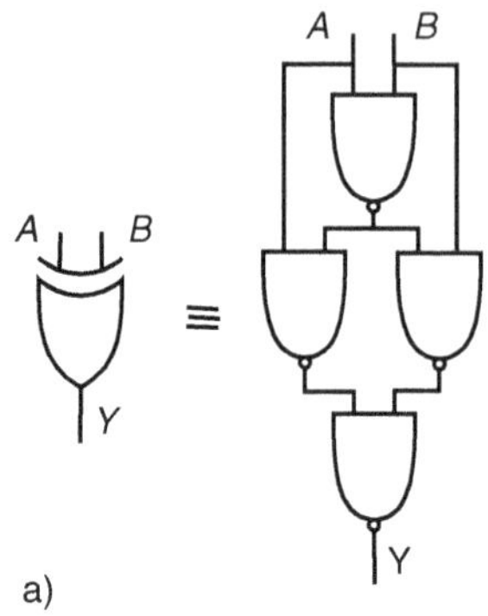

a)

A	B	Y
0	0	0
0	1	1
1	0	1
1	1	0

b)

Fig. 6.4 XOR gate in terms of NAND gates.

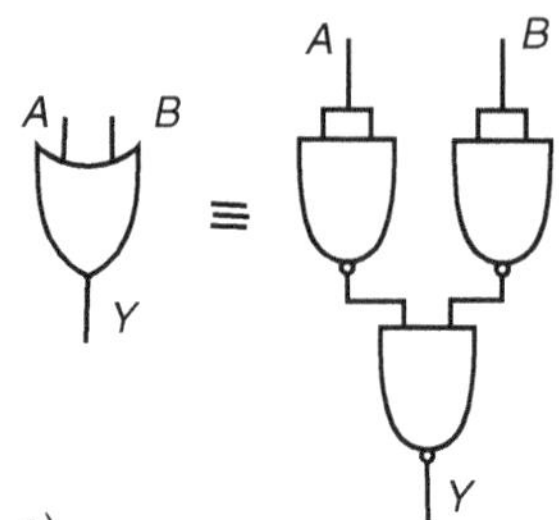

a)

A	B	Y
0	0	0
0	1	1
1	0	1
1	1	1

b)

Fig. 6.5 OR gate in terms of NAND gates.

If we are to use logic to represent numbers, we will need to represent a single number by several logic states. For example, a four-bit number A will need four logic states $A3$, $A2$, $A1$ and $A0$ (where $A0$ is the number of 1s, $A1$ is the number of 2s, $A2$ is the number of 4s and $A3$ is the number of 8s, i.e. the number $A_0 + 2A_1 + 4A_2 + 8A_3$ in decimal form). If we need to add this to another number B with the bits $B3$, $B2$, $B1$ and $B0$, then we can achieve this with combinatorial logic. We need to add the corresponding bits of both numbers but need to account for any carry from the previous bit and to produce any carry that might be needed for the next bit. This can be achieved with the one-bit adder shown in Figure 6.6a. For a complete addition, we will need to combine several of these one-bit adders as shown in Figure 6.6b. More complex operations, such as multiplication, will require us to perform a sequence of operations and this brings us to *sequential logic*. In sequential logic, the output depends not only on the input at the current time but also on past input and outputs of the circuit (see Figure 6.7). For example, multiplication can be regarded as addition which is repeated several times over.

An example of sequential logic is the JK flip-flop shown in Figure 6.8a. This device is able to maintain the state of its output Q ($\overline{Q}$ being the complementary state of Q) until suitably reset, the reset being enabled on the rising edge of a clock signal (the table shows that changes that can take place). Further, this device can act as a store for a single bit of data (the state of Q) and several of these devices can be used to

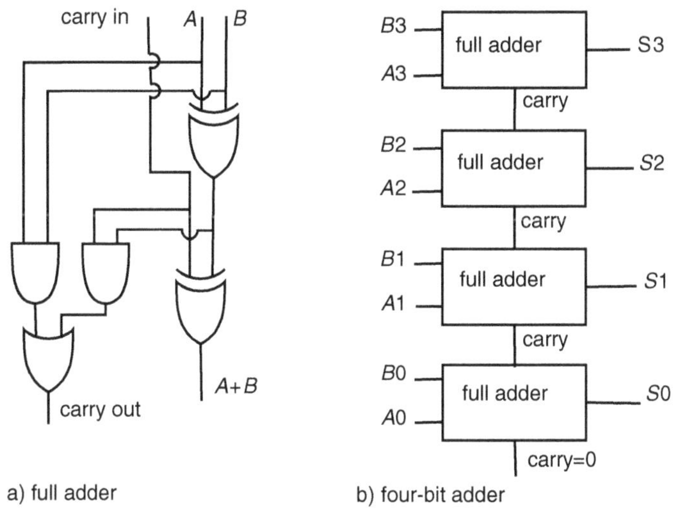

Fig. 6.6 Full adder and four-bit adder.

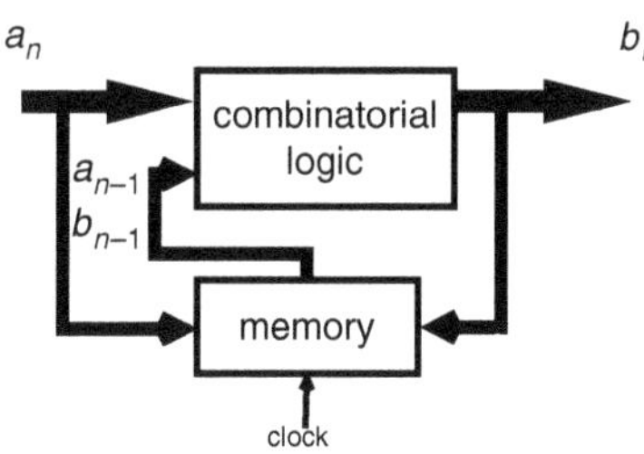

Fig. 6.7 Typical sequential logic.

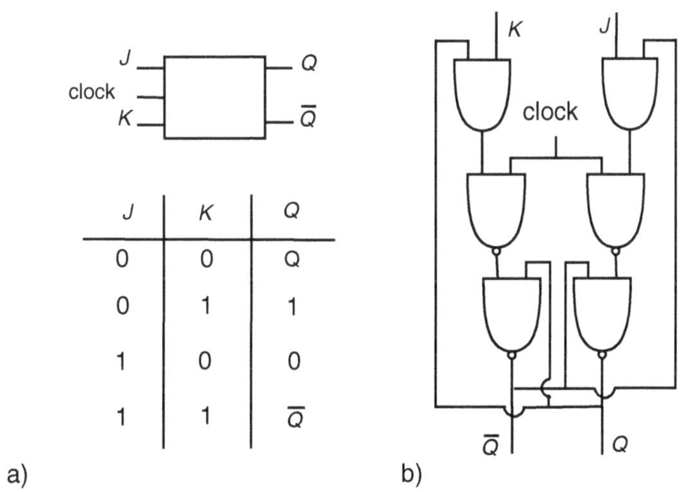

Fig. 6.8 A JK flip-flop.

store numbers in the binary system. Indeed such memory is an essential element in sequential logic. Another important element is the clock signal, often supplied by a quartz crystal oscillator, which is used to progress the signal through the logic. If a sequence of numbers is to be processed through several stages of combinatorial logic, the clock pulses will trigger the passage from one stage to the next. If a radio signal has been converted to a sequence of numbers, the various functions of a radio receiver will now become arithmetic operations. For example, amplification becomes multiplication by a fixed number and mixing becomes the multiplication of the sequences representing two different signals. Obviously, a typical radio will require many stages of sequential logic and this can become extremely complex. The logic, however, can be constructed as a combination of a vast number of simple logic elements such as NAND gates. This has been made easier with the advent of *field-programmable gate arrays* (FPGAs) . These are integrated circuits that contain millions of simple gates that can have their connections programmed and hence be made to perform the function of the desired software radio. These devices have revolutionised software radio since the operator can change the nature of the radio by simply reprogramming the FPGA.

6.2 Digital Signal Processing

We have already seen the convenience of analysing analogue RF signals in the complex domain and the same is also true for digital signals. In the complex domain, a real harmonic signal $s(t) = S\cos(2\pi ft)$ can be represented in terms of the complex harmonic signals $\exp(\mathrm{j}2\pi ft)$ and $\exp(-\mathrm{j}2\pi ft)$ as $s(t) = S(\exp(\mathrm{j}2\pi ft) + \exp(-\mathrm{j}2\pi ft))/2$, i.e. the real signal contains equal components on the frequencies f and $-f$. In general, a signal $s(t)$ will be a linear combination of complex harmonic signals at a variety of frequencies. Importantly, for a real signal, the distribution of positive frequency components will be a mirror image of those at negative frequencies (this is illustrated in Figure 6.9). Consequently, it is sufficient to represent a signal $s(t)$ by its positive frequency components $s_+(t)$. Signal $s_+(t)$ can be split into its real and imaginary parts, i.e. $s_+(t) = s^I(t) + \mathrm{j}s^Q(t)$ where s^I and s^Q are known as the *in-phase* and *quadrature* components respectively (note that the original real signal will be $2s^I$). In the case of a simple harmonic signal $s(t) = S\cos(2\pi ft)$, we have an in-phase component

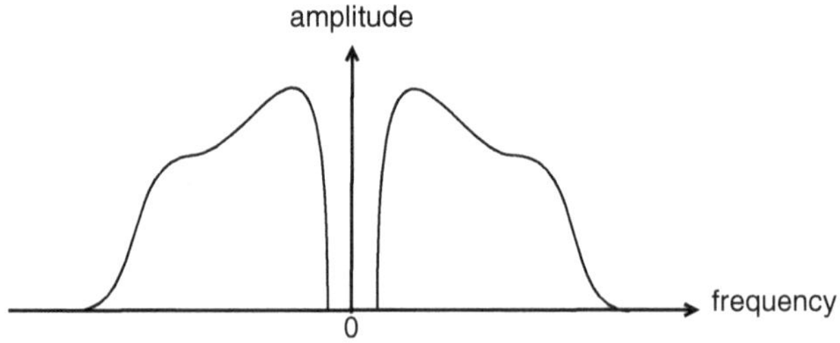

Fig. 6.9 The frequency content of a real signal.

$s^I(t) = S\cos(2\pi ft)/2$ and quadrature component $s^Q(t) = S\sin(2\pi ft)/2$. From this it can be seen that the in phase component is obtained from the quadrature component by a $\pi/2$ shift in phase. Likewise, for a general signal, the in-phase component is obtained from the quadrature component by a $\pi/2$ phase shift for each of its harmonic components.

In the digital domain, an RF signal will be represented by a time-ordered sequence of binary numbers, the numbers having been sampled from the original RF signal. The sampling will need to have been carried out at a rate that gives an adequate representation of the signal and so we will need to determine the limitations on this rate. To this end, we consider a signal $s(t)$ that has been sampled (see Figure 6.10) at a rate F for a finite number N of samples ($T = \frac{1}{F}$ is the time interval between samples). Let $s_0, s_1, s_2, \ldots, s_{N-1}$ be the samples and form the linear combination

$$\frac{1}{N}\sum_{k=0}^{N-1} s_k \sum_{n=0}^{N-1} \exp\left(\mathrm{j}\frac{2\pi n}{N}(l-k)\right) \tag{6.1}$$

where l is an arbitrary integer. Noting the identity

$$\sum_{n=0}^{N-1} \exp(\mathrm{j}\frac{2\pi n}{N}(l-k)) = \frac{1-\exp(\mathrm{j}2\pi(l-k))}{1-\exp\left(\mathrm{j}\frac{2\pi}{N}(l-k)\right)} \tag{6.2}$$

(this is obtained from the geometric sum $\sum_{i=0}^{N-1} a^i = (1-a^N)/(1-a)$ with $a = \exp(\mathrm{j}2\pi n(l-k)/N)$), we will find that this sum takes the value N when $l = k$ and is zero for other integral combinations of l and k. As a consequence, (6.1) will evaluate to s_l and then

$$s_l = \sum_{n=0}^{N-1}\left(\frac{1}{N}\sum_{k=0}^{N-1} s_k \exp(-\mathrm{j}\frac{2\pi kn}{N})\right)\exp\left(\mathrm{j}\frac{2\pi ln}{N}\right) \tag{6.3}$$

on changing the order of summation. We can now make the following approximation to the signal $s(t)$

$$s(t) \approx \frac{1}{N}\sum_{n=0}^{N-1} a_n \exp(\mathrm{j}2\pi n\Delta F t), \tag{6.4}$$

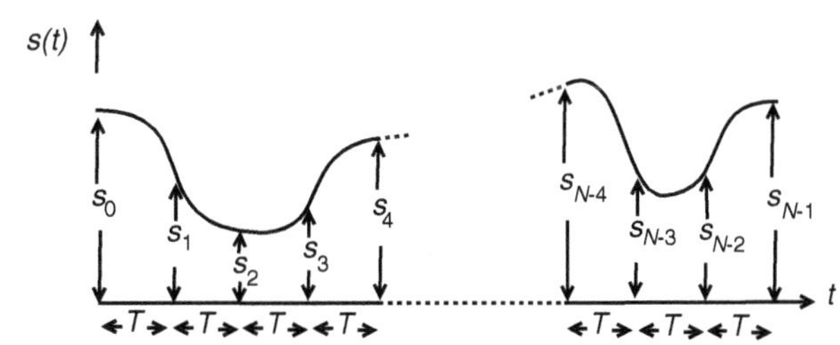

Fig. 6.10 The sampling of an analogue signal.

where $\Delta F = \frac{F}{N}$ and the coefficients $a_0, a_1, \ldots, a_{N-1}$ are given by

$$a_n = \sum_{k=0}^{N-1} s_k \exp\left(-\mathrm{j}\frac{2\pi kn}{N}\right). \tag{6.5}$$

This gives an exact representation of the signal at the sample times $0, T, 2T, 3T, \ldots, (N-1)T$ and interpolated values at other times. The coefficients $a_0, a_1, a_2, a_3, \ldots, a_{N-1}$ are the magnitude of the signal at frequencies 0, ΔF, $2\Delta F$, $3\Delta F$, ..., $(N-1)\Delta F$ respectively and together are known as the *discrete Fourier transform* (DFT). We now need to consider the limitations of the discrete representation of a signal and its spectral content. To this end, consider the complex harmonic signal $s(t) = \exp(\mathrm{j}2\pi ft)$ and form its DFT

$$\begin{aligned} a_n &= \sum_{k=0}^{N-1} \exp\left(-\mathrm{j}2\pi k\frac{n}{N}\right)\exp\left(\mathrm{j}2\pi k\frac{f}{F}\right) \\ &= \frac{1-\exp\left(-\mathrm{j}2\pi N\left(\frac{n}{N}-\frac{f}{F}\right)\right)}{1-\exp\left(-\mathrm{j}2\pi\left(\frac{n}{N}-\frac{f}{F}\right)\right)} \\ &= \exp\left(-\mathrm{j}\pi(N-1)\left(\frac{n}{N}-\frac{f}{F}\right)\right)\frac{\sin\left(\mathrm{j}\pi N\left(\frac{n}{N}-\frac{f}{F}\right)\right)}{\sin\left(\mathrm{j}\pi\left(\frac{n}{N}-\frac{f}{F}\right)\right)}. \end{aligned} \tag{6.6}$$

For frequencies that are a multiple of F/N (i.e. $f = m\Delta F$ where m is an integer), we have that $a_m = 1$ and that $a_n = 0$ when $n \neq m$. This means that the DFT has picked out the correct frequency content of the signal. However, when the frequency f is a non-integral multiple of F/N, the DFT will spread the energy around all the spectral coefficients. This is known as *spectral leakage* and is a particular problem will the DFT. Fortunately, the leakage will only be significant for coefficients associated with frequencies close to f. Clearly, ΔF sets the lower limit to the frequency resolution of the DFT.

We have seen that the sample rate and data length set a lower limit on frequency resolution, but the question arises as to what is the maximum frequency that can be resolved. It will be noted that the harmonic signal $s(t) = \exp(\mathrm{j}2\pi((f+mF)t)$ will have the same DFT as $s(t) = \exp(\mathrm{j}2\pi ft)$ when m is an integer. As a consequence the DFT will not distinguish between them, and their reconstructions (the inverse DFT) will be identical, i.e. both signals will be in the frequency range $[0, F]$. This phenomenon is known as *aliasing* and means that a DFT has the potential to *fold* signal content with frequencies outside the interval $[0, F]$ onto signals with frequency content within the interval. This is illustrated in Figure 6.11, which shows three signals with distinctly different frequency content but with the same DFT. Figure 6.12 shows an example of a signal with frequency content outside the range $[0,F]$. The signal can be pictured as two separate signals, one with frequency content within the interval $[0,F]$ and one with all its frequency content outside this range. The DFT, however, will fold the signal with frequencies outside $[0,F]$ onto one with content inside this interval and, when the signals

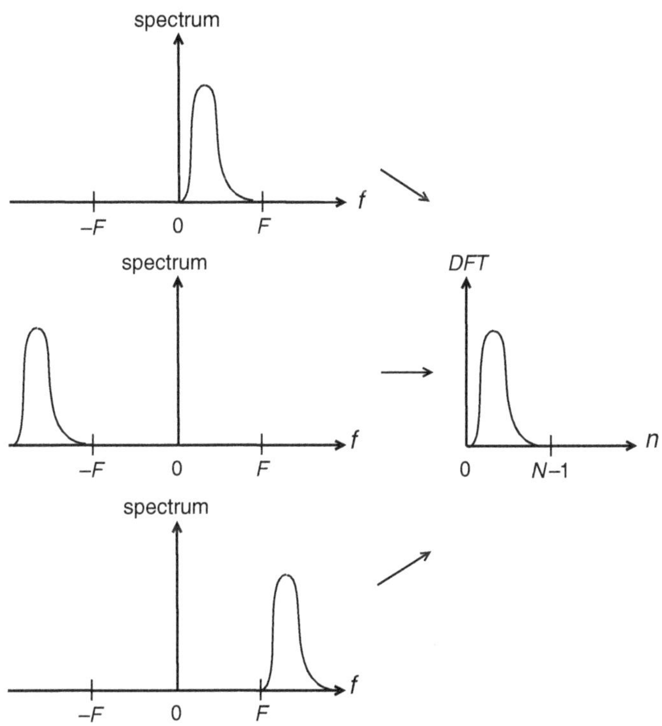

Fig. 6.11 Three signals with different spectral content, but identical DFT due to aliasing.

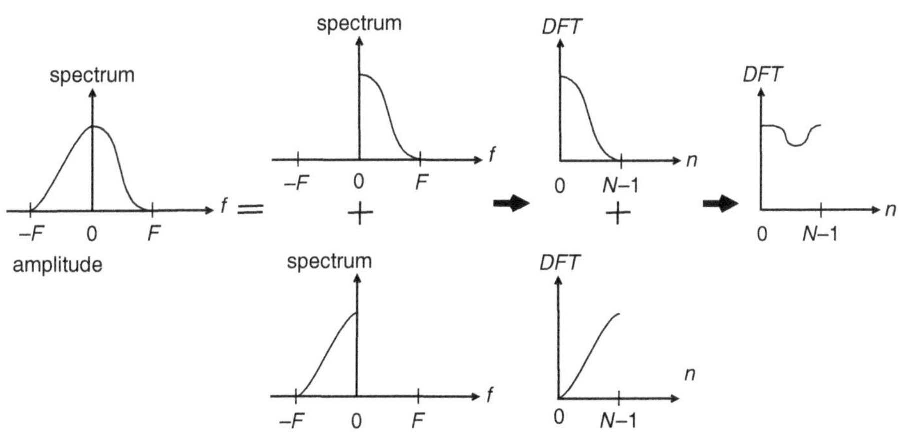

Fig. 6.12 Spectral contamination caused by aliasing.

are recombined, they will give a completely erroneous picture of frequency content. In order for the DFT to properly resolve the frequency content of the signal, the signal will need to be band-limited with the bandwidth dictated by the sampling rate. For a real signal, there will be both positive and negative frequencies and so the bandwidth needs to prevent the negative frequencies folding onto the positive frequencies and this leads to the *Nyquist sampling theorem* (Nyquist, 1928), i.e.

The sample rate F will need to be at least twice the maximum frequency present if aliasing is to be avoided.

As a consequence of this, it is normal to use an *anti-aliasing filter* before sampling in order to remove any signal components at frequencies above $F/2$.

As we have seen earlier, filtering is an important function in analogue RF signal processing and this obviously remains true in digital processing. The DFT suggests a possible approach to such filtering. Consider N consecutive samples be taken from a digitised signal and consider their DFT. If we want to pass the signal through a band-pass filter, lower and upper band edges $P\Delta F$ and $Q\Delta F$ respectively, we could do this by reconstructing the consecutive samples from the DFT coefficients a_P to a_Q alone, i.e.

$$s_l^F = \frac{1}{N}\sum_{n=P}^{Q} a_n \exp\left(\mathrm{j}2\pi\frac{ln}{N}\right), \tag{6.7}$$

where s^F denotes the filtered sequence. The coefficients a_n can be eliminated from (6.7) by means of (6.5) to yield

$$s_l^F = \sum_{k=0}^{N-1} s_k c_{k-l}, \tag{6.8}$$

where

$$c_m = \frac{1}{N}\sum_{n=P}^{Q} \exp\left(-\mathrm{j}2\pi\frac{mn}{N}\right), \tag{6.9}$$

i.e. the band-pass samples can be formed as linear combinations of the original samples. This suggests that, in the case of an infinite sequence, we form a filtered sequence by applying a process of the form of (6.8). (This is exemplified by a moving average that is used to smooth data, essentially a low-pass filtering process.) The filtered sequence (s_0^F, s_1^F, s_2^F, s_3^F, ...) is generated according to

$$s_M^F = \sum_{k=-L}^{N-1-L} s_{k+M} c_k \tag{6.10}$$

for suitable L (note that the output sequence is delayed by $N-1-L$ samples). The filter is implemented as the sequential circuit shown in Figure 6.13 which consists of a combination of adder, multiplier and delay elements. The signal samples enter on the left and are moved through at a rate of one stage per sample interval; the filtered sequence then exits to the right. The output of a delay block (labelled Z^{-1}) consists of its input value at the previous step of the process and therefore will involve some sort of memory. It is important to note that the above filter is based upon ideas that come from the filtering of a finite-length sequence of data, i.e. each output value is derived from a truncated sequence of input data. Unfortunately, this *windowing* process has the effect of

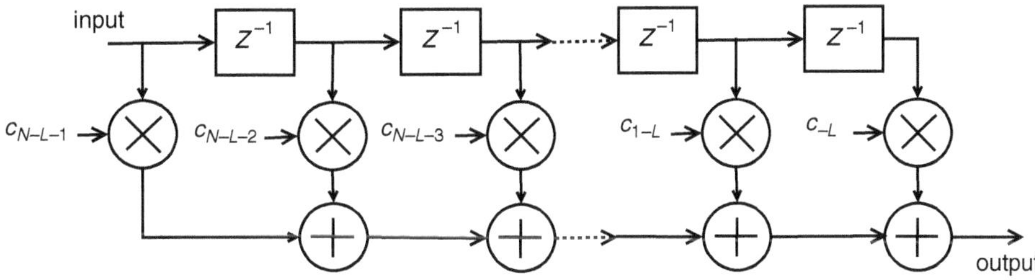

Fig. 6.13 A finite-length discrete signal filtering system.

spreading the spectrum of the output sequences outside the desired filter bounds (another example of spectral leakage). Fortunately, however, we can reduce this effect by making the window more benign. This can be achieved by making the filter coefficients (i.e. the c_k) taper towards zero as k approaches both 0 and N (to minimise the effects of the taper, it is advisable to choose L close to $N/2$).

Digital signal processing can be computationally expensive and this can make it hard to maintain the kind of real time throughput that is required by modern radio systems. One way of improving matters is to eliminate any unnecessary data processing. Fortunately, the Nyquist sampling theorem comes to the rescue as it tells us that we need only use a sampling rate that is twice the bandwidth. Consequently, providing we obey this limitation, we can reduce the computational requirement by only keeping every M^{th} sample, a process that is known as *down sampling*. To avoid any possibility of aliasing, such a process is often proceeded by a low-pass filter with bandwidth consistent with the new sampling rate (the combination is known as a *decimation filter*). If we need to change a signal back to a greater sample rate, we can repeat each sample M times, but we must then follow this process by a low-pass filter in order to remove any high-frequency component that is introduced by this interpolation process (this total process is known as an *interpolation filter*).

The computational expense of such processing has led to some clever ideas for filtering and one of the most important of these is the cascaded integrator comb (CIC) filter (Hogenauer, 1981). Figures 6.14a and 6.14b show an *integrator filter* and a *comb filter* respectively. The amplitude response of the integrator in terms of frequency is $1/2|\sin(\pi f/F)|$ and so it acts like a low-pass filter (F is the sampling rate). On the other hand, the comb filter has the amplitude response $2|\sin(\pi Mf/F)|$ and acts as a multiple notch filter that removes frequencies that are integral multiples of frequency F/M (i.e. it behaves like a comb). If we combine these filters in the manner of Figure 6.15, we obtain a filter with the frequency response that is also shown in this Figure 6.15 (the amplitude response is $|\sin(2\pi fRM/2F)/\sin(2\pi f/2F)|^N$). It will be noted that the filter does not contain any computationally expensive processes (only delays, additions and changing the sign of a number) and so can form the basis of an extremely efficient filter. One downside, however, is that the number of bits in the samples grows during passage through the filter by an amount of $N\log_2(RM)$. However, this growth can usually be handled by judicious truncation at various stages of the filter.

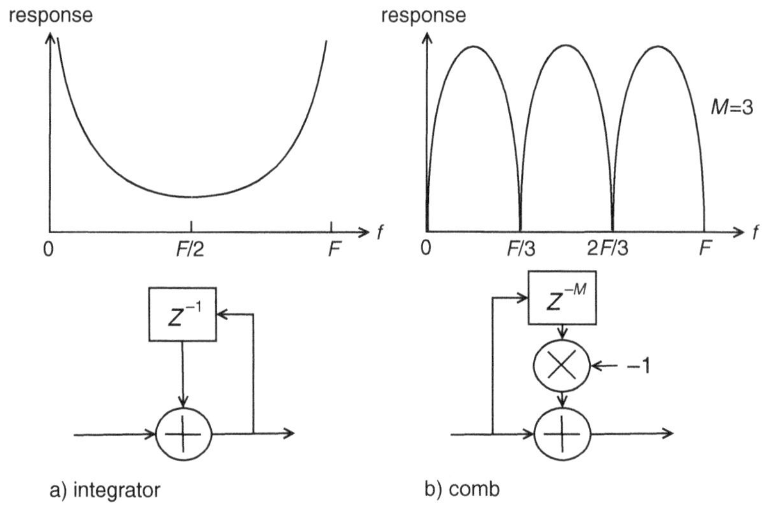

Fig. 6.14 Integrator and comb filters with their frequency response.

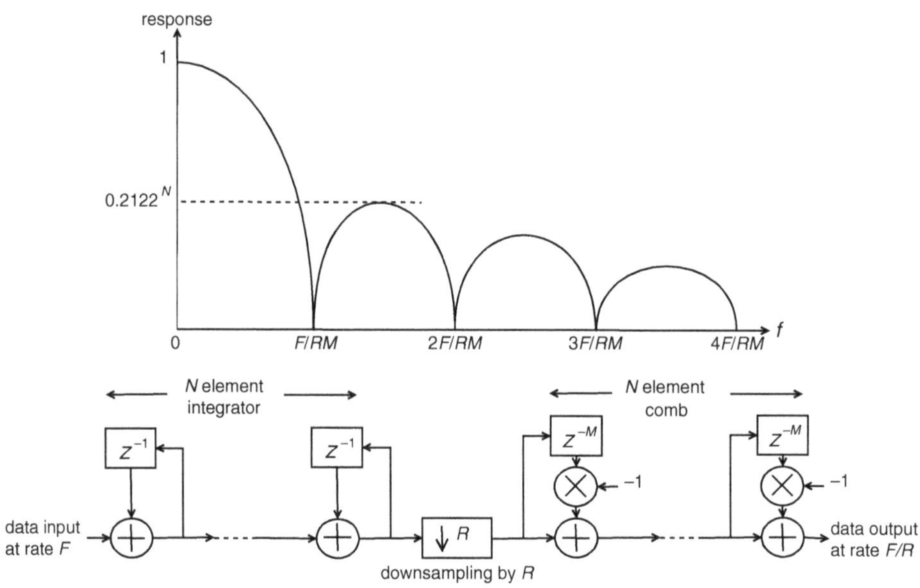

Fig. 6.15 CIC filter with its frequency response.

6.3 Analogue-to-Digital and Digital-to-Analogue Converters

Key to the realisation of software radio is the ability to convert an RF signal into a sequence of numbers through an appropriate *analogue-to-digital converter* (ADC). The most basic ADC is the threshold detector, an example of which is shown in example of Figure 6.16. This converter can be used to detect whether a pulse is present and can

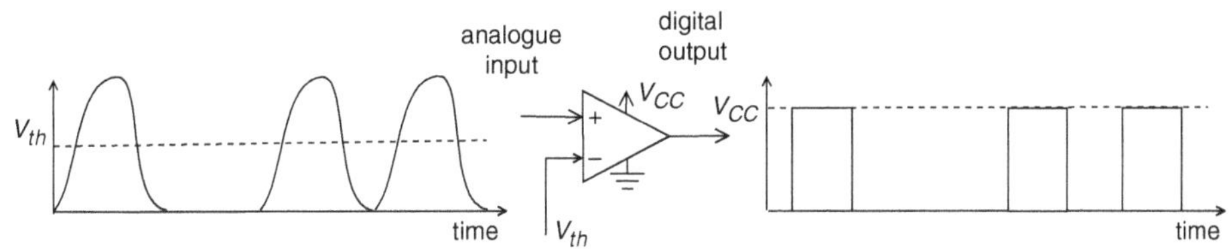

Fig. 6.16 A simple one-bit analogue-to-digital converter.

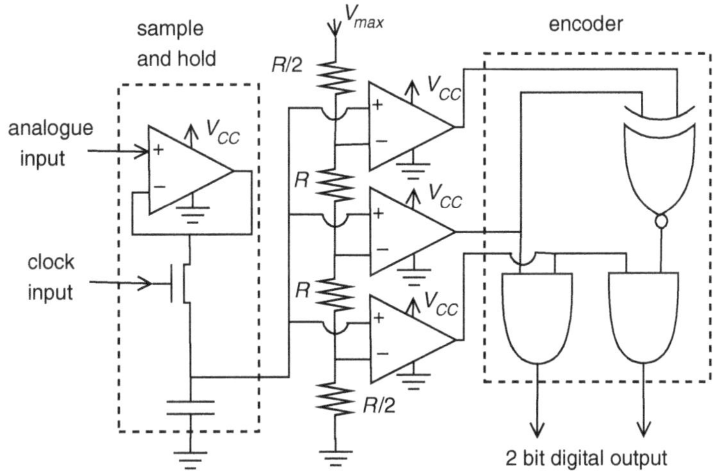

Fig. 6.17 A two-bit analogue-to-digital flash converter.

therefore be used to detect certain kinds of digital modulation. If the input voltage is above a certain threshold, its level is raised to that of the supply. However, if the input voltage is below the threshold, the output voltage will be that of the ground. Such a converter can only detect two levels of amplitude (i.e. it can provide one bit of data). For signals more complex than simple pulses, we will require a much greater amplitude resolution. Figure 6.17 shows a simple *flash converter* that can detect four levels of amplitude (i.e. it can provide two bits of data). At the input there is a *sample-and-hold* system to ensure that the samples have a constant amplitude during the acquisition period. Digitisation is achieved through a multi-threshold comparator and the logic circuits encode the output as a two-bit binary number. A two-bit converter is too crude (see Figure 6.18) for most software radio, but this simple converter demonstrates the general principles.

The conversion of an analogue signal, into one with only a finite number of fixed levels of voltage, will obviously lead to some inaccuracy and this is known as *quantisation noise* (see Figure 6.18). Let the sample levels have spacing Δ then, for a sampling interval length T, the signal will fluctuate up to $\pm\Delta/2$ about the mean. Assuming that the quantisation voltage error v_{qn} varies linearly between the quantisation levels, its mean square will be given by

$$\overline{v_{\text{qn}}^2} = \frac{\Delta^2}{12} = \frac{V_{\text{max}}^2 2^{-2b}}{12}, \tag{6.11}$$

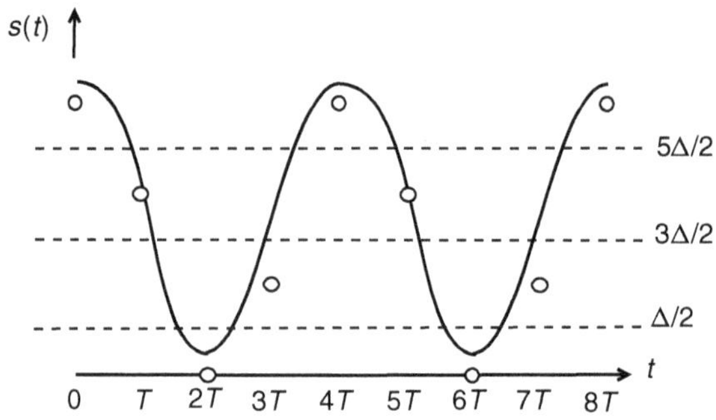

Fig. 6.18 Two-bit digitisation.

where b is the bit number resolution of the ADC and $V_{\rm max}$ is the voltage range. The power of the *quantisation noise* will therefore be

$$N_{qn} = \frac{V_{\rm max}^2 2^{-2b}}{12R_{\rm in}}, \tag{6.12}$$

where $R_{\rm in}$ is the input resistance of the ADC. $V_{\rm max}$ will be the largest peak-to-peak voltage that ADC can accommodate and so the largest possible SNR will be

$$\begin{aligned} SNR &= \frac{(V_{\rm max}/2)^2}{2R_{\rm in}} / N_{\rm qn} = \frac{3}{2} 2^{2b} \\ &= (6.02b + 1.75) \text{ dB}. \end{aligned} \tag{6.13}$$

This is the *dynamic range* of the converter. If the quantisation noise is spread uniformly across the *Nyquist bandwidth* ($F/2$), the SNR will increase by factor $F/2B$ when the ADC output is processed through a digital filter with bandwidth B (quantity $10\log_{10}(F/2B)$ is known as the processing gain). However, if the input consists of relatively few signals, the quantisation noise can become highly correlated and cause a considerable number of high-amplitude *spurs* (see Figure 6.19a). Fortunately, this can be alleviated by adding some noise at the input to the ADC (a process known as *dithering*) in order to assist the decorrelation of the quantisation noise and hence reduce the spurs (see Figure 6.19b). Such noise can be added into a section of the Nyquist bandwidth that is to be removed by subsequent processing, e.g. where there are no signals of interest.

As with sequential digital electronics in general, the clock signal is an important element in analogue-to-digital conversion and its stability is an important issue. A quartz crystal oscillation is usually used for stability, but imperfections can still arise in the signal from a practical oscillator. The resulting fluctuations in sample interval are known as *aperture jitter* and cause distortions that translate into additional noise that increases with frequency f. For high-resolution converters, this noise will dominate and the largest possible SNR will take the form

$$SNR = 20\log_{10}\left(\frac{1}{2\pi f \Delta\tau}\right) \text{dB}, \tag{6.14}$$

where $\Delta\tau$ is the rms value of the sample interval fluctuations.

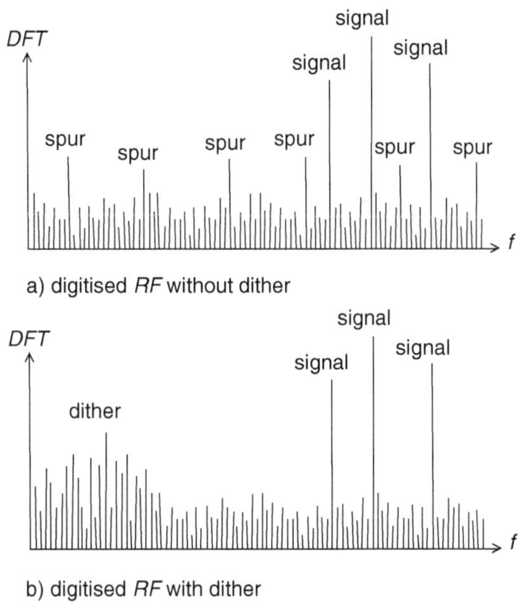

Fig. 6.19 Sampled signals with quantisation noise and spurs.

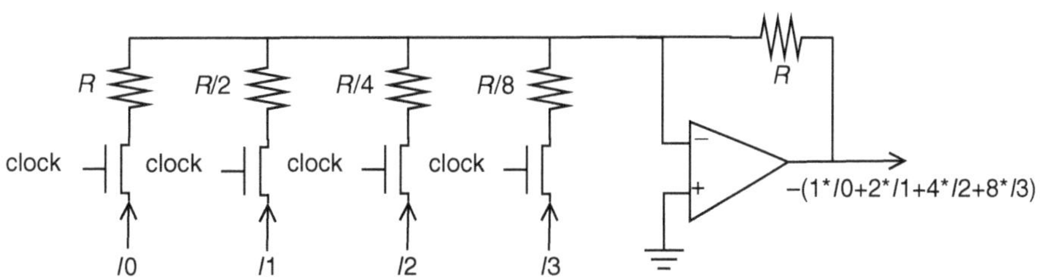

Fig. 6.20 Simple digital-to-analogue converter.

Once converted into digital form, the analogue signal can be processed by a computer or perhaps some programmable logic. Often, however, we will need to convert the processed signal back into analogue form (this is certainly the case for audio broadcast signals). To achieve this, we need a digital-to-analogue converter (DAC), a simple example being shown in Figure 6.20. This circuit adds suitable multiples of the input digits and produces an output that is proportional to the number represented by the digits.

We now consider the option of increasing the sample rate above that suggested by the Nyquist theorem, i.e. *over-sampling*. This might seem wasteful, but we will see that it does have some merit. We have already seen that the effective SNR is controlled by the bit number resolution of the ADC and Figure 6.21 illustrates a signal and accompanying quantisation noise. If we increase the sample rate by a factor k (i.e. sample at frequency kF) we obtain the same SNR, but now the noise is spread out over a wider frequency

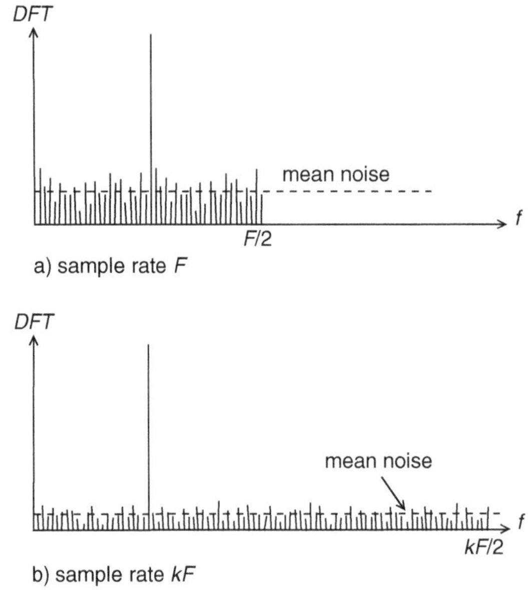

Fig. 6.21 The effect of over-sampling.

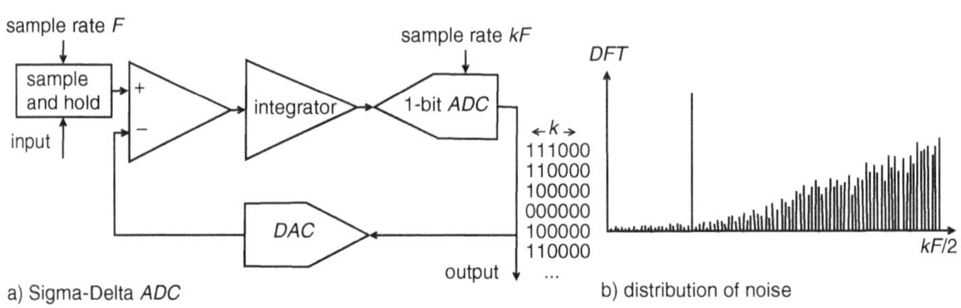

Fig. 6.22 Sigma-delta ADC.

range. Consequently, if we filter out the extra frequency range, we will obtain an increased SNR. This fact is used in what is known as a *sigma-delta* ($\Sigma\Delta$) ADC.

Figure 6.22a shows the circuit of a basic $\Sigma\Delta$ ADC. The signal is fed into a sample-and-hold circuit that operates at the desired sampling rate and then enters a comparator where the output of the converter (initially zero) is passed through a DAC and subtracted from the sampled input. This is then fed into a one-bit ADC with suitable threshold and the process repeated. The output will hence consist of a sequence of constant values (notionally 1s) until their combination has reached the level of the input signal and will thereafter be zero until k samples of the ADC have been achieved. The sequence of 1s and 0s will then constitute the digitised signal where the number of 1s is the value of the sample. The integrator before the ADC has the effect of moving the quantisation noise to the higher frequencies (see Figure 6.22b). Consequently, when the output is converted to a more conventional form by counting the number of 1s in the sequence of k digits, this noise will be averaged out. With the $\Sigma\Delta$ approach to ADCs, it is possible to make very simple high-resolution linear converters. However, their bandwidth is limited by the need to over-sample and so there is a trade-off between bandwidth and resolution.

6.4 Digital Receiver and Transmitter Architecture

In software radio, the various functions within an analogue radio become arithmetic operations performed on a sequence of samples representing the RF signal. Amplification becomes a multiplication by the voltage gain, mixing becomes the multiplication of two signal sequences and filtering becomes a weighted moving average. Essentially, the radio architecture remains the same, it is just the realisation that changes. The digital circuits can be realised as programs within a computer or, more often as not, constructed out of the logic within an FPGA.

Most software radio receivers are basically of the direct-conversion variety, i.e. they convert directly to baseband. Figure 6.23 is an example of such a receiver. It is essentially an analogue receiver with the analogue-to-digital conversion performed at the outputs of the mixers. We first analyse a simple cosine RF signal $s^{\mathrm{RF}} = S^{\mathrm{RF}}\cos(2\pi f_{\mathrm{RF}}t)$ and local oscillator signal $s^{\mathrm{LO}} = S^{\mathrm{LO}}\cos(2\pi f_{\mathrm{LO}}t)$. The in-phase signal output will be

$$\begin{aligned} s^{\mathrm{I}}(t) &= S^{\mathrm{RF}}S^{\mathrm{LO}}\cos(2\pi f_{\mathrm{LO}}t)\cos(2\pi f_{\mathrm{RF}}t) \\ &= \frac{S^{\mathrm{RF}}S^{\mathrm{LO}}}{2}(\cos(2\pi(f_{\mathrm{LO}}-f_{\mathrm{RF}})t)+\cos(2\pi(f_{\mathrm{LO}}+f_{\mathrm{RF}})t)) \end{aligned} \tag{6.15}$$

and the quadrature output

$$\begin{aligned} s^{\mathrm{Q}}(t) &= -S^{\mathrm{RF}}S^{\mathrm{LO}}\sin(2\pi f_{\mathrm{LO}}t)\cos(2\pi f_{\mathrm{RF}}t) \\ &= -\frac{S^{\mathrm{RF}}S^{\mathrm{LO}}}{2}(\sin(2\pi(f_{\mathrm{LO}}-f_{\mathrm{RF}})t)+\sin(2\pi(f_{\mathrm{LO}}+f_{\mathrm{RF}})t)). \end{aligned} \tag{6.16}$$

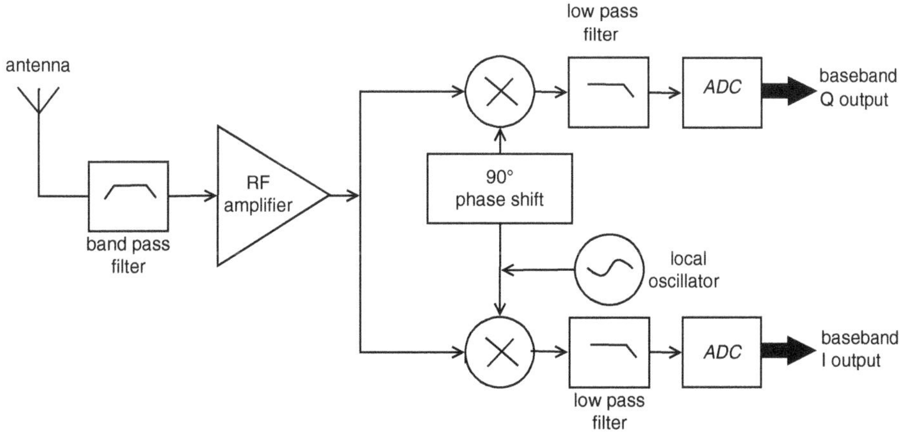

Fig. 6.23 A direct conversion receiver with baseband analogue-to-digital conversion.

The low-pass filters at the mixer outputs are used to remove the high-frequency components from the outputs and so the receiver will produce digitised copies of the input RF signal that have been translated downward in frequency by f_{LO}. However, one copy will be in phase with the original signal and the other will be $\pi/2$ out of phase (i.e. in quadrature). In general, the RF signal will be modulated and so contain components over a range of frequencies, all of which will be translated downward by f_{LO} with the quadrature components $\pi/2$ out of phase.

We now need to consider the demodulation of the I and Q outputs. Let s_n^{I} be an element of the sequence that represents the in-phase output and s_n^{Q} be an element of the sequence that represents the quadrature output. For an amplitude modulated signal, the sequence that represents the original modulation will have the terms $S_n^{\mathrm{demod}} = \left(|s_n^{\mathrm{I}}|^2 + |s_n^{\mathrm{Q}}|^2\right)^{\frac{1}{2}}$ and, for a phase modulated signal, the sequence will have terms $S_n^{\mathrm{demod}} = \arctan(s_n^{\mathrm{Q}}/s_n^{\mathrm{I}})$. Frequency is the rate of change of phase and so frequency demodulation will consist of differentiating the phase demodulated output. In the case of discrete samples, however, this will consist of forming the difference sequence

$$S_n^{\mathrm{demod}} = \frac{1}{T}\left(\arctan(s_n^{\mathrm{Q}}/s_n^{\mathrm{I}}) - \arctan(s_{n-1}^{\mathrm{Q}}/s_{n-1}^{\mathrm{I}})\right). \tag{6.17}$$

In the case of single sideband modulated signals, the architecture of Figure 6.23 is ideal. We now further process the quadrature output by a digital filter that shifts the phase by $-\pi/2$ at positive frequencies and by $\pi/2$ at negative frequencies (this is known as a *Hilbert transform* filter). Then, for a simple cosine RF signal, the output will be the digitised form of

$$\begin{aligned} s^{\mathrm{H}}(t) &= -\frac{S^{\mathrm{RF}}S^{\mathrm{LO}}}{2}\cos\left(2\pi(f_{\mathrm{LO}} - f_{\mathrm{RF}})t\right) \quad f_{\mathrm{RF}} > f_{\mathrm{LO}} \\ &= \frac{S^{\mathrm{RF}}S^{\mathrm{LO}}}{2}\cos\left(2\pi(f_{\mathrm{LO}} - f_{RF})t\right) \quad f_{\mathrm{RF}} < f_{\mathrm{LO}}. \end{aligned} \tag{6.18}$$

(The Hilbert transform has essentially multiplied positive frequency content by $-\mathrm{j}$ and negative frequency content by j.) We further note that the in-phase output will be the digitised form of

$$s^{\mathrm{I}}(t) = \frac{S^{\mathrm{RF}}S^{\mathrm{LO}}}{2}\cos\left(2\pi(f_{\mathrm{LO}} - f_{\mathrm{RF}})t\right). \tag{6.19}$$

In order to demodulate SSB signals, we will need to choose f_{LO} to be the carrier frequency; then the addition of $s^{\mathrm{I}}(t)$ and $s^{\mathrm{H}}(t)$ will eliminate signals above f_{LO} and subtraction will eliminate signals below. Consequently, we form output sequence $S_n^{\mathrm{demod}} = s_n^{\mathrm{I}} + s_n^{\mathrm{H}}$ for LSB and $S_n^{\mathrm{demod}} = s_n^{\mathrm{I}} - s_n^{\mathrm{H}}$ for USB (note that the terms s_n^{H} represent the quadrature samples s_n^{Q} after processing through a discrete version of the Hilbert transform filter).

The goal of software radio is to move the ADC closer and closer to the antenna until the function of the analogue electronics is to merely filter and amplify the incoming signal (these analogue functions are still necessary). In a fully digital receiver, the mixer function is performed by multiplying each term s_n^{RF} of the digitised RF signal by the corresponding term of a digital local oscillator sequence s_n^{LO}. A typical direct digitising

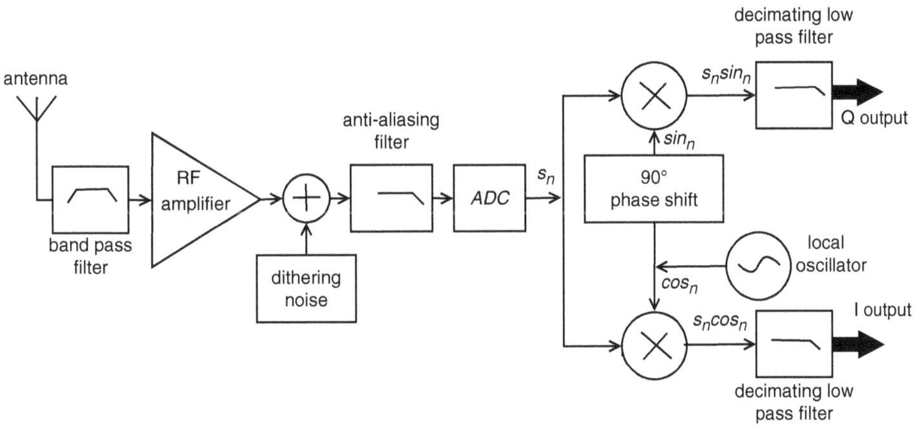

Fig. 6.24 A direct digitising receiver.

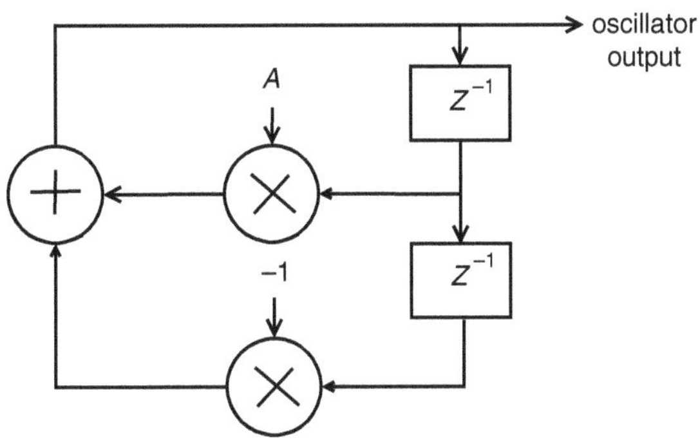

Fig. 6.25 A digital oscillator ($A = 2\cos(2\pi f_{LO}/F)$).

receiver is shown in Figure 6.24. It will be noted the mixers on the I and Q channels are followed by *decimating filters* since the sampling rate requirement at this stage will often be much lower than that of the original RF signal (according to the Nyquist theorem it will be twice the baseband bandwidth). This helps reduce the computational demand of subsequent processing. Obviously, in order to achieve the mixing in the digital domain, we will also need to generate discrete samples corresponding to the local oscillator signal. We can generate a cosine LO sequence by starting with $s_0^{LO} = 1$ and $s_1^{LO} = A\cos(2\pi f_{LO}/F)$ and then generating subsequent values through the recurrence relation $s_n^{LO} = 2\cos(2\pi f_{LO}/F)s_{n-1}^{LO} - s_{n-2}^{LO}$ where f_{LO} is the required oscillator frequency and F is the sample rate. The oscillator can be realised through the circuit of Figure 6.25 where the memory of the delays will be initially seeded with s_0^{LO} and s_1^{LO} (s sine sequence can be generated by as suitably delayed cosine sequence).

The major limitation of the above procedures is the availability of ADCs with suitable dynamic range and high enough sampling rate. Analogue down conversion before digitisation is one possibility, but another is to use what are commonly known as

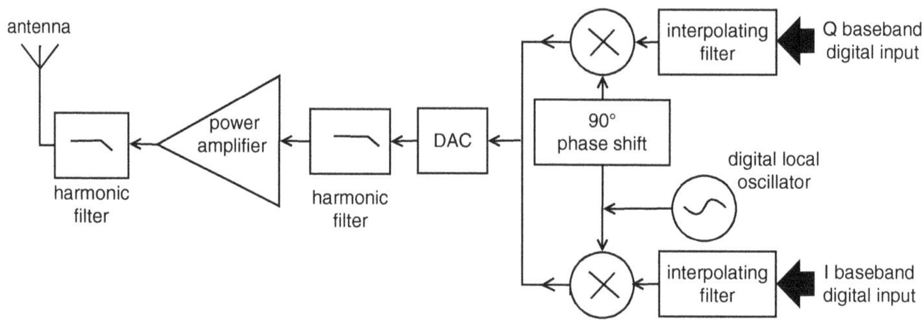

Fig. 6.26 A direct digitising transmitter.

under-sampling techniques. As we have noted previously, sampling a signal at frequency F will cause components at frequencies above $F/2$ to fold onto lower frequencies. In the case that these lower frequencies are unoccupied, we will have down conversion without the use of mixers. However, such an approach will require the use of a band-pass anti-aliasing filter (see Figure 6.24) with bandwidth less than the Nyquist bandwidth of the ADC. In addition, the sample and hold circuits in the ADC must have a response time that is fast enough to handle the input signal.

In the case of transmission, the architectures are once again based on those in the analogue domain. A typical architecture for a direct digitising transmitter is shown in Figure 6.26. We generate the desired modulation by forming the appropriate baseband I and Q sequences. In the case of SSB, we form these sequences from digitised audio with the quadrature input processed through a Hilbert transform filter. We have a digitised form of the phasing method. The audio will usually be sampled at a low rate due to its low frequency and so we use an *interpolating filter* to convert the samples into an equivalent sequence with the same the rate as those of the RF signals (this is achieved by interpolating between the low-rate samples). The output of the digital SSB generator is fed into a DAC after which it is filtered in order to remove any unwanted harmonics. Consequently, the output of the filter is the desired analogue RF signal which can then be amplified to the appropriate power level.

6.5 Conclusion

In the current chapter we have shown that many of the traditional functions of a radio can be performed within the digital domain. Indeed, the digitisation process can now almost reach the radio antenna. However, at the input of the receiver, and at the output of a transmitter, analogue filtering and amplification will always be necessary. Whilst we can now generate, and detect, radio signals with great fidelity, this is not much use if our ability to launch, propagate and capture radio waves is poor. As a consequence, an understanding of transmission lines, antennas and propagation is crucial to the design and implementation of an effective radio system. These topics are the subject of the next three chapters.

7 Transmission Lines

In most radio systems, the radio signal spreads out from the source and can service many receivers. The downside is that the power of the radio waves falls away as the inverse square of distance. Transmission lines, however, are the means by which electromagnetic waves are directly transferred between transmitter and receiver with very little loss in power. It is obvious that such transmissions are needed under special circumstances and, in particular, in transferring power between a radio and its antenna. Long before the advent of radio, however, transmission lines were a key element in communications through the telegraph. The telegraph transmitted information as electromagnetic waves along a transmission line that consisted of a wire strung above the ground between poles. Telegraph systems were developed, around the year 1837, by William Fothergill Cooke and Charles Wheatstone in the UK and, independently, by Samuel Morse in the US. In the following chapter we will develop the basic theory of transmission lines and also look at some of the radio techniques that are a direct result of this theory.

7.1 Transmission-Line Theory

Some common types of transmission lines are depicted in Figure 7.1. A *coaxial transmission line* is shown in Figure 7.1a and consists of a central conductor that is surrounded by a cylindrical conductor (the space between the conductors can often be filled by a dielectric). Figure 7.1b shows a *parallel-wire transmission line*, i.e. two parallel conductors. The major thing to note is that the conductors possess inductance and that there will be capacitance between the conductors. Consequently, a simple model of the form shown in Figure 7.2 can be used. This model consists of short segments in which the conductors are replaced by inductors and a capacitor is added to represent the capacitance between the conductors ($L1$ and $L2$ are the inductances per unit length and C is the capacitance per unit length). The circuit represents a segment of transmission line that has length dz over which the voltage between the conductors changes from V to $V+\mathrm{d}V$ and the current through the conductors from I to $I+\mathrm{d}I$. (The return current in the lower conductor has the same magnitude as that in the upper conductor, but in the opposite direction.) We can apply the Kirchhoff circuit laws to each section of transmission line and obtain

$$\mathrm{d}\mathcal{V} = -\frac{\partial \mathcal{I}}{\partial t}L1\mathrm{d}z - \frac{\partial \mathcal{I}}{\partial t}L2\mathrm{d}z \text{ and } \mathrm{d}\mathcal{I} = -\frac{\partial \mathcal{V}}{\partial t}C\mathrm{d}z, \tag{7.1}$$

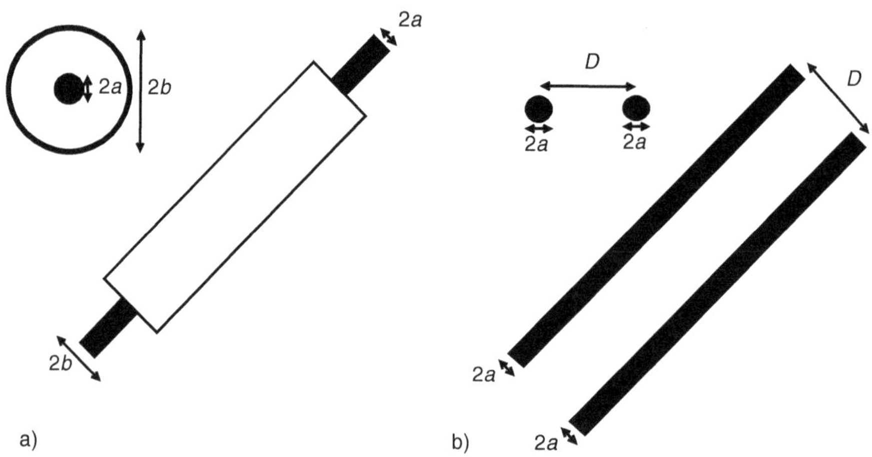

Fig. 7.1 Coaxial and parallel-wire transmission lines.

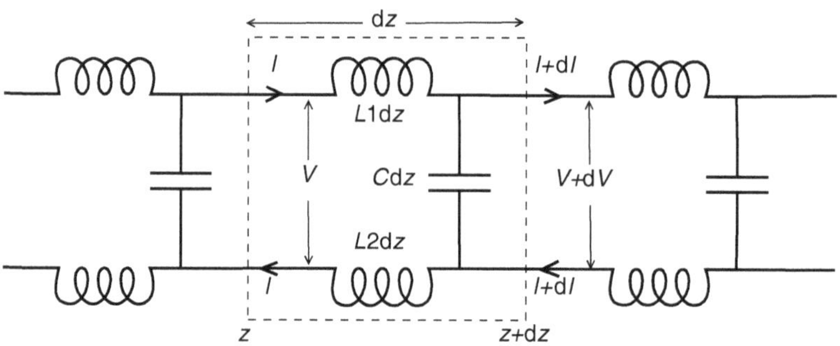

Fig. 7.2 Circuit model of a transmission line.

which can then be rewritten as

$$\frac{\partial \mathcal{V}}{\partial z} = -L\frac{\partial \mathcal{I}}{\partial t} \text{ and } \frac{\partial \mathcal{I}}{\partial z} = -C\frac{\partial \mathcal{V}}{\partial t}, \tag{7.2}$$

where $L = L1 + L2$ (i.e. all the inductance can be lumped into one inductor). From (7.2), we have that $\mathcal{V}$ and $\mathcal{I}$ both satisfy the same wave equation

$$\frac{\partial^2 \mathcal{V}}{\partial z^2} - \frac{1}{c^2}\frac{\partial^2 \mathcal{V}}{\partial t^2} = 0 \text{ and } \frac{\partial^2 \mathcal{I}}{\partial z^2} - \frac{1}{c^2}\frac{\partial^2 \mathcal{I}}{\partial t^2} = 0, \tag{7.3}$$

where $c = 1/\sqrt{LC}$. It is clear from the above equations that energy travels down a transmission line in the form of a wave, albeit a one-dimensional wave. The above wave equations have the solutions

$$\mathcal{V} = \mathcal{V}^+(z - ct) + \mathcal{V}^-(z + ct) \text{ and } \mathcal{I} = \mathcal{I}^+(z - ct) + \mathcal{I}^-(z + ct). \tag{7.4}$$

It should be noted, however, that $\mathcal{V}$ and $\mathcal{I}$ are not independent solutions of these wave equations and must also satisfy relations (7.2). These relations will then imply that

$$\mathcal{V}^{\pm} = \pm Z_0 \mathcal{I}^{\pm}, \tag{7.5}$$

where $Z_0 = \sqrt{L/C}$ is known as the *characteristic impedance* of the transmission line (this relationship has the characteristics of Ohm's law). As a consequence, we find that

$$\mathcal{I} = \frac{1}{Z_0}\left(\mathcal{V}^{+}(z-ct) - \mathcal{V}^{-}(z+ct)\right). \tag{7.6}$$

For a coaxial transmission line $L = \mu \ln(b/a)/2\pi$ and $C = 2\pi\epsilon/\ln(b/a)$. In the case of a parallel-line transmission line, $L = \mu\cosh^{-1}(D/2a)/\pi$ and $C = \pi\epsilon/\cosh^{-1}(D/2a)$. Another type of transmission line is the *parallel-strip line* shown in Figure 7.3a. This line is almost a parallel-plate capacitor and so $C = \epsilon w/h$. The inductance per unit length is given by $L = \mu h/w$ and so speed of propagation is that of the medium between the plates. A related transmission line is the *microstrip* (see Figure 7.3b). This is extremely useful as it can be easily incorporated into a printed circuit design. Providing that $h \ll W$ the expressions for C and L are well approximated by those for the parallel-strip line. Outside this limit, however, the design of a microstrip becomes far more complex.

Real transmission lines will suffer losses and the model of Figure 7.4 provides a more realistic representation of the transmission line. The resistances per unit length ($R1$

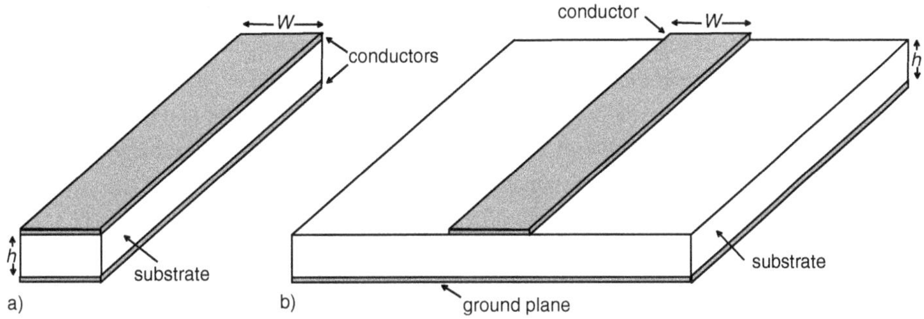

Fig. 7.3 Parallel-strip and microstrip transmission lines.

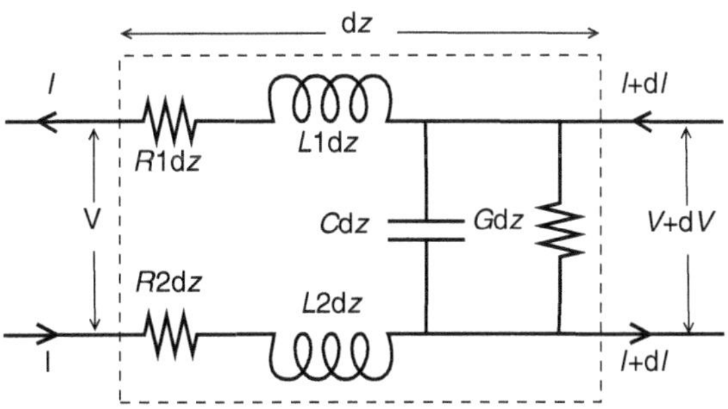

Fig. 7.4 Realistic model of a transmission line.

and $R2$) represent the ohmic losses in the conductors and the conductance (the inverse of resistance) per unit length G represents the losses in the dielectric separating the conductors. The transmission line equations will then become

$$\frac{\partial \mathcal{V}}{\partial z} = -R\mathcal{I} - L\frac{\partial \mathcal{I}}{\partial t} \text{ and } \frac{\partial \mathcal{I}}{\partial z} = -G\mathcal{V} - C\frac{\partial \mathcal{V}}{\partial t}, \tag{7.7}$$

where $R = R1 + R2$. We have from (7.7) that $\mathcal{V}$ and $\mathcal{I}$ both satisfy the same equations

$$\frac{\partial^2 \mathcal{V}}{\partial z^2} - \frac{1}{c^2}\frac{\partial^2 \mathcal{V}}{\partial t^2} - (RC + GL)\frac{\partial \mathcal{V}}{\partial t} - GR\mathcal{V} = 0$$

$$\text{and } \frac{\partial^2 \mathcal{I}}{\partial z^2} - \frac{1}{c^2}\frac{\partial^2 \mathcal{I}}{\partial t^2} - (RC + GL)\frac{\partial \mathcal{I}}{\partial t} - GR\mathcal{I} = 0. \tag{7.8}$$

The above form of equation is known as *telegrapher's equation* since it describes the propagation of Morse code pulses down a telegraph line, one of the first major uses of transmission lines. Indeed, the equation was first developed by Oliver Heaviside around 1885 in order to explain such propagation. The telegrapher's equation is difficult to solve and so we merely quote the solution for a right-travelling harmonic wave at frequency ω, i.e.

$$\mathcal{V} = V_0 \exp(-t(R/L + G/C)/2) \sin\left(\omega t - \frac{\sqrt{\omega^2 + \alpha^2}}{c} z\right), \tag{7.9}$$

where $\alpha = (R/L - G/C)/2$ and V_0 is the amplitude when $t = 0$. The first thing to note is that the amplitude of the wave reduces as it propagates, i.e. energy is dissipated in the conductance and resistance of the line. Just as important, however, is that the speed of the wave is given by $v = c\omega/\sqrt{\omega^2 + \alpha^2}$ which, it will be noted, is frequency-dependent. Normally, the wave of interest will consist of a modulated harmonic wave (the pulses of Morse code for instance) and therefore it will be a mixture of components at many different frequencies. Since these components will travel at different speeds, the signal will lose integrity over time (see Figure 7.5) with a loss of information. This phenomenon, known as *dispersion*, is every bit as problematic as signal loss for long-distance telegraph lines. The solution, suggested by Heaviside, was to make a cable such that $R/G = L/C$, i.e. a cable free of dispersion.

Fig. 7.5 The effect of dispersion.

7.2 The Termination of Transmission Lines

We consider the situation where a wave is travelling down a transmission line to a resistive load R where it is then reflected back down the transmission line (see Figure 7.6). The incoming wave will have the form $\mathcal{V}^{+}(z-ct)$ and the reflected wave the form $\mathcal{V}^{-}(z+ct)$. At the load, Ohm's law will imply $\mathcal{V}=R\mathcal{I}$ and this, together with (7.6), will then imply that

$$\mathcal{V}^{+}(l-ct)+\mathcal{V}^{-}(l+ct)=\frac{R}{Z_0}\left(\mathcal{V}^{+}(l-ct)-\mathcal{V}^{-}(l+ct)\right). \tag{7.10}$$

Rearranging this expression

$$\mathcal{V}^{-}(l+ct)=\frac{R-Z_0}{R+Z_0}\mathcal{V}^{+}(l-ct)=\Gamma\mathcal{V}^{+}(l-ct), \tag{7.11}$$

where $\Gamma=(R-Z_0)/(R+Z_0)$ is known as the *reflection coefficient*. It should be noted that, when the load R is equal to the characteristic impedance Z_0, $\Gamma=0$ and there is no reflection. In this situation, the incoming wave is totally absorbed by the load and the load is said to be *matched* to the transmission line.

We now look at the case of a harmonic wave travelling down the transmission line, i.e.

$$\mathcal{V}=\Re\{V\exp(\mathrm{j}\omega t)\} \text{ and } \mathcal{I}=\Re\{I\exp(\mathrm{j}\omega t)\}, \tag{7.12}$$

then

$$V=V^{+}\exp(-\mathrm{j}\beta z)+V^{-}\exp(\mathrm{j}\beta z) \text{ and } I=\frac{1}{Z_0}\left(V^{+}\exp(-\mathrm{j}\beta z)-V^{-}\exp(\mathrm{j}\beta z)\right), \tag{7.13}$$

where $\beta=\omega/c$ is the *propagation constant* (subscript + refers to the incoming wave and − refers to the reflected wave). At the load, the time-harmonic Ohm's law will imply that $V=IZ_L$ and so

$$V^{+}\exp(-\mathrm{j}\beta l)+V^{-}\exp(\mathrm{j}\beta l)=\frac{Z_L}{Z_0}\left(V^{+}\exp(-\mathrm{j}\beta l)-V^{-}\exp(\mathrm{j}\beta l)\right) \tag{7.14}$$

from which

$$V^{-}=\frac{Z_L-Z_0}{Z_L+Z_0}\exp(-2\mathrm{j}\beta l)V_{+}=\Gamma_{\text{in}}V^{+} \tag{7.15}$$

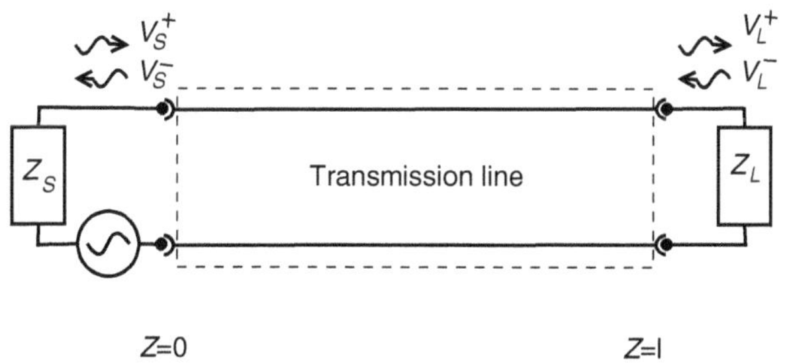

Fig. 7.6 A terminated transmission line.

(we have implicitly assumed that the source is matched to the transmission line and so all of the reflected wave is absorbed by the source). Γ_{in} is the reflection coefficient looking into the transmission line (i.e. where $z = 0$) and it is obvious that $\Gamma_{in} = \Gamma_L \exp(-2j\beta l)$ where $\Gamma_L = (Z_L - Z_0)/(Z_L + Z_0)$ is the reflection coefficient looking into the load. We will now have that

$$V = V^+ \left(\exp(-j\beta z) + \Gamma_L \exp(-2j\beta l) \exp(j\beta z)\right) \tag{7.16}$$

and

$$I = \frac{V^+}{Z_0} \left(\exp(-j\beta z) - \Gamma_L \exp(-2j\beta l) \exp(j\beta z)\right). \tag{7.17}$$

From (7.16), we have that

$$|V| = |V_L^+||1 + \Gamma_L \exp(2j\beta(z - l))| \tag{7.18}$$

and from which we see that, along the transmission line, the voltage amplitude varies from a minimum of $V_{\text{min}} = |V_L^+|(1 - |\Gamma_L|)$ to a maximum of $V_{\text{max}} = |V_L^+|(1 + |\Gamma_L|)$. An important measure of the degree to which a transmission line is matched to a load is the *voltage standing-wave ratio* (VSWR) which is defined to be

$$\text{VSWR} = \frac{V_{\text{max}}}{V_{\text{min}}} = \frac{1 + |\Gamma_L|}{1 - |\Gamma_L|}. \tag{7.19}$$

There will be a perfect match for a VSWR of 1 and the higher the value of VSWR the greater the mismatch. VSWR is not a full description of the load, but the magnitude of the reflection coefficient can be derived from it.

The reflection coefficient Γ provides an alternative description of a device when it is connected to other devices through a transmission line. The reflection coefficient is related to the impedance of the device through $Z = Z_0(1 + \Gamma)(1 - \Gamma)$ where Z_0 is the characteristic impedance of the transmission line. The resistive part R of the impedance can be expressed in terms of the reflection coefficient as

$$R = Z_0 \Re \left\{ \frac{1 + \Gamma}{1 - \Gamma} \right\} = \frac{1 - |\Gamma|^2}{|1 - \Gamma|^2}. \tag{7.20}$$

From the above expression, it is clear that the resistance of the device will be positive when $|\Gamma| < 1$ and negative when $|\Gamma| > 1$. As we have seen in Chapter 5, a device that exhibits negative resistance has the ability to cause oscillations and is thus unstable. Passive devices are inherently stable, but active devices can often exhibit instability. Hence, to ensure the stability of a device we need to have a reflection coefficient with $|\Gamma| < 1$. If we load an unstable device with a passive impedance Z_L, the resulting system will oscillate if $Z_L + Z = 0$ (this is the condition that allows an arbitrary current to flow in the circuit), i.e. $R_L + R = 0$ and $X_L + X = 0$. The device resistance will need to be sufficiently negative to cancel the effect of the load resistance and the oscillation frequency will be determined by $X_L + X = 0$. If $R_L + R < 0$, the amplitude of oscillation will rise until the inherent nonlinearities of the device have increased the resistance R to a point where $R_L + R = 0$.

7.3 Transmission-Line Resonators

From (7.16) and (7.17), the impedance looking into a transmission line of length l, terminated by a load Z_L, is given by

$$Z_{\text{in}} = \frac{V(0)}{I(0)} = Z_0 \frac{1 + \Gamma_L \exp(-2\text{j}\beta l)}{1 - \Gamma_L \exp(-2\text{j}\beta l)}, \tag{7.21}$$

where $\Gamma_L = (Z_L - Z_0)/(Z_L + Z_0)$. This can be rearranged into

$$Z_{\text{in}} = Z_0 \frac{Z_L \cos(\beta l) + \text{j}Z_0 \sin(\beta l)}{Z_0 \cos(\beta l) + \text{j}Z_L \sin(\beta l)} \tag{7.22}$$

and, from the above expression, it can be seen that a length of transmission line can be used to transform an impedance. There are two important special cases, the *open-circuit transmission line* ($Z_L = \infty$) and the *short-circuit transmission line* ($Z_L = 0$). For both cases, the impedance looking into the transmission line is entirely reactive with

$$Z_{\text{in}} = -\text{j}Z_0 \cot(\beta l) \tag{7.23}$$

for the open-circuit line and

$$Z_{\text{in}} = \text{j}Z_0 \tan(\beta l) \tag{7.24}$$

for the short-circuit line. Figure 7.7 shows how the reactance varies with line length for a given frequency ω (the length is measured in terms of wavelength $\lambda = 2\pi c/\omega$ on the line). It will be noted that the reactances for both cases vary between capacitative reactance and inductive reactance. Importantly, for lengths less than a quarter wavelength, the open-circuit line has capacitative reactance (it is negative) and the short-circuit line has inductive reactance (it is positive). Both lines have a resonance at a quarter wavelength, the open circuit exhibiting a series resonance and the short circuit exhibiting a parallel resonance.

From the above discussion, it is clear that an open-circuit transmission line behaves like a series resonant tuned circuit with a resonance at a frequency ω_0 for which the length

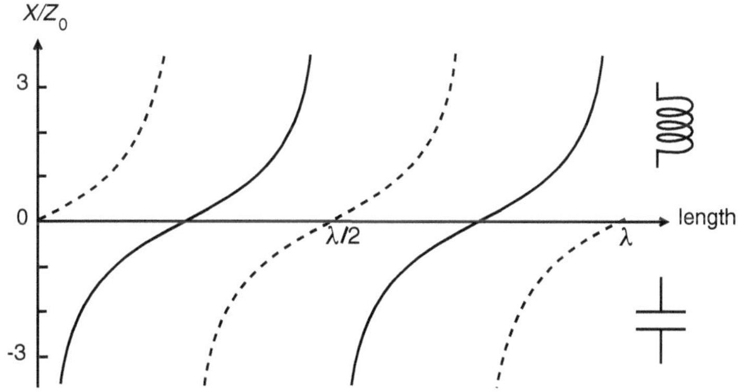

Fig. 7.7 Impedance of an open-circuit (unbroken-line) and a short-circuit (broken-line) transmission line (TL).

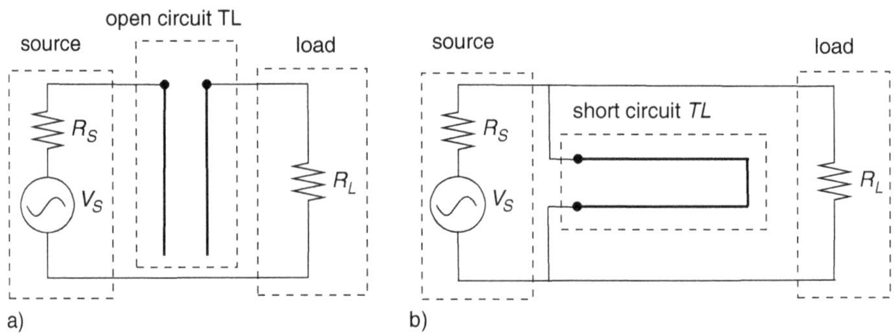

Fig. 7.8 Filtering using open-circuit and short-circuit quarter-wave transmission lines.

l corresponds to a quarter wavelength ($l = \pi c/2\omega$). The impedance of this resonator is given by $Z = -\mathrm{j}Z_0 \cot\left(\frac{\pi}{2}\frac{\omega}{\omega_0}\right)$. If we use the resonator in the configuration shown in Figure 7.8a, the voltage drop across the load resistor is then given by

$$V_L = V_S \frac{R_L}{R_S + R_L + Z} = V_S \frac{R_L}{R_S + R_L} \frac{1}{1 - \mathrm{j}\frac{Z_0}{R_S + R_L} \cot\left(\frac{\pi}{2}\frac{\omega}{\omega_0}\right)} \tag{7.25}$$

and, around the resonance at ω_0, this can be approximated by

$$V_L = V_S \frac{R_L}{R_S + R_L} \frac{1}{1 - 2\mathrm{j}Q\frac{\delta\omega}{\omega_0}}, \tag{7.26}$$

where $Q = \pi Z_0/4(R_L + R_S)$ and $\delta\omega = \omega - \omega_0$. It is possible to make transmission-line resonators with very high unloaded Q (much higher than is possible with inductor- and capacitor-based tuned circuits) and so the above tuned circuit can have an extremely sharp response if the source and/or load resistances are large. Figure 7.9 shows the insertion gain of the filter for various values of Q (solid line for a Q of 5, broken line for a Q of 2 and dotted line for a Q of 1). As to be expected, the filtering action is sharper for the higher Q. However, unlike inductor and capacitor combinations, it will be noted that the transmission-line resonators have a whole series of resonances (these occur at ω_0, $3\omega_0$, $5\omega_0$, etc.). Consequently, in using transmission-line resonators one needs to ensure that these higher-frequency responses are not an issue.

A short-circuited transmission line behaves like a parallel combination of an inductor and a capacitor with a resonance at a frequency ω_0 for which the line is a quarter wavelength. The impedance of such a resonator is given by $Z = \mathrm{j}Z_0 \tan\left(\frac{\pi}{2}\frac{\omega}{\omega_0}\right)$. If we use the resonator in the configuration shown in Figure 7.8b, the voltage drop across the load resistor is given by

$$V_L = V_S \frac{ZR_L}{R_S Z + R_L Z + R_S R_L} = V_S \frac{R_L}{R_S + R_L} \frac{1}{1 - \mathrm{j}\frac{R_S \| R_L}{Z_0} \cot\left(\frac{\pi}{2}\frac{\omega}{\omega_0}\right)}. \tag{7.27}$$

Around resonance at ω_0, the behaviour of V_L can be approximated by (7.26), but with $Q = (\pi R_S \parallel R_L)/4Z_0$. Further, like the open-circuit transmission line, the short-circuit line will have additional resonances at $3\omega_0$, $5\omega_0$, etc.

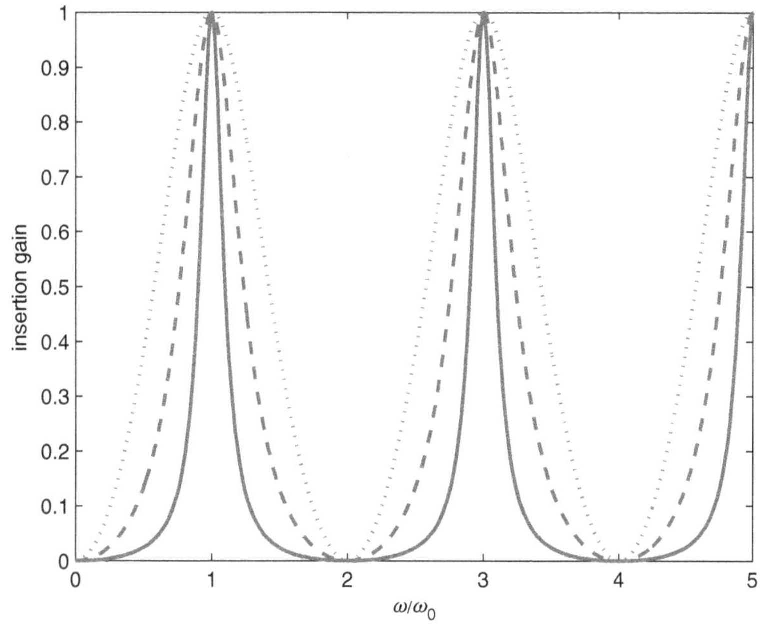

Fig. 7.9 Insertion gain for an open-circuit transmission line with $Q = 5$ (solid line), $Q = 2$ (broken line) and $Q = 1$ (dotted line).

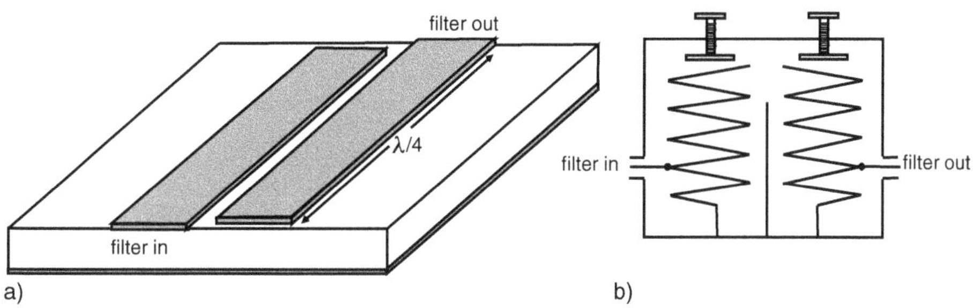

Fig. 7.10 Coupled transmission-line filters (microstrip and helical).

In Chapter 3 we saw that series and parallel resonant circuits could be combined to form complex band-pass filters. It is therefore possible to build transmission line versions of these filters by replacing the series and parallel resonant circuits of these filters by open- and short-circuit quarter-wavelength transmission lines. An alternative filter is the coupled microstrip filter shown in Figure 7.10a. In this filter we have two quarter-wavelength resonators that are close enough for power to be transferred through their mutual capacitance and inductance. Using such an approach, effective filters can be made for frequencies above 500 MHz, but below these frequencies the physical size of a quarter-wavelength line can become an issue. The physical size can be reduced by slowing down the wave on the transmission line. In the case of the microstrips, this can be done by increasing the permittivity of the substrate dielectric (substrates with $\epsilon = 10$

are certainly available). Another approach, however, is to increase the inductance of the conductors. This is the case in the helical filter shown in Figure 7.10b. This filter consists of two transmission-line resonators, each consisting of a metal chamber containing a central conductor that is a helix; this helical shape increases the inductance of the central conductor. The two windings are then coupled through a window between the chambers (note the adjustable capacitance at the top of a chamber that allows the filter frequency to be fine-tuned). In order to ensure that the filter input and output present the correct impedance to the source and load, these are fed into a tap at a suitable point along the helix.

Another application of transmission lines is the matching of a source Z_S to a load Z_L. In the case that the source and load have real impedances, this is simply achieved using a quarter-wave transmission line. From (7.22), we have that the impedance looking into the loaded line will be $Z_{\text{in}} = Z_0^2/Z_L$. Consequently, if we choose $Z_0 = \sqrt{Z_L Z_S}$, we will have that $Z_{\text{in}} = Z_S$, i.e. we have matched the load to the source. In the case of general source and load impedances, the matching is far more difficult. Consider the situation shown in Figure 7.11. We will consider the problem in terms of the admittances $Y_L = 1/Z_L$ and $Y_S = 1/Z_S$ (the admittance of parallel impedances is additive). From (7.22), the admittance transforms along the transmission line according to

$$Y_{in} = Y_0 \frac{Y_L \cos(\beta l) + jY_0 \sin(\beta l)}{Y_0 \cos(\beta l) + jY_L \sin(\beta l)}, \tag{7.28}$$

where $Y_0 = 1/Z_0$. For a matching to occur, we need the admittances Y_{in} and Y_S to be conjugate and so first choose the length l such that $\Re\{Y_{\text{in}}\} = \Re\{Y_S\}$. Then, if $\Im\{Y_{\text{in}}\} \neq -\Im\{Y_S\}$, we manufacture a stub reactance X such that

$$\Im\{Y_{\text{in}}\} - \frac{1}{X} = -\Im\{Y_S\}. \tag{7.29}$$

Although we have shown the stub as open circuit, it could equally well be a shorted stub. At the end of the day, however, it is best to choose the type of stub that has the shortest length that gives the desired reactance X.

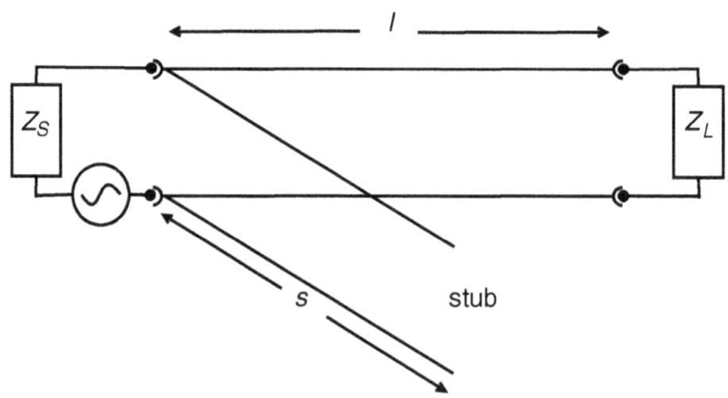

Fig. 7.11 Stub matching a source to a load.

7.4 Scattering Matrices

One of the problems with the description of a circuit in terms of voltages and currents is that the measurement of these quantities becomes more and more problematic as frequency rises. In particular, the probes used to measure these quantities can severely disturb the operation of the circuit. We have hitherto described our circuits in terms of impedances and sources, but we have already seen that a reflection coefficient Γ can provide an alternative description of the load (the impedance can be related to the reflection coefficient through the relation $Z = Z_0(1+\Gamma)/(1-\Gamma)$). In the impedance description of RF electronics, we analyse circuits in terms of sources (both current and voltage) and impedances that are modelled by Ohm's law, i.e.

$$V = IZ, \tag{7.30}$$

where V is the voltage drop across an impedance Z through which a current I flows. In the reflection coefficient description of RF circuits, we deal with sources and devices that are connected through transmission lines. We describe the circuit in terms of voltage waves that travel back and forth along the transmission lines and model a device through its reflection coefficient, i.e.

$$V^r = \Gamma V^i, \tag{7.31}$$

where V^i is the voltage wave incident at the impedance, V^r is the voltage wave reflected at the impedance and Γ is the reflection coefficient at the impedance. It should be noted, however, that this description depends on the characteristic impedance Z_0 of the transmission lines and that Γ can change if Z_0 changes. The Smith chart is a graphical representation of the reflection coefficient and is particularly useful for illustrating the behaviour of an impedance with respect to variations of a parameter such as frequency. The vertical axis represents the imaginary component of the reflection coefficient and the horizontal axis the real component.

Figure 7.13a shows curves of constant $|\Gamma|$ plotted on a Smith chart, Figure 7.13b curves of constant resistance (normalised on Z_0) and Figure 7.13c curves of constant reactance (normalised on Z_0). Figure 7.14 shows a complex tuned circuit with its Smith chart. The variation with frequency shows two resonant frequencies (these are points where the reactance is zero). There is a series resonance at 7.16 MHz where the value of resistance is around $R = 50\,\Omega$ and parallel resonance at 10 MHz where the resistance is large.

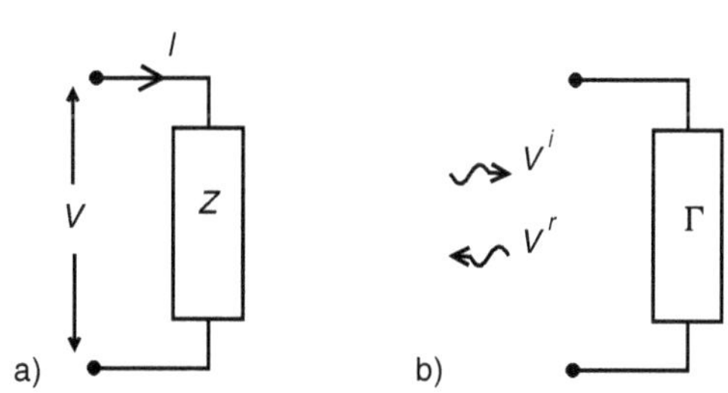

Fig. 7.12 Impedance and reflection coefficient models of a one-port device.

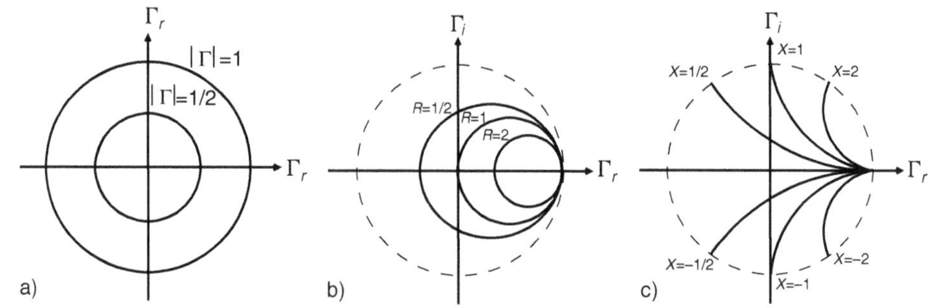

Fig. 7.13 Smith-charts representation of the reflection coefficient $\Gamma = \Gamma_r + j\Gamma_i$, resistance R and reactance X.

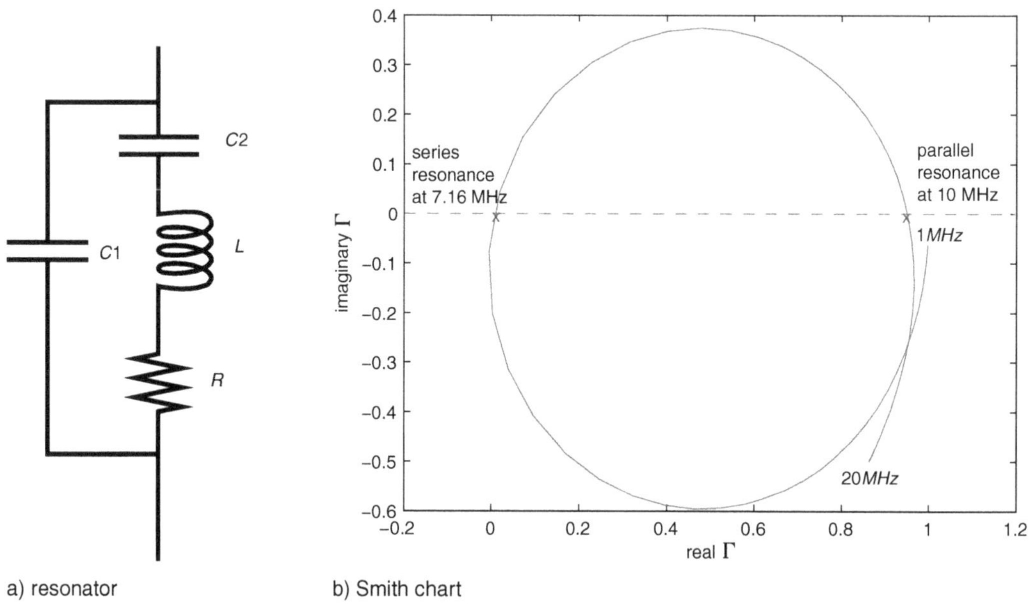

Fig. 7.14 Smith chart of a resonator with both parallel and series resonances ($C1 = C2 = 50\,\text{pF}$, $L = 10\ \mu\text{H}$ and $R = 50\,\Omega$).

Impedance is a concept that applies to a one port device, but can be extended to multiport devices through the concept of an *impedance matrix*. An amplifier is an example of a two-port device, the input and the output of the amplifier being the two ports. In the world of voltage and current, the currents flowing into the ports are related to the voltages applied to the ports through an impedance matrix Z. Figure 7.15a illustrates the situation for a two-port device. The relationship between the voltages V_1 and V_2 at the ports and the currents I_1 and I_2 flowing into these ports is then given by

$$V_1 = Z_{11}I_1 + Z_{12}I_2$$
$$V_2 = Z_{21}I_1 + Z_{22}I_2, \tag{7.32}$$

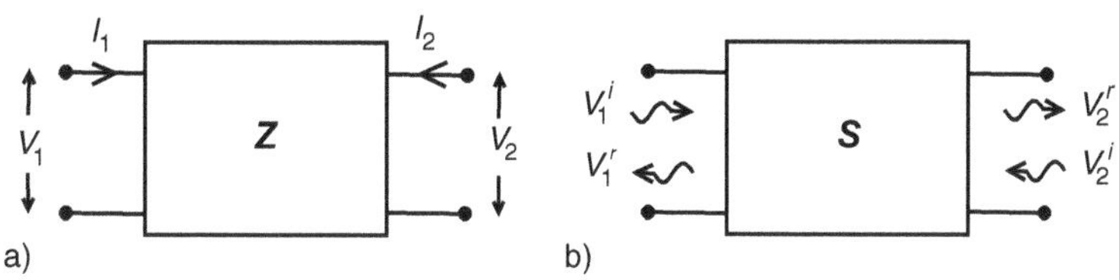

Fig. 7.15 Impedance-matrix and scattering-matrix models of a two-port device.

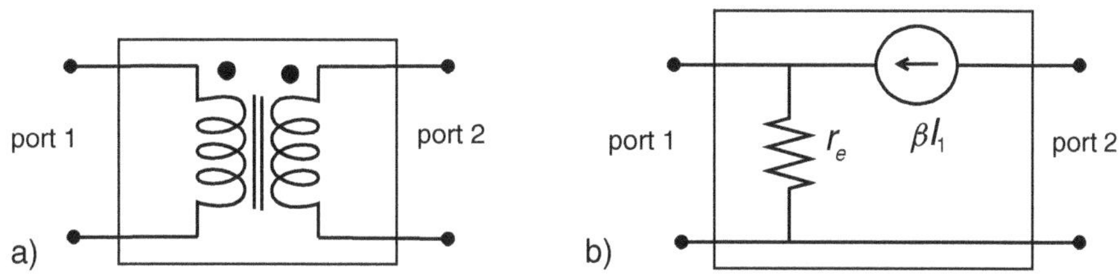

Fig. 7.16 Examples of two-port networks.

where Z_{11}, Z_{12}, Z_{21} and Z_{22} are the coefficients of the impedance matrix. In the language of matrices

$$\mathbf{V} = Z\mathbf{I}, \tag{7.33}$$

where

$$\mathbf{V} = \begin{pmatrix} V_1 \\ V_2 \end{pmatrix}, \quad \mathbf{I} = \begin{pmatrix} I_1 \\ I_2 \end{pmatrix} \quad \text{and} \quad Z = \begin{pmatrix} Z_{11} & Z_{12} \\ Z_{21} & Z_{22} \end{pmatrix}.$$

We have already seen such a relationship of the form of (7.32) when studying transformers in Chapter 3 (see Figure 7.16a). In this case, the impedance matrix is given by

$$Z = \begin{pmatrix} \mathrm{j}\omega L_{11} & \mathrm{j}\omega L_{12} \\ \mathrm{j}\omega L_{21} & \mathrm{j}\omega L_{22} \end{pmatrix}$$

where $L_{12} = L_{21}$ is the mutual impedance between the transformer windings and L_{11} and L_{22} are the self impedance of these windings. (A more general schematic of a transformer is used in Figure 7.16a with the dots indicating 'in phase' terminals and the bars indicating the presence of an iron core.)

In a similar fashion to impedance, the concept of a reflection coefficient can be extended to multiport devices through the *scattering matrix*. In the scattering-matrix approach, the incident and reflected voltage waves at the ports of a device are related to each other through the scattering matrix. Figure 7.15b illustrates the situation for a two-port device. The relationship between the incident waves V_1^i and V_2^i and the reflected waves V_1^r and V_2^r is then given by

$$\begin{aligned} V_1^r &= S_{11}V_1^i + S_{12}V_2^i \\ V_2^r &= S_{21}V_1^i + S_{22}V_2^i, \end{aligned} \tag{7.34}$$

where S_{11}, S_{12}, S_{21} and S_{22} are the coefficients of the scattering matrix. In the language of matrices

$$\mathbf{V}^r = S\mathbf{V}^i, \tag{7.35}$$

where

$$\mathbf{V}^i = \begin{pmatrix} V_1^i \\ V_2^i \end{pmatrix}, \quad \mathbf{V}^r = \begin{pmatrix} V_1^r \\ V_2^r \end{pmatrix} \quad \text{and} \quad S = \begin{pmatrix} S_{11} & S_{12} \\ S_{21} & S_{22} \end{pmatrix}.$$

The above description might seem incomplete (there is no mention of current) but we must remember that, in the scattering-matrix world, devices are connected through transmission lines. As a consequence, the current and voltage waves are related through $V_1^i = Z_0 I_1^i$, $V_1^r = -Z_0 I_1^r$, $V_2^i = -Z_0 I_2^i$ and $V_2^r = Z_0 I_2^r$, where Z_0 is the characteristic impedance of the transmission lines. In the same way that the reflection coefficient and impedance of a one-port device are related, the scattering matrix and the impedance matrix of a multiport device are also related (a reflection coefficient is merely scattering matrix for a one-port device). The relation is given by $S = (Z - Z_0 I)(Z + Z_0 I)^{-1}$ where I is the unit matrix. For an ideal transformer, this then implies that

$$S = \frac{1}{\mathrm{j}\omega(L_{11} + L_{22}) + Z_0} \begin{pmatrix} \mathrm{j}\omega(L_{11} - L_{22}) - Z_0 & 2\mathrm{j}\omega L_{12} \\ 2\mathrm{j}\omega L_{21} & \mathrm{j}\omega(L_{22} - L_{11}) - Z_0 \end{pmatrix}. \tag{7.36}$$

We now consider the transformer shown in Figure 7.16a to be ideal, i.e. $V_1 = nV_2$ and $I_1 = -I_2/n$, where $n = N_1/N_2$ is the turns ratio. Since $V_1 = V_1^i + V_1^r$ and $V_2 = V_2^i + V_2^r$, we have that

$$V_1^r - nV_2^r = nV_2^i - V_1^i. \tag{7.37}$$

Further, since $I_1 = V_1^i/Z_0 - V_1^r/Z_0$ and $I_2 = -V_2^i/Z_0 + V_2^r/Z_0$ (note that the incident wave at port 2 is now leftward-travelling), we have that

$$V_1^r + \frac{V_2^r}{n} = \frac{V_2^i}{n} + V_1^i. \tag{7.38}$$

Adding (7.37) to n^2 times (7.38), we obtain that

$$V_1^r(n^2 + 1) = (n^2 - 1)V_1^i + 2nV_2^i \tag{7.39}$$

and subtracting (7.37) from (7.38), we obtain that

$$V_2^r(n + \frac{1}{n}) = 2V_1^i + (\frac{1}{n} - n)V_2^i. \tag{7.40}$$

Bringing this all together, we have a scattering matrix of the form

$$S = \begin{pmatrix} \frac{n^2-1}{n^2+1} & \frac{2n}{n^2+1} \\ \frac{2n}{n^2+1} & -\frac{n^2-1}{n^2+1} \end{pmatrix}. \tag{7.41}$$

In the limit that the transformer is strongly coupled, and the windings have large reactance, (7.36) reduces to the above expression.

Now consider the S matrix of the BJT shown in Figure 7.16b. The current flowing through r_e will be $(1+\beta)I_1$ and so $V_1 = r_e(1+\beta)I_1$. At the input, we have $V_1 = V_1^i + V_1^r$

and $I_1 = V_1^i/Z_0 - V_1^r/Z_0$. As a consequence, $V_1 = V_1^i + V_1^r = r_e(1+\beta)(V_1^i - V_r)/Z_0$ and from this

$$V_1^r = \frac{(\beta+1)r_e - Z_0}{(\beta+1)r_e + Z_0} V_1^i. \tag{7.42}$$

Comparing this with (7.34) we have that $S_{12} = 0$ and

$$S_{11} = \frac{(\beta+1)r_e - Z_0}{(\beta+1)r_e + Z_0}. \tag{7.43}$$

If we load port 2 with the characteristic impedance Z_0, there will be no reflection from this load and hence $V_2^i = I_2^i = 0$. The current flowing through the load will be βI_1 from which $V_2^r = V_2 = \beta I_1 Z_0$ and so

$$V_2^r = \beta(V_1^i - V_1^r). \tag{7.44}$$

Eq. (7.34) will also imply that $V_2^r = S_{21}V_1^i$ and so this, together with (7.44), will imply that

$$S_{21} = \beta\left(1 - \frac{V_1^r}{V_1^i}\right) = \beta(1 - S_{11}). \tag{7.45}$$

Finally, looking into port 2, the current source will present an open circuit (i.e. infinite impedance) and so $S_{22} = 1$. Bringing this all together, we have a scattering matrix of the form

$$S = \begin{pmatrix} \frac{(\beta+1)r_e - Z_0}{(\beta+1)r_e + Z_0} & 0 \\ \frac{2\beta Z_0}{(\beta+1)r_e + Z_0} & 1 \end{pmatrix}. \tag{7.46}$$

For the simple devices above, we have the luxury of being able to obtain an analytic expression for the S matrix. For more complex devices, however, the S matrix is usually obtained through measurements. Such measurements are important due to the fact that most RF devices are now characterised by their manufacturers in terms of S parameters. To do the necessary measurements, we need a device which we can sample the incoming and reflected voltage waves at the device ports, a device known as a *directional coupler*.

Figure 7.17 shows a coupler that is based around two transformers with turns ratio of n. Ports 3 and 4 are loaded by the impedances Z_0 (the characteristic impedance of the transmission lines). As a consequence, the left-hand transformer causes a current $(V^i - V^r)/nZ_0$ to flow into the line joining ports 3 and 4, half of which flows into the left load and half into the right load. The right-hand transformer will generate a voltage $(V^i + V^r)/n$ in the line which causes an additional current $(V^i + V^r)/2nZ_0$ to flow into the left load and $-(V^i + V^r)/2nZ_0$ into the right load. As a consequence, we will have voltages V^i/n and $-V^r/n$ across the left and right loads respectively. We now have a device which, if placed in line, allows us to sample left- and right-travelling waves. Figure 7.18 shows the setup for measuring the S parameters of a two-port device. Since the output port is loaded with the characteristic impedance Z_0, we will have $V_2^i = 0$ and so $V_1^r = S_{11}V_1^i$ and $V_2^r = S_{21}V_1^i$. Consequently, we can obtain S_{11} and S_{21} by measuring the input and output voltage waves. To measure S_{12} and S_{22}, we simply reverse the DUT (device under test).

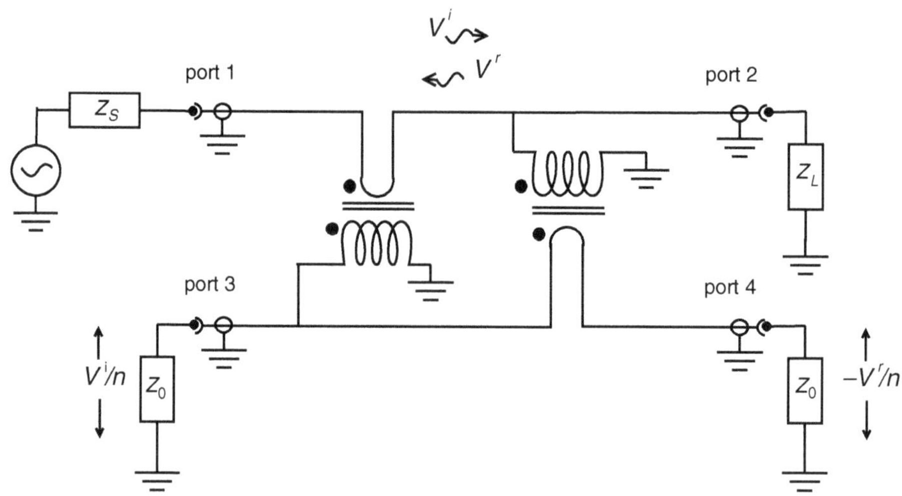

Fig. 7.17 Directional coupler.

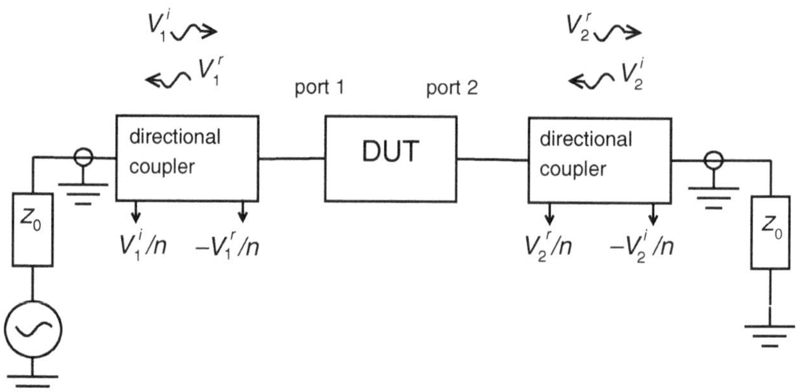

Fig. 7.18 The measurement of S parameters.

An amplifier is an important example of a two-port device and we now look at how its S parameters relate to its gain and stability. We first consider the device to be loaded at the output and so $V_2^i = \Gamma_L V_2^r$. Then, from (7.34),

$$\begin{aligned} V_1^r &= S_{11} V_1^i + S_{12} \Gamma_L V_2^r \\ V_2^r &= S_{21} V_1^i + S_{22} \Gamma_L V_2^r \end{aligned} \tag{7.47}$$

and eliminating V_2^r between these equations,

$$\frac{V_1^r}{V_1^i} = S_{11} + \frac{S_{21} S_{12} \Gamma_L}{1 - S_{22} \Gamma_L}. \tag{7.48}$$

Consequently, the reflection coefficient looking in the device input, i.e. V_1^r / V_1^i, is given by

$$\Gamma_{\text{in}} = S_{11} + \frac{S_{21} S_{12} \Gamma_L}{1 - S_{22} \Gamma_L}. \tag{7.49}$$

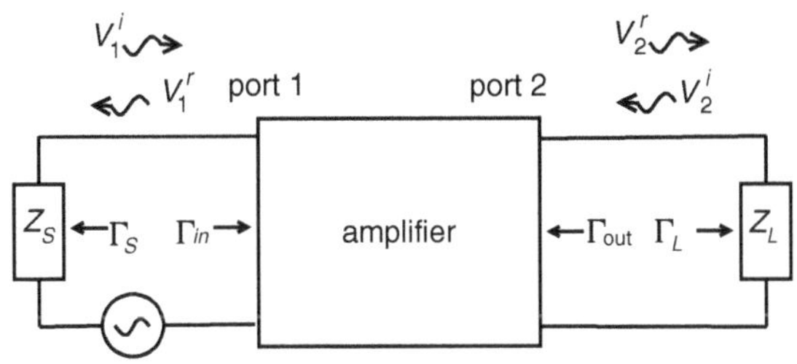

Fig. 7.19 Reflection coefficients for a loaded two-port device.

In a similar fashion, the reflection coefficient looking in the device output, i.e. V_2^r/V_2^i, is given by

$$\Gamma_{\text{out}} = S_{22} + \frac{S_{12}S_{21}\Gamma_S}{1 - S_{11}\Gamma_S} \tag{7.50}$$

As we have seen, in the case of a one-port device, we need the reflection coefficient at a device port to be less than one in order to ensure stability. In the case of our two-port device, this will require that $\Gamma_{\text{in}} < 1$ and $\Gamma_{\text{out}} < 1$. Obviously, if there is feedback (i.e. if $S_{12} \neq 0$), there is possibility of instability. However, even if $S_{12} = 0$, there is still the possibility of instability through $S_{22} > 1$ and/or $S_{11} > 1$. As for a one-port active device, this can arise through negative resistance. As can be seen from (7.49) and (7.50), the stability of a device will depend on the source and load (both assumed to be stable, i.e. $|\Gamma_S| < 1$ and $|\Gamma_L| < 1$). Consequently, it is sometimes possible to stabilise a potentially unstable amplifier with the correct choice of load and/or source. Obviously, it would be preferable to have a device that is stable for all sources and loads. Such a device is said to be *unconditionally stable* and this will be the case if

$$K = \frac{1 - |S_{11}|^2 - |S_{22}|^2 + |\Delta|^2}{2|S_{12}||S_{21}|} > 1, \tag{7.51}$$

where $\Delta = S_{11}S_{22} - S_{21}S_{21}$ and K is known as the Rollett stability factor.

We now consider the *transducer power gain* of the amplifier, i.e. the power delivered to the load divided by the maximum power available from the source. Firstly, note that the voltage at port 1 is given by $V_1 = V_S Z_{\text{in}}/(Z_{\text{in}} + Z_S)$ where Z_{in} is the input impedance of the device (Z_S and Z_{in} act as a voltage divider). We note that $V_1 = V_1^r + V_1^i = V_1^i(1+\Gamma_{\text{in}})$ and that $Z_{\text{in}} = Z_0(1+\Gamma_{\text{in}})/(1-\Gamma_{\text{in}})$. We therefore have that

$$V_1^i(1+\Gamma_{\text{in}}) = V_S\frac{Z_{\text{in}}}{Z_{\text{in}} + Z_S} = V_S\frac{Z_0(1+\Gamma_{\text{in}})}{Z_0(1+\Gamma_{\text{in}}) + Z_S(1-\Gamma_{\text{in}})}. \tag{7.52}$$

We now note that $Z_L = Z_0(1+\Gamma_S)/(1-\Gamma_S)$ and, substituting this into (7.52), we obtain that

$$V_1^i = \frac{V_S}{2}\frac{1-\Gamma_S}{1-\Gamma_S\Gamma_{\text{in}}}. \tag{7.53}$$

From (7.47), the voltage wave arriving at the load will be

$$V_2^r = V_1^i\frac{S_{21}}{1-S_{22}\Gamma_L} = \frac{V_S}{2}\frac{1-\Gamma_S}{1-\Gamma_S\Gamma_{\text{in}}}\frac{S_{21}}{1-S_{22}\Gamma_L} \tag{7.54}$$

and so the power absorbed by the load will be

$$P_L = \frac{1}{8Z_0}\frac{|V_S|^2|1-\Gamma_S|^2 S_{21}^2(1-|\Gamma_L|^2)}{|1-\Gamma_S\Gamma_{\text{in}}|^2|1-S_{22}\Gamma_L|^2} \tag{7.55}$$

i.e. the difference between the power arriving at the load and that reflected. The maximum power will be delivered by the source when there is a conjugate match between the source and the device input. Consequently, the maximum available power will be $P_{\text{av}} = |V_S|^2/8R_S$ where

$$R_S = \Re\{Z_S\} = \Re\left\{Z_0\frac{1+\Gamma_S}{1-\Gamma_S}\right\} = Z_0\frac{1-|\Gamma_S|^2}{|1-\Gamma_S|^2}. \tag{7.56}$$

As a consequence, the transducer gain will be given by

$$G_T = \frac{P_L}{P_{\text{av}}} = \frac{(1-|\Gamma_S|^2)S_{21}^2(1-|\Gamma_L|^2)}{|1-\Gamma_S\Gamma_{\text{in}}|^2|1-S_{22}\Gamma_L|^2}. \tag{7.57}$$

It is clear that the gain is maximum when $\Gamma_L = \Gamma_S = 0$, i.e. when input and output are matched for zero reflection.

7.5 Transmission-Line Transformers

From our various examples of transmission lines, we see that there are two distinct groups. The first group, known as unbalanced lines, use the ground (or a ground plane) for the return current and the second group, known as balanced lines, use an identical conductor that is quite separate from the ground. Telegraph, microstrip and coaxial lines are examples of unbalanced lines. Parallel-wire and parallel-strip lines are examples of balanced lines. We can regard balanced lines as pairs of unbalanced lines with the return currents through the ground cancelling each other out. If we consider the situation of Figure 7.20a, the currents flowing through the two conductors of the transmission line will be equal, provided that the loads are balanced (i.e. $Z_1 = Z_2$). There will be no flow down to the ground and this connection can be removed without any effect. If the loads are unbalanced (i.e. $Z_1 \neq Z_2$), however, the currents through the conductors will be unequal (i.e. $I_1 \neq I_2$) and there will be a net flow into the ground. Unfortunately, this will have a highly undesirable impact upon the operation of the transmission line. With a balanced load the radiation from the two conductors will cancel since they are closely spaced (the spacing is usually much less than a wavelength) and $I_2 = I_1$. With an unbalanced load, however, $I_1 \neq I_2$ and there will be net radiation from transmission line.

In general, the currents on a transmission line can be divided into a differential mode (equal magnitude currents I_d on the conductors in opposite directions) and a common mode (equal magnitude currents I_c on the conductors in the same direction). It is the common mode currents I_c that cause the radiation and which need to be removed for the proper operation of a transmission line (i.e. the transference of energy with minimum losses).

Now consider the case of an unbalanced transmission line that is connected to a balanced load (see Figure 7.20b). By its construction, the inside of the coaxial line can only support the differential mode. There is, however, a route for a common mode current I_c along the outside of the transmission line and this will result in radiation. From our

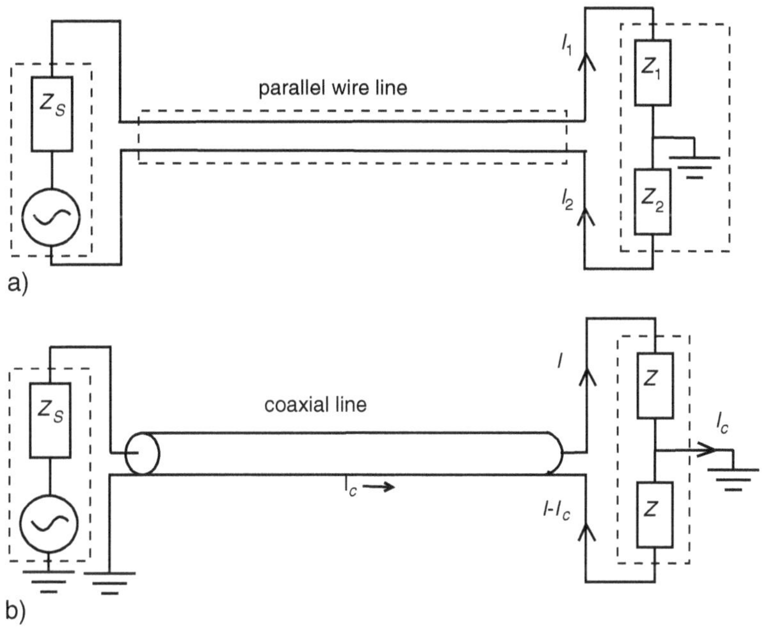

Fig. 7.20 Balanced and unbalanced loads.

Fig. 7.21 Transmission line BALUN transformer using a coaxial transmission line.

above considerations, we see that the common mode currents arise if we try to connect a balanced transmission line to an unbalanced load, or a balanced load (a dipole antenna is an example of such a load) to an unbalanced transmission line. The solution to this problem is what is known as a BALUN (balanced-to-unbalanced) transformer. Such a transformer is shown in Figure 7.21 and consists of a short section of transmission line (usually less than $1/10$ of a wavelength) that is wound on a low-loss iron core (a ferrite or iron-dust toroid is usually preferred). The common-mode current will now encounter a large reactance and so will be choked off by the BALUN. Through such a device, an unbalanced load can be connected to a balanced transmission line or a balanced load to an unbalanced transmission line.

A BALUN does not transform impedance and its main purpose is to provide a blockage to all but the differential mode. This property, however, can be used to make a transformer that does transform impedance. Consider the circuit of Figure 7.22 in which there is a transmission line that only allows the differential mode. We will calculate the average power P_L dissipated in the load R_L. From Kirchhoff's voltage law we will have that $v_S = v_1 + R_S(i_1 + i_2)$ and $v_1 + v_2 = i_2 R_L$. If the length of the transmission line l is much smaller than a wavelength, then $i_2 \approx i_1$ and $v_2 \approx v_1$. Consequently, $v_S \approx v_1 + 2R_S i_2$ and $v_1 \approx i_2 R_L/2$. Eliminating V_1, we then obtain that $i_2 \approx 2V_S/(R_L + 4R_S)$ and so power

$$\begin{aligned} P_L &= i_2^2 R_L \\ &\approx \frac{2v_S^2 R_L}{(4R_S + R_L)^2} \end{aligned} \tag{7.58}$$

will be dissipated in the load R_L (note that this expression is frequency-independent). Solving $\partial P_L/\partial R_L = 0$, we find that this power will be the maximum when $R_L \approx 4R_S$ (i.e. the transformer matches source R_S to load $4R_S$). The important thing to note is that we now have a 4:1 transformer that works at frequencies for which the transmission line is much shorter than a wavelength (less than $\lambda/10$). The lower frequency limit is set by the requirement that the inductance of the line be large enough to choke off the common mode. At the high-frequency end, the range can be extended by choosing a transmission-line impedance that is the geometric average of the source and load (i.e. $Z_0 = 2R$). Further examples of transformers that are based on the BALUN are shown in Figure 7.24. Figure 7.24a shows a 9:1 broadband transformer and Figure 7.24b shows a power combiner whose output is the sum of the two input signals (note that the input sources need to have twice the impedance of the load).

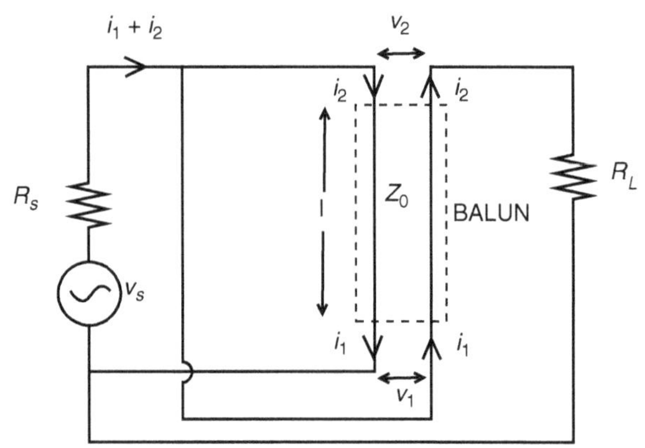

Fig. 7.22 Circuit of a transmission-line transformer.

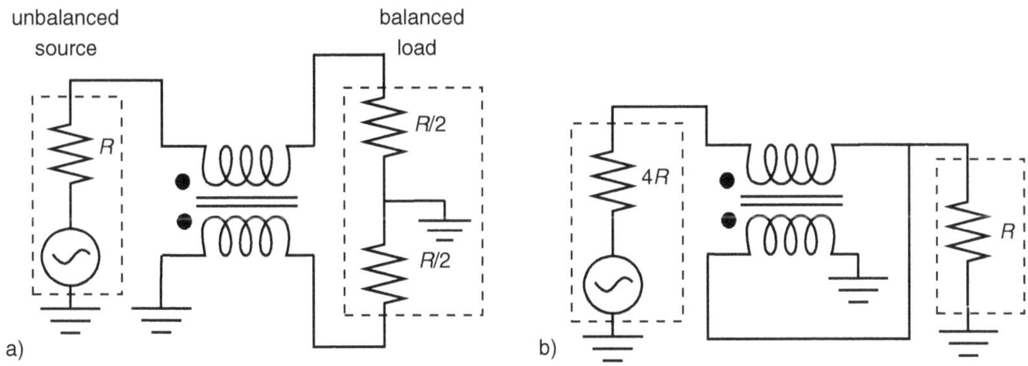

Fig. 7.23 A BALUN and a 4:1 broadband transformer.

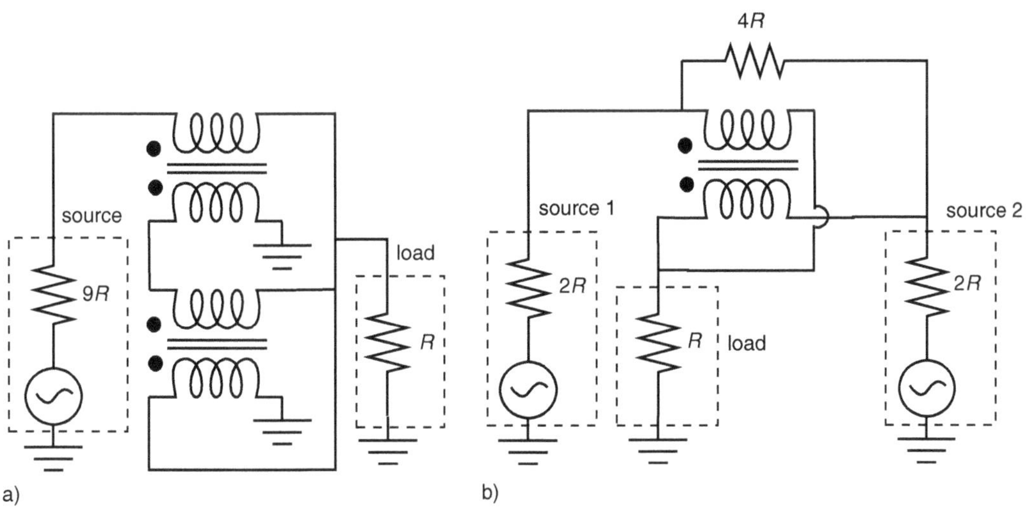

Fig. 7.24 A 9:1 broadband transformer and a combiner.

7.6 Conclusion

In the present chapter, we have studied transmission lines. These are the means by which RF power can be efficiently transferred over long distances and were the basis of the telegraph, the major means of communication before radio. We have found transmission lines have properties that allow us to use them to manufacture high-quality components, especially suitable for frequencies in the GHz range. In radio, the electromagnetic energy spreads out from an antenna that is usually fed by a transmission line, i.e. an antenna is the connector that joins a one-dimensional propagation medium to the three-dimensional propagation medium that is free space. In the next chapter we will study antennas in considerable detail.

8 Antennas

We have already seen that antennas are important component of a radio system and, in Chapter 2, have already discussed the most basic of all antennas, the dipole. The radio electronics merely produce an oscillating voltage and it is the function of the antenna to turn this voltage into the desired radio wave. Consequently, the efficiency with which an antenna performs this function is of great importance to radio technology. In the following chapter we will further develop the theory of antennas and, in particular, will look at the factors that control the performance of these devices. Further, we will study how different forms of antenna can meet the requirements of different radio technologies. We will look at electrically small antennas for situations where size is important (in personal communications for example), broadband antennas for situations where multiple frequencies are to be used (in communications via the ionosphere for example) and array antennas for situations where the radiation needs to be steered in a particular direction (in radar for example).

8.1 The Electric Dipole

In Chapter 2 we introduced the idea of a dipole antenna, an antenna which consisted of a metallic rod with an RF signal driven into its centre. An electric field develops between the two sides of the antenna and so it is often known as an *electric dipole*. Obviously, this antenna must be driven by a radio transmitter and the power will be transferred from the transmitter to the antenna through a transmission line. The dipole is a balanced load and so, from the considerations of the previous chapter, it needs to be fed by a balanced transmission line (if not a BALUN will need to be used). We will assume that the antenna is fed by a parallel-wire transmission line and then the dipole antenna can be pictured as a transmission line that has been opened out (see Figure 8.1) to connect into free space. The speed of propagation c on parallel wires does not depend on the distance between the wires and so will be the same when the wires are opened out (this will be the speed of light in free space). Consequently, the current on the dipole will satisfy a wave equation with wave speed c. Let the arms of the antenna have length l (see Figure 8.2a) and assume the antenna is driven by a time-harmonic current source at the dipole centre ($\mathcal{I}_0(t) = \Re\{I_0 \exp(j\omega t)\}$). Noting that the current will be zero at the end of each dipole arm, the current will be distributed on the dipole according to

$$\mathcal{I}_0(t,z) = \Re\left\{ I_0 \exp(j\omega t) \frac{\sin(\beta(l-|z|))}{\sin \beta l} \right\}, \tag{8.1}$$

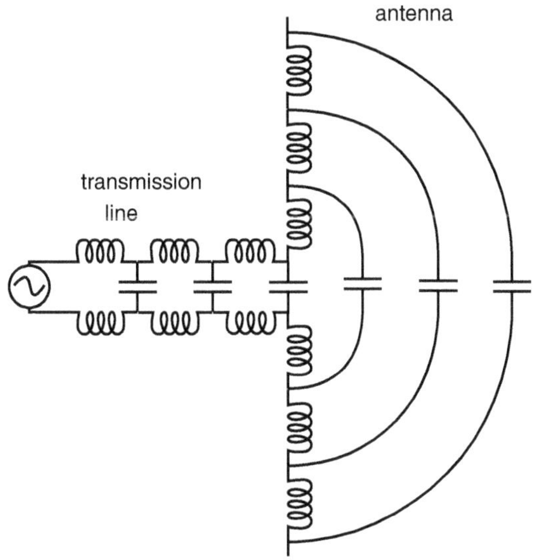

Fig. 8.1 Circuit model of a dipole.

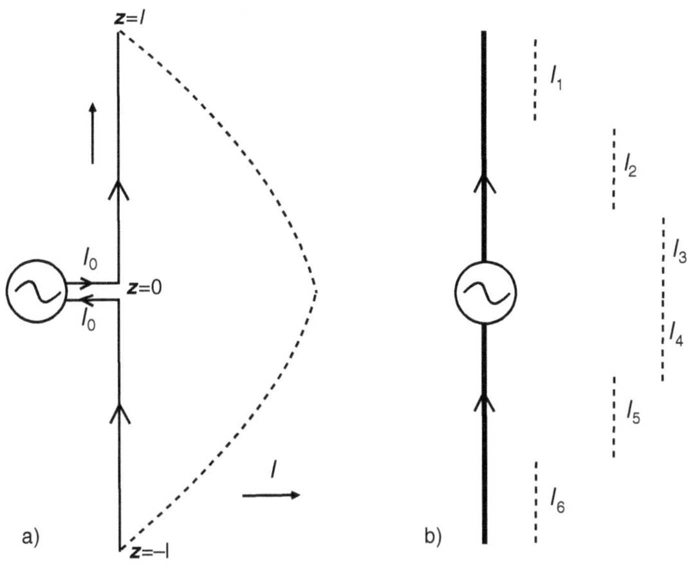

Fig. 8.2 Current distribution along a dipole and its approximation by ideal dipoles.

where $\beta = 2\pi/\lambda$ (note that the complex amplitude of current on the dipole current is given by $I = I_0 \sin(\beta(l - |z|))/\sin\beta l$).

In Chapter 2 we considered an *ideal dipole*, a harmonically oscillating current that is uniformly distributed along the length of the dipole. At large distances from this dipole, the field is given by $\mathcal{E}_t = \Re\{\exp(\mathrm{j}\omega t)E_t\}$, where

$$E_t(R,\theta) = \mathrm{j}\omega\mu_0 I \Delta z \sin\theta \frac{\exp(-\mathrm{j}\beta R)}{4\pi R} \tag{8.2}$$

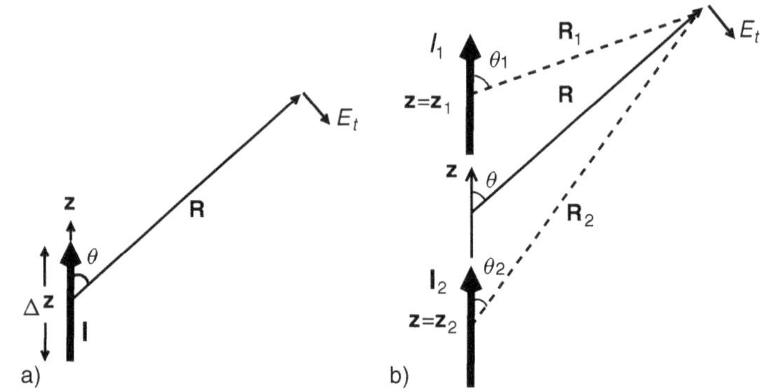

Fig. 8.3 Single and dual ideal dipole elements.

in which I is the complex amplitude of the current, Δz is the dipole length, R is the distance from the dipole and θ is the angle between the dipole axis and the radial vector to the observation point (see Figure 8.3a). The field is a spherical wave that travels outwards from the dipole. In the locality of the observation point P, the field will behave like a plane wave which propagates in the radial direction. The electric field is transverse to this direction (in the plane of the dipole and point P) with magnitude $\mathcal{E}_t$ and the magnetic field is orthogonal to both the electric field and propagation direction with magnitude $\mathcal{B}_t = \mathcal{E}_t/c$. (Note that the electric field, magnetic field and propagation direction form an orthogonal triad that obeys the right-hand screw rule.) For a general ideal dipole with current I in the direction of unit vector $\hat{\mathbf{I}}$, the electric field will be given by $\boldsymbol{\mathcal{E}} = \Re\{\exp(\mathrm{j}\omega t)\mathbf{E}\}$ where

$$\mathbf{E}(R,\theta) = \mathrm{j}\omega\mu_0 I\Delta l\hat{\mathbf{R}} \times (\hat{\mathbf{R}} \times \hat{\mathbf{I}})\frac{\exp(-\mathrm{j}\beta R)}{4\pi R}, \tag{8.3}$$

with $\hat{\mathbf{R}}$ a unit vector in the radial direction and Δl the dipole length. On noting the vector identity $\mathbf{A} \times (\mathbf{B} \times \mathbf{C}) = \mathbf{BA} \cdot \mathbf{C} - \mathbf{CA} \cdot \mathbf{B}$, we can put this in the more meaningful form

$$\mathbf{E}(R,\theta) = \mathrm{j}\omega\mu_0 I\Delta l(\hat{\mathbf{R}}\hat{\mathbf{R}} \cdot \hat{\mathbf{I}} - \hat{\mathbf{I}})\frac{\exp(-\mathrm{j}\beta R)}{4\pi R}, \tag{8.4}$$

where $\hat{\mathbf{R}}\hat{\mathbf{R}} \cdot \hat{\mathbf{I}} - \hat{\mathbf{I}}$ is a vector in a direction transverse to $\hat{\mathbf{R}}$ having magnitude $\sin\theta$ where θ is the angle between vectors $\hat{\mathbf{R}}$ and $\hat{\mathbf{I}}$ (note that $\hat{\mathbf{I}} - \hat{\mathbf{R}}\hat{\mathbf{R}} \cdot \hat{\mathbf{I}}$ is essentially $\hat{\mathbf{I}}$ with its component in the $\hat{\mathbf{R}}$ direction removed).

The field of a general current distribution can be made up of a combination of ideal dipoles (the field of a set of dipoles is simply the sum of the fields of the individual dipoles). Consider the two elements (both of length Δz) shown in Figure 8.3b, then the field will be given by

$$E_t(R,\theta) = \mathrm{j}\omega\mu_0\Delta z\left(I_1 \sin\theta_1 \frac{\exp(-\mathrm{j}\beta R_1)}{4\pi R_1} + I_2 \sin\theta_2 \frac{\exp(-\mathrm{j}\beta R_2)}{4\pi R_2}\right). \tag{8.5}$$

From the law of cosines for triangles, we will have that

$$R_1 = \sqrt{R^2 + z_1^2 + 2|z_1|R\cos\theta}$$

$$\text{and} \quad R_2 = \sqrt{R^2 + z_2^2 + z_1^2 - 2|z_2|R\cos\theta}. \tag{8.6}$$

Since we are considering the field far away from the antenna we will have $|z_1| \ll R$. Consequently, on noting that $\sqrt{1+x} \approx 1+x/2$ for small x, $R_1 = R\sqrt{1+z_1^2/R^2+2|z_1|\cos\theta/R}$ can be approximated by $R_1 \approx R+|z_1|\cos\theta$ and, in a similar fashion, $R_2 \approx R-|z_2|\cos\theta$. By the sine rule for triangles, $\sin\theta_1/R = \sin\theta/R_1$ and so, to the leading order in $1/R$, $\theta_1 \approx \theta$ (in a similar fashion $\theta_2 \approx \theta$). We will now have that

$$E_t(R,\theta) \approx j\omega\mu_0\Delta z\,(I_1\exp(j\beta z_1\cos\theta) \\ +I_2\exp(j\beta z_2\cos\theta))\sin\theta\frac{\exp(-j\beta R)}{4\pi R}. \tag{8.7}$$

In general, for elements at $z_1, z_2, \ldots, z_N$ with currents $I_1, I_2, \ldots, I_N$,

$$E_t(R,\theta) \approx \mathrm{j}\omega\mu_0\Delta z\left(\sum_{i=1}^{N} I_i\exp(\mathrm{j}\beta z_i\cos\theta)\right)\sin\theta\frac{\exp(-\mathrm{j}\beta R)}{4\pi R}. \tag{8.8}$$

We can approximate the dipole as a series of current elements (see Figure 8.2b) and then, in the limit that the number of current elements tends to infinity, the sum will become an integral, i.e.

$$E_t(R,\theta) = \mathrm{j}\omega\mu_0\left(\int_{-l}^{l} I(z)\exp(\mathrm{j}\beta z\cos\theta)dz\right)\sin\theta\frac{\exp(-\mathrm{j}\beta R)}{4\pi R}. \tag{8.9}$$

For a general dipole

$$I(z) = I_0\frac{\sin(\beta(l-|z|))}{\sin\beta l} \tag{8.10}$$

and so

$$E_t(R,\theta) = 2\mathrm{j}\eta_0 I_0 g(\theta)\sin\theta\frac{\exp(-\mathrm{j}\beta R)}{4\pi R}, \tag{8.11}$$

where

$$g(\theta) = \frac{cos(\beta l\cos\theta) - \cos(\beta l)}{\sin^2\theta\sin\beta l}. \tag{8.12}$$

Two special cases are the *short dipole* ($l \ll \lambda$) for which $g(\theta) = \beta l/2$ and the *half-wave dipole* ($l = \lambda/4$) for which $g(\theta) = \cos(\frac{\pi}{2}\cos\theta)/\sin^2\theta$.

We consider the average power P_{rad} that is radiated by a dipole antenna. From (2.37), when the antenna is driven by a time-harmonic current, this is

$$P_{\mathrm{rad}} = \frac{1}{2\eta_0}\int_S |E_t|^2\mathrm{d}S. \tag{8.13}$$

Then, from (2.38),

$$P_{\mathrm{rad}} = \frac{1}{2\eta_0}\int_0^{\pi} |E_t|^2 2\pi R^2\sin\theta\mathrm{d}\theta \tag{8.14}$$

when the surface S is a sphere of radius R. In the case of an ideal dipole with length Δz, this can be integrated to yield $P_{\mathrm{rad}} = 40\pi^2\Delta z^2|I_0|^2/\lambda^2$. Further, for a general dipole

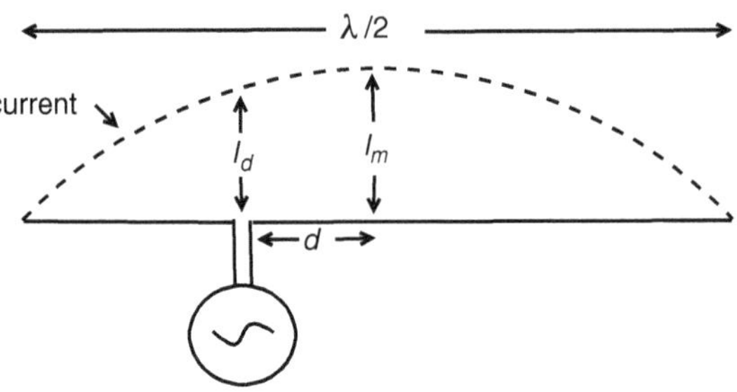

Fig. 8.4 The offset fed dipole.

with current distribution 8.10,

$$P_{\text{rad}} = \frac{\eta_0 |I_0|^2}{4\pi} \int_0^{\pi} g(\theta)^2 \sin^3\theta \mathrm{d}\theta. \tag{8.15}$$

For the case of a short dipole, the integral can be performed analytically to yield $P_{\text{rad}} = 40\pi^2 l^2 |I_0|^2/\lambda^2$. In general, the integral needs to be integrated numerically and, in the case of a half-wave dipole, this leads to the result $P_{\text{rad}} = 1.218\lambda|I_0|^2/4\pi$. If all the power that is supplied to the antenna is radiated, the source of the antenna will see a load resistance $R_{\text{rad}} = 2P_{\text{rad}}/I_0^2$. Consequently, for the ideal dipole $R_{\text{rad}} = 80\pi^2\Delta z^2/\lambda^2$ ohms, for the short dipole $R_{\text{rad}} = 80\pi^2 l^2/\lambda^2$ ohms and for the half-wave dipole $R_{\text{rad}} \approx 73$ ohms. It turns out that dipoles can also exhibit some reactance X_A. For a half-wave dipole $X_A \approx 43$ ohms and for a short dipole $X_A = -120(\ln(l/a) - 1)/\beta l$ ohms where a is the radius of the dipole rod. If the half-wave dipole is slightly shortened (by about 5 per cent) it is possible to reduce X_A to zero.

Thus far, we have only considered dipoles that are fed at their centre. This does not need to be the case and a dipole can be fed at any point. However, the impedance that the dipole presents to a source will change with the position at which it is fed. For a half-wave dipole, the current at a point displaced distance d from the centre will be

$$I_d = I_m \cos\left(2\pi \frac{d}{\lambda}\right). \tag{8.16}$$

The radiated power P_{rad} will be the same wherever the antenna is fed and so, at distance d from the centre, the input resistance R_d will satisfy $P_{\text{rad}} = \frac{1}{2}R_d I_d^2 = \frac{1}{2}RI_m^2$ where R_m is the input resistance at the antenna centre. As a consequence,

$$R_d = R_m \frac{1}{\cos^2\left(2\pi\frac{d}{\lambda}\right)} \tag{8.17}$$

and from which we see that the resistance is minimum at the centre and increases towards the dipole ends. (Although an infinite resistance is predicted for a dipole end, in reality this will be a finite, but very large, value.)

For a perfectly conducting antenna, loss will only occur through radiation, i.e. the loss resistance R_{rad}. However, if the material of a dipole antenna is imperfectly conducting, power will be dissipated as heat in the antenna structure. For DC, the resistance of a wire of length Δz is given by $R = \Delta z/\pi a^2\sigma$ where σ is the conductivity

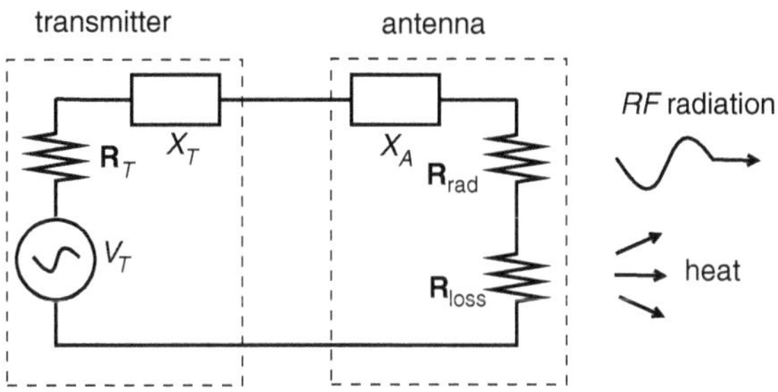

Fig. 8.5 Circuit model of a transmit antenna.

of the wire. At radio frequencies, however, things are not quite so simple due to a phenomenon known as the *skin effect*. Radio waves are heavily attenuated in metals and so current is unable to penetrate very far into the metal, being confined to a thin layer at the surface (thickness $\delta = \sqrt{2/\omega\mu_0\sigma}$). The resistance of a wire of length Δz will therefore be $R = \Delta z/2\pi a\delta\sigma$ and the average power lost in this wire will be $\Delta P_{\text{loss}} = (\Delta z/2\pi a\delta\sigma)|I|^2/2$. ($\sigma = 3.5\times 10^7$ siemens per metre in the case of aluminium, a common material used in the construction of antennas.) Summing such losses across the dipole, and taking the limit $\Delta z \to 0$, we obtain that

$$P_{\text{loss}} = \int_{-l}^{l} \frac{|I|^2}{4\pi a\delta\sigma} \text{d}z. \tag{8.18}$$

For a short dipole $P_{\text{loss}} = lI_0^2/6\pi a\delta\sigma$ and for a half-wave dipole $P_{\text{loss}} = lI_0^2/4\pi a\delta\sigma$. The loss resistance of the dipole will be $R_{\text{loss}} = 2P_{\text{loss}}/|I_0|^2$ which, for a short dipole, is given by $R_{\text{loss}} = l/3\pi a\delta\sigma$ and for a half-wave dipole by $R_{\text{loss}} = l/2\pi a\delta\sigma$. The antenna and transmitter can now be modelled as the circuit of Figure 8.5 and from this it can be seen that the maximum power will be transferred from the transmitter to the antenna when $X_A = -X_T$ and $R_T = R_{\text{rad}} + R_{\text{loss}}$ i.e. when there is a conjugate match.

The fact that some of the power that is fed to the antenna will be lost as heat on the antenna body, and not radiated, brings us to the concept of *antenna efficiency*. This is a measure of how well the antenna radiates the power that it accepts and is defined by

$$\text{efficiency} = \frac{P_{\text{rad}}}{P_{\text{rad}} + P_{\text{loss}}} = \frac{R_{\text{rad}}}{R_{\text{rad}} + R_{\text{loss}}}. \tag{8.19}$$

It is clear that, for an efficient antenna, the radiation resistance will need to be much larger than the loss resistance. Although this will usually be the case for an antenna such as a half-wave dipole, this will not be the case for antennas that are short in comparison to a wavelength. It will be noted that, for a short dipole, $R_{\text{loss}} \propto l$ and that $R_{\text{rad}} \propto l^2$. This is bad news as it implies that the efficiency will tend to zero as the length of the dipole tends to zero. The basic message is that small antennas can be very inefficient.

8.2 The Magnetic Dipole

We have already seen that loops were used as receiving antennas in the early days of radio. If a harmonic EM wave impinges on such a loop, Faraday's law implies that the magnetic field $\mathcal{B}$ of the wave will induce an open circuit voltage $V = \text{j}\omega A\mathcal{B}\cdot\mathbf{n}$ where A

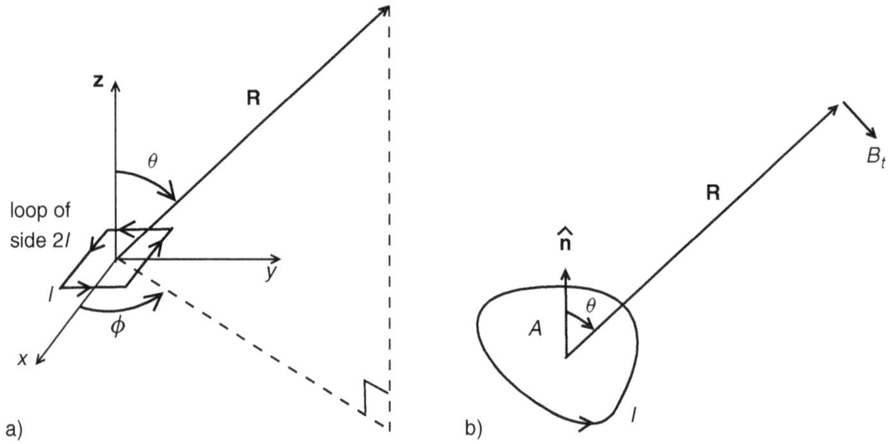

Fig. 8.6 Magnetic dipole antenna.

is the area of the loop and $\mathbf{n}$ is a unit vector perpendicular to the loop (we assume the loop has dimensions much less than a wavelength). If, on the other hand, we use the loop for transmission, we will drive a current I into the loop and, in the case of a rectangular loop, we can model this as four electric dipoles of length $2l$ (see Figure 8.6a).

For observation points that are a large distance away from the rectangular loop, the difference between a dipole that constitutes a side and one located at the origin is merely a phase factor (as we have seen in our considerations of the electric dipole). The sides that are parallel to the x axis will generate a transverse field $\mathbf{E}^x$ that lies in the plane of the x axis and the radial direction. In a similar fashion, the sides that are parallel to the y axis will generate a transverse field $\mathbf{E}^y$ that lies in the plane of the y axis and the radial direction. From (8.4), we will have that

$$\mathbf{E}^x(R,\theta,\phi) = \mathrm{j}\omega\mu_0 2lI(\hat{\mathbf{R}}\hat{\mathbf{R}}\cdot\hat{\mathbf{x}}-\hat{\mathbf{x}})\left(\exp(-\mathrm{j}\beta l\hat{\mathbf{R}}\cdot\hat{\mathbf{y}})-\exp(\mathrm{j}\beta l\hat{\mathbf{R}}\cdot\hat{\mathbf{y}})\right)\frac{\exp(-\mathrm{j}\beta R)}{4\pi R} \tag{8.20}$$

and

$$\mathbf{E}^y(R,\theta,\phi) = \mathrm{j}\omega\mu_0 2lI(\hat{\mathbf{R}}\hat{\mathbf{R}}\cdot\hat{\mathbf{y}}-\hat{\mathbf{y}})\left(\exp(\mathrm{j}\beta l\hat{\mathbf{R}}\cdot\hat{\mathbf{x}})-\exp(-\mathrm{j}\beta l\hat{\mathbf{R}}\cdot\hat{\mathbf{x}})\right)\frac{\exp(-\mathrm{j}\beta R)}{4\pi R}. \tag{8.21}$$

(Note that, in terms of the polar coordinates θ and ϕ, $\hat{\mathbf{R}} = \sin\theta\cos\phi\hat{\mathbf{x}}+\sin\theta\sin\phi\hat{\mathbf{y}}+\cos\theta\hat{\mathbf{z}}$.) We will assume the lengths of the sides are much less than a wavelength and, by means of the approximation $\exp x \approx 1+x$, the total field ($\mathbf{E}=\mathbf{E}^x+\mathbf{E}^y$) will reduce to

$$\begin{aligned}\mathbf{E}(R,\theta,\phi) &= \omega\mu_0\beta AI\left((\hat{\mathbf{R}}\hat{\mathbf{R}}\cdot\hat{\mathbf{x}}-\hat{\mathbf{x}})\hat{\mathbf{R}}\cdot\hat{\mathbf{y}}-(\hat{\mathbf{R}}\hat{\mathbf{R}}\cdot\hat{\mathbf{y}}-\hat{\mathbf{y}})\hat{\mathbf{R}}\cdot\hat{\mathbf{x}}\right)\frac{\exp(-\mathrm{j}\beta R)}{4\pi R}\\ &= \omega\mu_0\beta AI\hat{\mathbf{z}}\times\hat{\mathbf{R}}\frac{\exp(-\mathrm{j}\beta R)}{4\pi R}.\end{aligned} \tag{8.22}$$

For a general plane loop, with area A and unit normal $\hat{\mathbf{n}}$, the electric field will be

$$\mathbf{E}(R,\theta,\phi) = \omega\mu_0\beta AI\hat{\mathbf{n}}\times\hat{\mathbf{R}}\frac{\exp(-\mathrm{j}\beta R)}{4\pi R}. \tag{8.23}$$

Further, at large distances from the source, the field will behave as a plane wave travelling in the direction of $\hat{\mathbf{R}}$ and so $\mathbf{B} = -\hat{\mathbf{R}} \times \mathbf{E}/c$. As a consequence,

$$\mathbf{B}(R,\theta,\phi) = \mu_0\beta^2 AI\hat{\mathbf{R}} \times (\hat{\mathbf{R}} \times \cdot\hat{\mathbf{n}})\frac{\exp(-j\beta R)}{4\pi R}. \tag{8.24}$$

From (8.24) and (8.3), it can be seen that the magnetic field for a loop is proportional to the electric field of a short dipole in the direction $\hat{\mathbf{n}}$. Because of this, a current loop is often known as a *magnetic dipole*. By analogy with the electric dipole, there is only a transverse component of magnetic field (see Figure 8.6b) which can be written as

$$B_t(R,\theta) = -\mu_0\beta^2 AI\sin\theta\frac{\exp(-\mathrm{j}\beta R)}{4\pi R}, \tag{8.25}$$

where θ is the angle between the loop normal and the radial vector to the observation point (see Figure 8.6b)

We would now like to calculate the average power radiated by a magnetic dipole and note that, from (8.13) and the relationship between $\mathbf{E}$ and $\mathbf{B}$, that

$$P_{\text{rad}} = \frac{c^2}{\eta_0}\int_S |B_t|^2 \mathrm{d}S. \tag{8.26}$$

From (8.25) and (8.26) we obtain that $P_{\text{rad}} = 40\beta^4 A^2|I|^2$ and hence that the radiation resistance will be given by $R_{\text{rad}} = 20\beta^4 A^2$. As far as reactance is concerned, the loop is an inductance with $X_A = \omega\mu_0 L(\ln(4L/\pi a) - 1.75)/2\pi$ where L is the length of wire in the loop. The loss resistance of the loop is given by $R_{\text{loss}} = L/2\pi a\delta\sigma$.

8.3 Reciprocity

The transmission and reception of radio signals can be viewed as a problem in circuit theory with the ports of the transmit and receive antennas regarded as those of a two-port network (see Figure 8.7a), i.e. we can view the problem in terms of an impedance matrix Z with

$$\begin{aligned} V_1 &= Z_{11}I_1 + Z_{12}I_2 \\ V_2 &= Z_{21}I_1 + Z_{22}I_2. \end{aligned} \tag{8.27}$$

We have already seen, in the case of transmission lines, that a propagation medium can be modelled as a combination of series inductors, shunt capacitors and resistors (both series and shunt). Indeed, a general propagation medium can also be modelled as a network of these passive circuit elements and for such a network we will always have that $Z_{12} = Z_{21}$. Consider port 1 driven by current I_1 and then the open circuit ($I_2 = 0$) voltage V_{21} in terminal 2 will be given by $V_{21} = Z_{21}I_1$. Now consider port 2 driven by current I_2 and then the open circuit ($I_1 = 0$) voltage V_{12} in terminal 1 will be given by $V_{12} = Z_{12}I_2$. Since $Z_{12} = Z_{21}$, we will therefore have that

$$\frac{V_{12}}{V_{21}} = \frac{I_2}{I_1} \text{ or } V_{12}I_1 = V_{21}I_2. \tag{8.28}$$

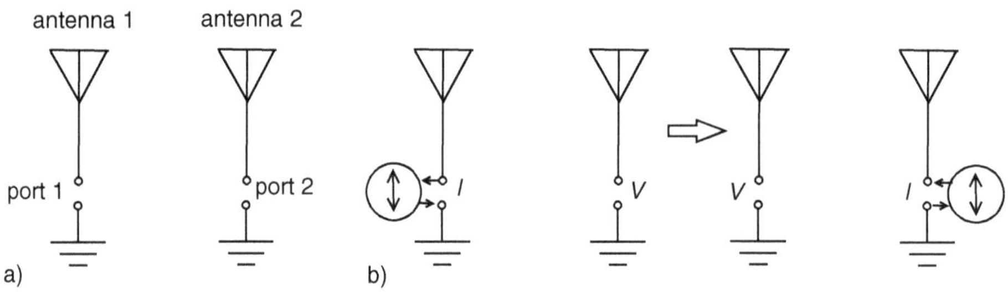

Fig. 8.7 Reciprocity of antennas.

If the ports are driven, in turn, by the same current I, (8.28) implies that the corresponding open circuit will exhibit the same voltage V (see Figure 8.7b). This is known as *reciprocity* and has important consequences for antennas. In particular, it implies that we can infer the two-way performance of an antenna combination from its one-way performance.

In general, the electric field of a radiating source will behave as

$$\mathbf{E} = \mathrm{j}\omega\mu_0 I \mathbf{h}_{\text{eff}} \frac{\exp(-\mathrm{j}\beta R)}{4\pi R}, \tag{8.29}$$

where $h_{\text{eff}} = |\mathbf{h}_{\text{eff}}|$ has the dimensions of length and is known as the *effective length* of the antenna. It is clear that $\mathbf{h}_{\text{eff}}$ can depend on the direction of observation (e.g. in the case of a short dipole $\mathbf{h}_{\text{eff}} = l\hat{\mathbf{R}} \times (\hat{\mathbf{R}} \times \hat{\mathbf{t}})$ and, in the case of a small loop, $\mathbf{h}_{\text{eff}} = -\mathrm{j}\beta A\hat{\mathbf{n}} \times \hat{\mathbf{R}}$). We will consider a two-port system in which one port is provided by an antenna with general effective length $\mathbf{h}_{\text{eff}}$ and the other by a small loop antenna with area A and unit normal $\hat{\mathbf{n}}$. From Faraday's law, the complex amplitude of the open circuit voltage in the terminals of the loop will be

$$\begin{aligned} V &= \mathrm{j}\omega A\mathbf{B} \cdot \hat{\mathbf{n}} = -\frac{\mathrm{j}\omega A}{c}\hat{\mathbf{R}} \times \mathbf{E} \\ &= \frac{\omega^2 \mu_0 IA}{c}(\hat{\mathbf{R}} \times \mathbf{h}_{\text{eff}}) \cdot \hat{\mathbf{n}} \frac{\exp(-\mathrm{j}\beta R)}{4\pi R} \\ &= \frac{\omega^2 \mu_0 IA}{c}(\hat{\mathbf{n}} \times \hat{\mathbf{R}}) \cdot \mathbf{h}_{\text{eff}} \frac{\exp(-\mathrm{j}\beta R)}{4\pi R} \\ &= \mathbf{h}_{\text{eff}} \cdot \left(\omega\mu_0 \beta IA\hat{\mathbf{n}} \times \hat{\mathbf{R}} \frac{\exp(-\mathrm{j}\beta R)}{4\pi R}\right) \end{aligned} \tag{8.30}$$

on noting the vector identity $\mathbf{A} \cdot (\mathbf{B} \times \mathbf{C}) = \mathbf{B} \cdot (\mathbf{C} \times \mathbf{A}) = \mathbf{C} \cdot (\mathbf{A} \times \mathbf{B})$. Consequently, $V = \mathbf{h}_{\text{eff}} \cdot \mathbf{E}_{\text{loop}}$ where $\mathbf{E}_{\text{loop}}$ is the electric field of the loop (see (8.23)). Then, by the reciprocity theorem, we also know that this is the open-circuit voltage at the input of the antenna with effective length $\mathbf{h}_{\text{eff}}$ when a current I is driven into the loop antenna. In general, if an antenna of effective length $\mathbf{h}_{\text{eff}}$ is placed in a time-harmonic electric field $\mathbf{E}$, then an open-circuit voltage

$$V = \mathbf{h}_{\text{eff}} \cdot \mathbf{E} \tag{8.31}$$

will appear across its terminals. Once again, the transmitting capability of an antenna is related to its receiving capability. We can now form the circuit model of a receive

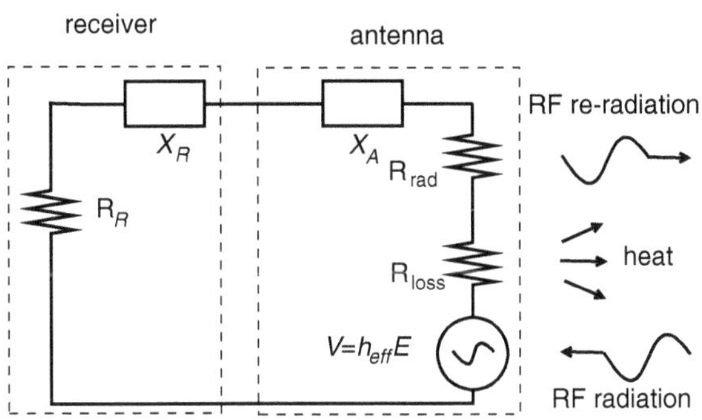

Fig. 8.8 Circuit model of a receive antenna.

antenna shown in Figure 8.8. It will be noted that the antenna has now become a source to the receiver circuit, but there are losses in this source due to heating of the antenna and the reradiation of some of the incoming radio wave. From the expression (8.31), if $\mathbf{h}^1_{\text{eff}}$ and $\mathbf{h}^2_{\text{eff}}$ are the effective lengths of two antennas, their mutual impedance will be

$$Z_{12} = \mathrm{j}\omega\mu_0 I \mathbf{h}^1_{\text{eff}} \cdot \mathbf{h}^2_{\text{eff}} \frac{\exp(-\mathrm{j}\beta R_{12})}{4\pi R_{12}}, \tag{8.32}$$

where R_{12} is the distance between the two antennas. Obviously, Z_{11} and Z_{22} will be given by the input impedances of the respective antennas and so we have the complete impedance matrix of a pair of antennas regarded as a two-port device.

8.4 Gain

We have seen that even the simplest of antennas has fields that are highly dependent upon the direction from which they are observed. A useful description of this property is what is known as the *directivity* of an antenna. This is the power radiated in a given direction when scaled upon the average power radiated in all directions. We can evaluate the gain over a sphere at a large distance R from the antenna, and then

$$D(\theta,\phi) = \frac{\bar{\mathbf{E}}(\theta,\phi) \cdot \mathbf{E}(\theta,\phi)}{2\eta_0} \frac{4\pi R^2}{P_{\text{rad}}}. \tag{8.33}$$

In the case that the antenna is a short dipole, the directivity will given by $D(\theta,\phi) = \frac{3}{2}\sin^2\theta$ and this will be the same for a small magnetic loop. In the case of a half-wave dipole, the directivity will be given by

$$D(\theta,\phi) = \frac{5}{3}\left(\frac{\cos(\frac{\pi}{2}\cos\theta)}{\sin\theta}\right)^2. \tag{8.34}$$

Directivity tells us something above the direction in which an antenna is sending the power, but it is incomplete as a measure of antenna performance. In particular, it does not tell us anything about the efficiency of an antenna. A more useful measure is known as *gain*. This is the power radiated in a given direction when scaled upon the power accepted by the antenna. The gain G can be related to the directivity D and efficiency *e* of the antenna through the relation $G = eD$. For an antenna with size of the order of a wavelength (a half-wave dipole for example), the gain is usually fairly close to the directivity. However, as we have already noted, *e* can be quite low for antennas that are very small in comparison to a wavelength. Due to reciprocity, gain will also provide a measure of how an antenna responds to an incoming signal when operating in the receive mode and we will see further evidence of this when we discuss aperture antennas.

Gain is often represented as a surface that is centred on some origin with the distance of the surface from the origin in a given direction being the value of gain in that direction. In the case of dipoles, the gain surface will be rotationally symmetric about the axis of the dipole and so can be represented by a slice through this surface. Figure 8.9 shows such a slice for a half-wavelength dipole and Figure 8.10 for a 5/4-wavelength dipole. It will be noted that the longer dipole is far more directional than the shorter one. In

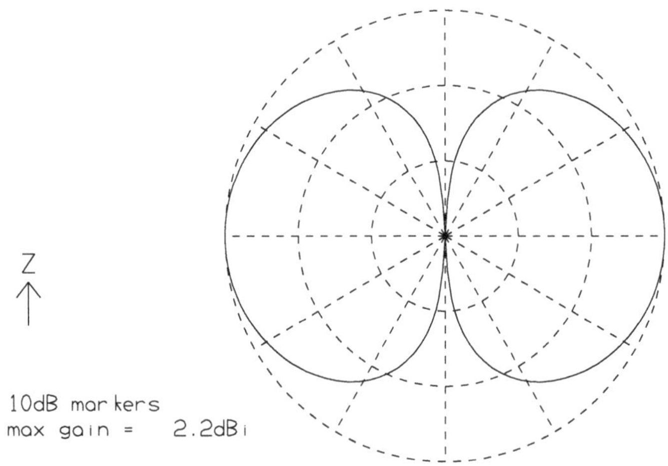

Fig. 8.9 Gain pattern of a half-wavelength dipole.

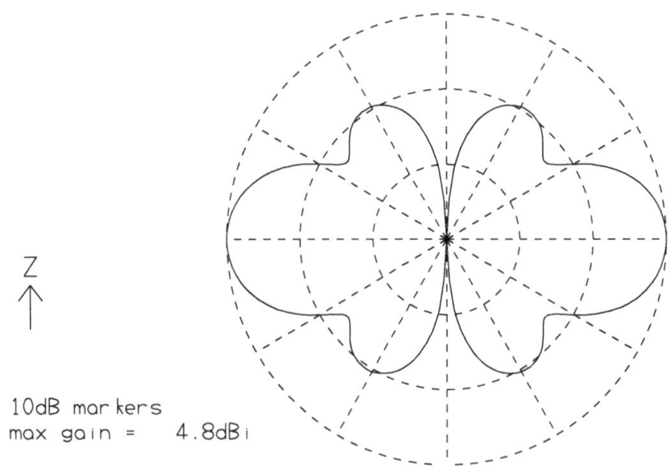

Fig. 8.10 Gain pattern of a 5/4-wavelength dipole.

the Figures, the gain G is expressed in terms of *decibels* (or dB for short), a logarithmic scale in which the value of the gain is given by $10\log_{10}G$ (note that the notation dBi is often used to signify dB with respect to an isotropic antenna). Such a scale is far more convenient when a large range of gain values needs to be considered.

8.5 The Monopole Antenna

Thus far, we have assumed that an antenna is completely isolated from other objects, but this is rarely the case. Objects closer than a wavelength can often have a significant impact, affecting both the gain pattern and the antenna input impedance. Consider a radiating current element that is placed above a perfectly conducting ground plane (see Figure 8.11), then there will be a reflection of some of the radiation by the plane (the mirror reflection that we have in the case of light). We could remove the plane and model its effect by an image current element and then the field in the upper space would be the same as the if the ground were still there.

Now consider a vertical rod with its end just above the conducting plane and driven against the plane (see Figure 8.12a). This is known as a monopole antenna and will constitute an unbalanced load to the source that drives it. The feeder will also need to be

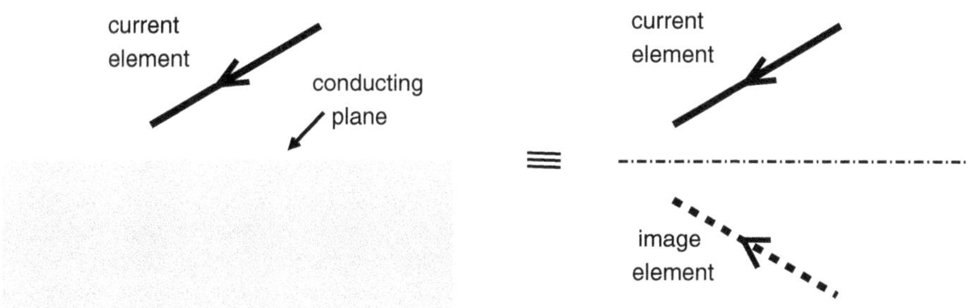

Fig. 8.11 Current element over a conducting plane and equivalent system that uses an image element.

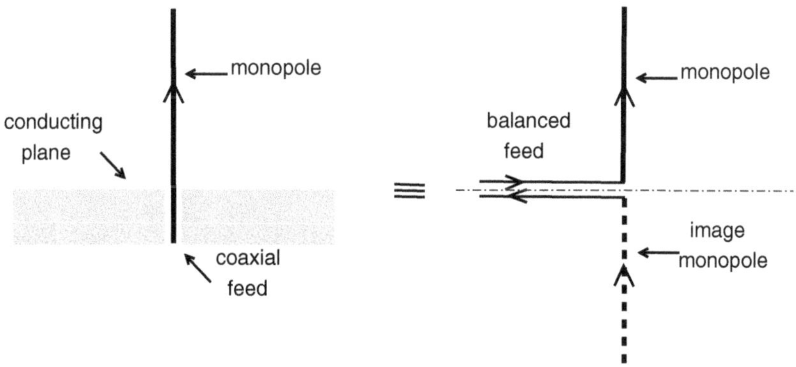

Fig. 8.12 Monopole and equivalent dipole system.

unbalanced and this will usually be a coaxial transmission line. Indeed, the monopole antenna can be regarded as the end of a coaxial line in which the centre conductor has become the monopole element and the outer conductor opened out to form the conducting plane. If we replace the conducting plane by an image of the monopole, this will constitute an effective dipole that yields the same field as the monopole in the upper space (see Figure 8.12b). Further, the coaxial feeder will now be replaced by a balanced feeder whose centre is at the same position as the conducting plane. As a consequence, the voltage in the monopole feed V_{mon} will be half that of the effective dipole V_{dip}. However, the current in the monopole I_{mon} will be the same as that in the dipole I_{dip}. The input impedance of the monopole Z_{mon} will therefore be

$$Z_{\text{mon}} = \frac{V_{\text{mon}}}{I_{\text{mon}}} = \frac{\frac{1}{2}V_{\text{dip}}}{I_{\text{dip}}} = \frac{1}{2}Z_{\text{dip}}, \tag{8.35}$$

where Z_{dip} is the dipole input impedance. Consequently, for a short monopole

$$Z_{\text{mon}} = R_{\text{rad}} + R_{\text{loss}} + \mathrm{j}X_A = 40\left(\frac{\pi l}{\lambda}\right) + \frac{l}{6\pi a\delta\sigma} - \mathrm{j}\frac{60}{\beta l}\left(\ln\left(\frac{l}{a}\right) - 1\right). \tag{8.36}$$

The directivity of a monopole will not be the same as that of the equivalent dipole. The radiation into the upper space will be the same, but there will be no radiation into the lower space and hence the total radiated power will be half that of the dipole. As a consequence, the directivity of a monopole will be twice that of the equivalent effective dipole.

In reality, the conducting plane will not be perfect and the image dipole will be distorted. In this case it is simpler to describe the effect of the conducting plane in terms of what is known as a *reflection coefficient*. The electric field of a short monopole (see Figure 8.13a) will be given by

$$E_t(R,\theta) = \mathrm{j}\omega\mu_0 Il(1+R_g)\sin\theta\frac{\exp(-\mathrm{j}\beta R)}{4\pi R}, \tag{8.37}$$

where R_g is the reflection coefficient of the conducting plane and R is the distance from the base of the monopole to the observation point. For a perfectly conducting plane we have $R_g = 1$ but, for the more realistic case of a finite conducting plane, the reflection coefficient R_g will depend on angle θ. For an observation point close to the ground,

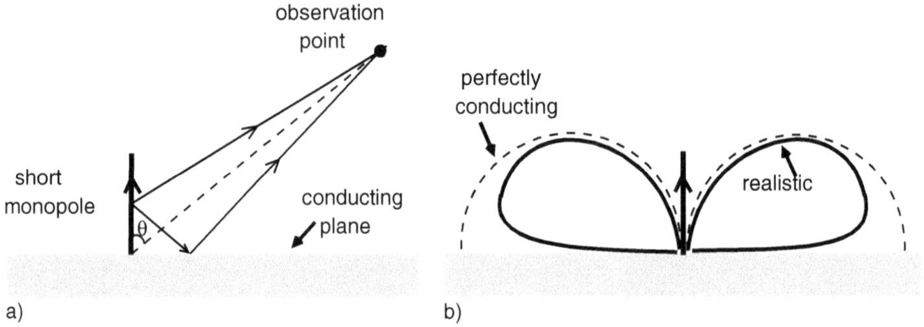

Fig. 8.13 A short monopole and its gain pattern.

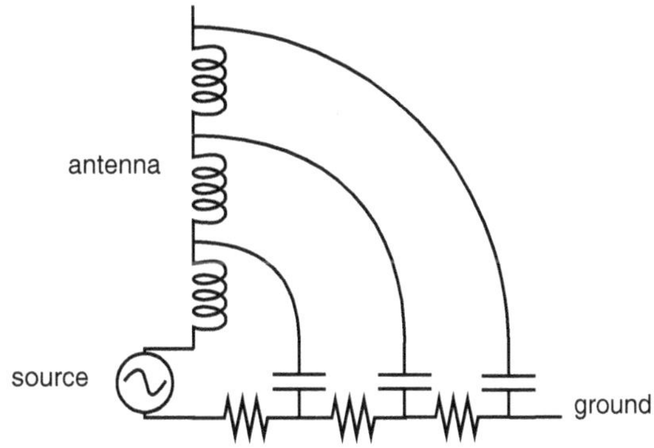

Fig. 8.14 Circuit model of a monopole over a lossy ground.

we will have $R_g \approx -1$ changing to $R_g \approx 1$ at the highest elevations. (Note that we will discuss the reflection coefficient further in the next chapter.)

The effect of a finitely conducting plane upon gain is to make this quantity zero at zero elevation (the elevation of maximum gain in the case of a perfectly conducting plane) and to reduce its value at higher elevations (see Figure 8.13b). Figure 8.14 shows a circuit model of a monopole over a finitely conducting plane. From this it will be seen that the return path for the capacitance is through the resistors that represent the finitely conducting ground. It is the loss in these resistances that reduces the efficiency of the monopole and hence its gain.

8.6 Reducing the Size of Antennas

At lower frequencies, such as those used in broadcasting, a wavelength can be physically large (several hundred metres) and antennas, such as a half-wave dipole, will be far too cumbersome for practical purposes. The simplest approach might be to use a short monopole together with a matching circuit that ensures maximum power is transferred to the antenna (see Figure 8.15a). Another approach might be to use a small resonant loop with suitable matching (see Figure 8.15b). However, as we have seen above, small antennas have a small radiation resistance that can be of the same order as the loss resistance and therefore can be very inefficient. On the other hand, antennas operated close to resonance, such as the half-wave dipole, tend to have a much larger radiation resistance and therefore a much higher efficiency.

If we look at the circuit model of the dipole (see Figure 8.1), we can see a possible approach to achieving a shorter resonant dipole. For a given frequency, we can lower the electrical wavelength $\lambda = c/\omega$ by lowering the propagation speed of the wave on the antenna. Treating the dipole as an opened-out transmission line, we have that $c = 1/\sqrt{LC}$ and so we will need to increase the inductance L, and/or the capacitance C, per unit length. We could increase L by replacing the straight wire with a helical winding (see Figure 8.16a), i.e. the dipole arms become solenoids.

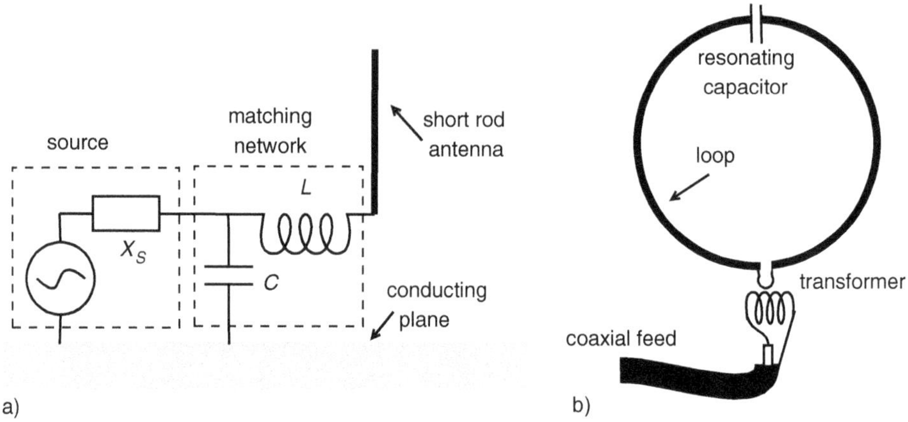

Fig. 8.15 Short monopole and small loop with matching networks.

We can improve on this by making the distribution of current along the wire more uniform and this is done by concentrating the additional inductance at a point further along each dipole arm (see Figure 8.16b). Since we have lumped all the inductance increase into a short section, the wave speed in this section is very low and most of the current drop will occur here. This means that most of the antenna length will support only high current with a consequent increase in effective length. The downside is that we need a lot more additional inductance the further we go out on the arm. (This follows from the fact that, as a parallel wire transmission widens, the capacitance gets weaker and the inductance stronger.) Consequently, to get the necessary level of inductance, the solenoid may need to be made of very thin wire and this will lead to a large increase is loss resistance.

An alternative is to increase the capacitance by means of what is known as a capacity hat (see Figure 8.16c). This can be a large conducting disc at the end of each arm, the disc being orthogonal to the arm. Such an approach overcomes the problem of the loss resistance in a lumped inductance. The downside is that the capacity hats can be quite cumbersome. Despite this, such an approach found favour in the early days of radio. The capacity hat is often made less cumbersome by replacing the conducting discs with a set of radial rods at the ends of the dipole arms.

Besides low efficiency, small antennas bring with them the problem of low bandwidth. The Q of an antenna will be given by $Q = |X_A/(R_{\text{rad}} + R_{\text{ohm}})|$ and this is related to bandwidth B through $B = \omega/Q$. We consider the case of a dipole (length $2l$) with a capacity hat (see Figure 8.16). For the case where $l \ll \lambda$, the current will be uniformly distributed and the antenna will behave as an ideal dipole. The radiation resistance of an ideal dipole is $R_{\text{rad}} = 80\beta^2 l^2$. Further, the reactance will be dominated by the capacitance of the capacity hats and will be given by $X_A = 2l\eta_0/A\beta$. As a consequence, $Q = 3\pi/Al\beta^3$ where A is the area of each capacity hat. In general, a result due to Wheeler (1947) states that Q will be limited by $Q \geq \alpha^{-3}\beta^{-3}$ where α is the radius of the smallest sphere that contains the antenna. This result sets a severe limit on the bandwidth of small antennas.

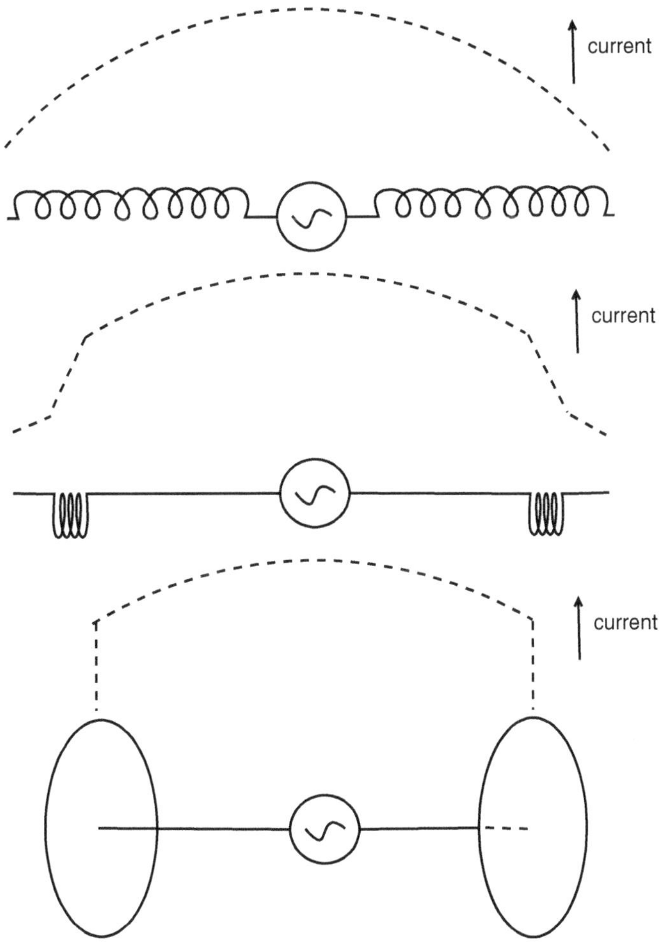

Fig. 8.16 Shortening a dipole antenna.

8.7 Broadband Antennas

A famous result of Rumsey (1966) states that:

If a structure upon scaling by $1/\tau$ is equal to itself, then it will have the same properties at frequencies ω and $\tau\omega$.

It is easiest to see this in the one-dimensional case, i.e. the transmission line (Jones, 1994). From (7.7), the time-harmonic equations for the current I and voltage V are

$$\frac{\mathrm{d}V}{\mathrm{d}z} = -\mathrm{j}\omega L(z)I \text{ and } \frac{\mathrm{d}I}{\mathrm{d}z} = -\mathrm{j}\omega C(z)V, \tag{8.38}$$

where z is the coordinate along the line axis. (Note that C and L are now position-dependent, as would be the case in a transmission line that is opened out into a dipole antenna.) We would like to know the conditions under which the current and voltage at position z, and frequency ω, will be the same as those at position $z' = z/\tau$, and frequency $\omega' = \tau\omega$, i.e.

$$V(z,\omega) = V(z',\omega') \text{ and } I(z,\omega) = I(z',\omega'). \tag{8.39}$$

Noting that $\mathrm{d}/\mathrm{d}z' = \tau\mathrm{d}/\mathrm{d}z$, will have from (8.38) that

$$\frac{\mathrm{d}V}{\mathrm{d}z'} = -\mathrm{j}\omega' L(\tau z')I \text{ and } \frac{\mathrm{d}I}{\mathrm{d}z'} = -\mathrm{j}\omega' C(\tau z')V. \tag{8.40}$$

For Eqs. (8.38) to (8.40) to be consistent we require

$$L(z') = L(\tau z') \text{ and } C(z') = C(\tau z'), \tag{8.41}$$

i.e. L and C should be invariant under the change of scale. As a consequence, the structure of the transmission line needs to be scale-invariant. Consider a parallel-wire transmission line that is opened out (see Figure 8.18), then the radiation from the separate lines will no longer totally cancel and the transmission line will act as an antenna. For the parallel-wire transmission line $L = \mu_0 \cosh^{-1}(D/2a)/\pi$ and $C = \pi\epsilon_0/\cosh^{-1}(D/2a)$, where D is the distance between the wires and a is the wire radius. Consequently, for the transmission to be invariant under the change of scale, we need D/a to be invariant under this change. The simplest way to achieve this is to choose D/a to be constant along the wire, i.e. the wires are conical in shape. The resulting antenna will then have the same properties at all frequencies.

The radiation will introduce loss into the transmission line and so it is better described by

$$\frac{\mathrm{d}V}{\mathrm{d}z} = -R(z)I - \mathrm{j}\omega L(z)I \text{ and } \frac{\mathrm{d}I}{\mathrm{d}z} = -\mathrm{j}\omega C(z)V, \tag{8.42}$$

where R is the radiation resistance of a unit length of the transmission line. Only current orthogonal to the line axis will contribute (radiation from the horizontal components still cancels) and so, from previous considerations of an ideal dipole, $R \approx 80\beta^2 \sin^2\gamma$ where γ is the angle between the wire axis and the z axis (γ is assumed to be small). The current distribution on the upper wire of the transmission line will now take the form

$$I(z) = I_0 \exp(-\mathrm{j}\beta z - \alpha z), \tag{8.43}$$

where z is now the distance along the wire (the current on the lower wire is $-I$). Assuming the radiation loss to be small, the propagation constant along the line will be $\beta \approx \omega\sqrt{LC} = \omega/c_0$, the impedance will be $Z_0 \approx \sqrt{L/C} = \eta_0/\pi \cosh^{-1}(D/2a)$ and the attenuation constant will be $\alpha \approx R/2Z_0 = \beta^2 \sin^2\gamma/3\cosh^{-1}(D/2a)$ (on noting that $\eta_0 = 120\pi$). The important thing to note is that the current will decay along the wire due to radiation and that, as a consequence, it is possible to truncate the antenna. Over a length $l_{\text{atten}} = 3\cosh^{-1}(D/2a)/\beta^2 \sin^2\gamma$ the current amplitude will drop by the factor $1/e \approx 0.36788$. Consequently, once we know the lowest frequency ω_L at which the antenna is required to operate, we can then truncate it at a length for which the missing section will carry negligible current for the frequencies above ω_L. We will then have an antenna with approximately the same properties for frequencies above ω_L.

When the wires spread out to form a pair of cones with a common axis (i.e. $\gamma = \pi/2$), we have what is known as a biconical dipole (see Figure 8.17). This dipole has an input

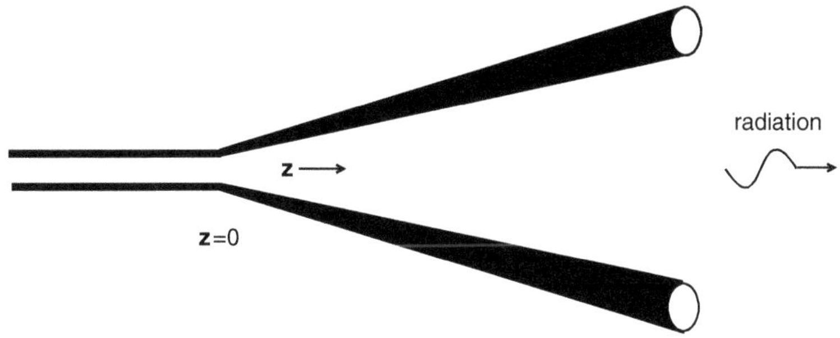

Fig. 8.17 Radiating transmission line.

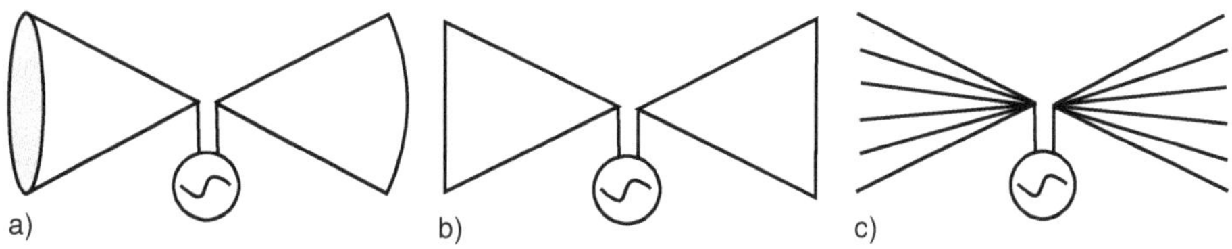

Fig. 8.18 Biconical and bow-tie antennas.

impedance $Z_{in} = 120\ln(\cot(\theta_h/2))$ where θ_h is the cone half angle. Once again, because of radiation losses, the current will be effectively zero after a distance of just over half a wavelength along each arm of the dipole. Consequently, the antenna radiation pattern will behave like a normal dipole up to a frequency where its length is just over a wavelength and thereafter its pattern will stabilise.

Another antenna that satisfies the above scaling condition is the infinite bow-tie dipole, the planar version of the biconical dipole. Once again, this can only be realised in its truncated form (see Figure 8.17b). A metal cone, or bow tie, can be rather cumbersome, but fortunately we can replace such structures by a series of rods that spread out from the feed (see Figure 8.17c for the case of a finite bow tie). In general, it is found that metal objects that contain gaps of less than $\lambda/10$ behave as if impervious to electromagnetic waves. Further, from the above considerations, currents are only significant out to a finite distance from the antenna feed. Consequently, providing the rods are sufficiently dense, the fact that they spread out from the feed does not stop them acting as if impervious.

It is also possible to form a conical monopole (see Figure 8.19a). As with the rod monopole, the input impedance is half that of the equivalent dipole and the gain is double. A further development of the monopole is the discone antenna. In this, the ground plane is replaced by a disc (see Figure 8.19b). As with the biconical dipole, it is possible to replace the cone, and the disc, by rods that spread out radially from the feed. In the case of the disc, it is usually sufficient for the rods to spread out to the same extent as the cone base.

A further variety of antenna that has broadband properties is the *travelling-wave antenna*. Consider a parallel-wire transmission line that has been opened out to form a

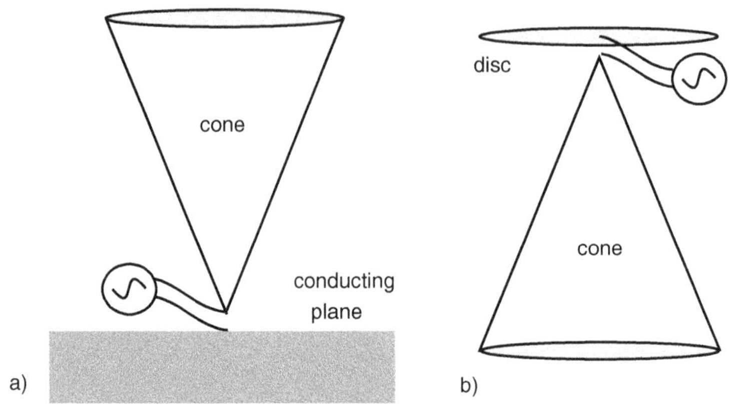

Fig. 8.19 Conical monopole and discone antennas.

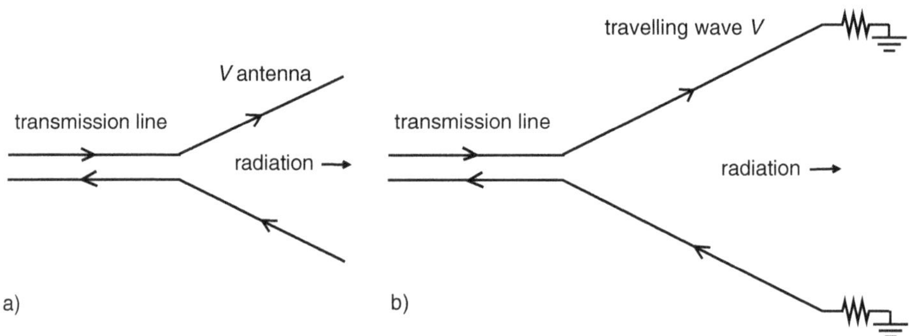

Fig. 8.20 V and travelling-wave V antennas.

V-shaped antenna (see Figure 8.20a), this will have a specific frequency at which it is resonant (i.e. it presents a real impedance to the transmission line) and away from this frequency the antenna will present some reactive impedance. From transmission-line theory, the resonance will occur at a frequency for which the arms of the V are approximately one quarter of a wavelength long. Away from this frequency, the waves travelling into the antenna will be reflected back at the antenna ends and cause a reactive impedance at the antenna input. In the case of a transmission line, we can prevent these reflections by placing a matched load across the end of the line and, in a travelling-wave antenna, we can do the same. As shown in Figure 8.20b, we load each arm of the antenna by connecting it to the ground through half the characteristic impedance (we can do this since the antenna is balanced). The antenna will then present a real impedance at all frequencies. The downside of such an antenna, however, is that it can be lossy at low frequencies. At the higher frequencies the decay in current due to radiation means that very little power will reach the loads. However, as the frequency drops, more and more power will reach the loads and so less and less will be radiated.

8.8 Array Antennas

So far we have only considered some fairly simple antenna elements, but it is possible to manufacture antennas with more complex gain patterns through arrays of such elements. Some insight into this process can be gained by considering a two-element array in the receive mode (remember the reciprocity between the behaviour of antennas in receive and transmit modes). Consider the configuration shown in Figure 8.21 and a harmonic wave of complex amplitude E_r arriving at the rear element from the rear. For an antenna with effective length h_{eff}, this will result in a voltage $h_{\text{eff}}E_r\exp(-\mathrm{j}\beta(L+S+D))$ at the combiner due the front element and a voltage $h_{\text{eff}}E_r\exp(-\mathrm{j}\beta L)$ arriving at the combiner due to the rear element (the signals travel path lengths $L+S+D$ and L respectively). The combined voltage at the output of the combiner will now be

$$V_r = h_{\text{eff}}E_r\exp(-\mathrm{j}\beta L)(\exp(-\mathrm{j}\beta(S+D))+1). \tag{8.44}$$

In a similar fashion, for a field E_f arriving at the front from the front,

$$V_f = h_{\text{eff}}E_f\exp(-\mathrm{j}\beta L)(\exp(-\mathrm{j}\beta S)+\exp(-j\beta D)). \tag{8.45}$$

A harmonic wave of amplitude E_s arriving from the side of the array will reach both elements simultaneously and this will result in a voltage

$$V_s = h_{\text{eff}}E_s\exp(-\mathrm{j}\beta L)(\exp(-\mathrm{j}\beta S)+1) \tag{8.46}$$

at the output of the combiner. We first make the feeds to the combiner of equal length (i.e. $S=0$) in order to generate a zero-phase increment between the elements. Then, if we make the separation between the antennas a half wavelength (i.e. $D=\lambda/2$), we will have $V_r = h_{\text{eff}}E_r\exp(-\mathrm{j}\beta L)(\exp(-\mathrm{j}\pi)+1)=0$, $V_f = h_{\text{eff}}E_f\exp(-\mathrm{j}\beta L)(1+\exp(-\mathrm{j}\pi)) = 0$ and $V_s = 2h_{\text{eff}}E_s\exp(-\mathrm{j}\beta L)$ (note that $\beta = 2\pi/\lambda$). In other words, we have zero response to the front and rear, but enhanced response to the sides. From the terminology of battleships, the array is known as a *broadside array* since, in the transmit mode, it fires off orthogonal to the axis of the array. Greater detail can be found in Figure 8.22 which

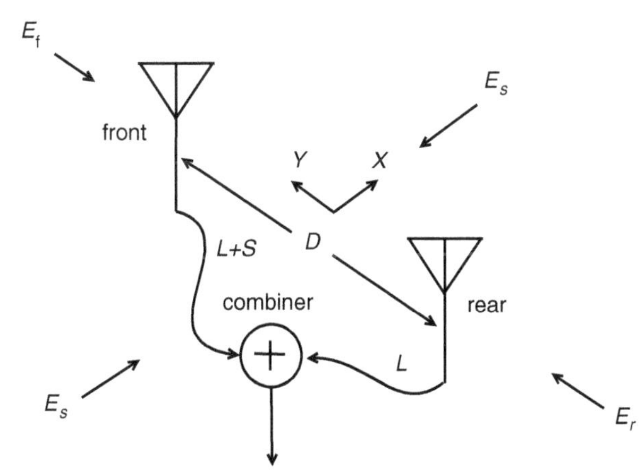

Fig. 8.21 Array of two elements in receive configuration.

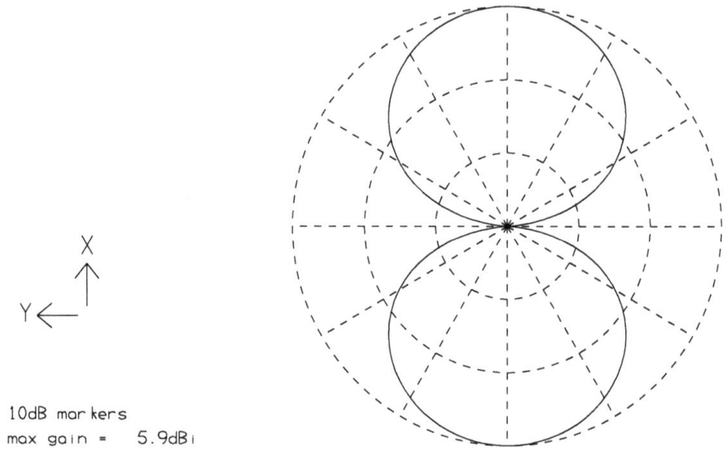

Fig. 8.22 Gain pattern of a two-dipole broadside array with $\lambda/2$ spacing.

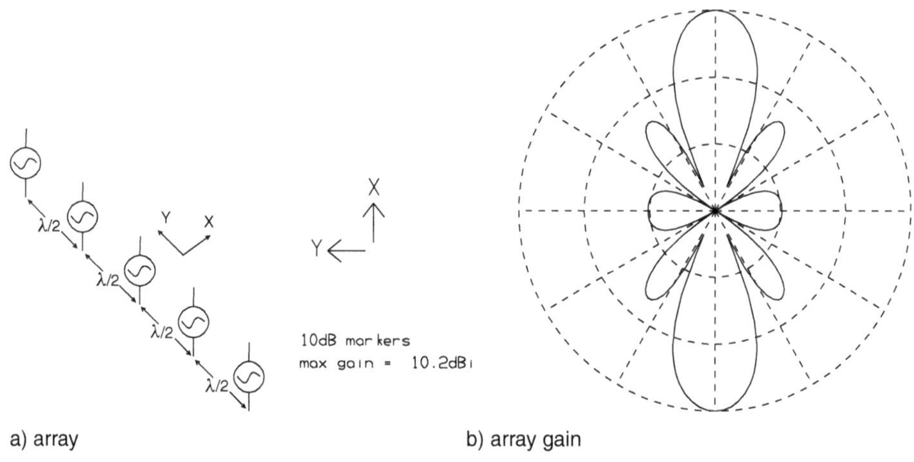

Fig. 8.23 Gain pattern of a five-dipole broadside array with $\lambda/2$ spacing.

shows the gain pattern derived from a numerical simulation of the array. Figure 8.23 shows the gain pattern for a broadside array with five elements ($D=\lambda/2$ and a zero phase increment between the elements). From this simulation, it will be noted that there are now several *side lobes* in addition to the main lobe and these can have significant amplitude. Importantly, it will be noted that there has been a significant narrowing of the main lobe with a consequent increase in gain. Obviously, if the angle through which the major part of the radiation flows becomes narrower, the density of power within this angle is greater. For a broadside array, the angle between the nulls either side of the main lobe will be λ/L radians where $L=ND$ is the effective length of the array, N is the number of elements and D is their spacing. Further, the gain will increases by the multiplicative factor $2ND/\lambda$. At this point one might be tempted to increase the spacing between the elements in order to make a narrower beam and increase the gain. However, this can lead to unwanted consequences as illustrated in Figure 8.24 for the case of a wavelength

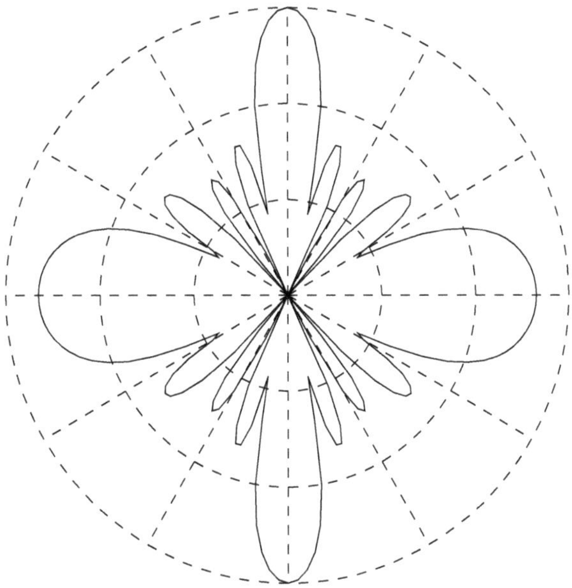

Fig. 8.24 The effect of increasing spacing to λ.

spacing between the elements. Although the beam has indeed narrowed and the gain increased, there is now strong lobes from either end of the array. These lobes, known as *grating lobes*, can give the array strong response in undesired directions. It turns out that $\lambda/2$ is the practical limit of element spacing if grating lobes are to be avoided.

For the two-element array, if we now make the feeds to the combiner differ in length by half a wavelength (i.e. $S = \lambda/2$), there will be a phase increment of π between the elements and we will find that $V_s = 0$, $V_f = 2h_{\text{eff}}E_f \exp(-\text{j}\beta(L+\lambda/2))$ and $V_r = 2h_{\text{eff}}E_r \exp(-\text{j}\beta L)$. We now have zero response to the sides and equal magnitude response to the fore and rear. From the terminology of battleships, the array is known as an *endfire array* since, in the transmit mode, it now fires off along the axis of the array. Greater detail can be found in Figure 8.25 which shows the gain pattern derived from a numerical simulation of the array. The important thing to note is that, by varying the phase increment between then elements from 0 to π, we have *steered* the array from broadside to endfire. As a consequence, although we could steer the main lobe by physically rotating a broadside array in the *xy* plane, we can also steer the array by changing the phase increment between the array elements. Besides adjusting the phase across the array, the distribution of current amplitude across the array can also be adjusted as a further means of manipulating the array gain pattern. Whilst the phase can be used to direct the array main lobe, the current amplitude can be used to control the level of side lobes and hence make a much more directionally selective antenna. Figure 8.26 shows the gain pattern of a five-element broadside array with a Dolph-Chebyshev taper (i.e. current amplitudes distributed in proportion to 1:2.41:3.14:2.41:1 across the array) and from which it will be noted that the side lobes have almost been eliminated. The downside, however, is that the main lobe is now slightly wider.

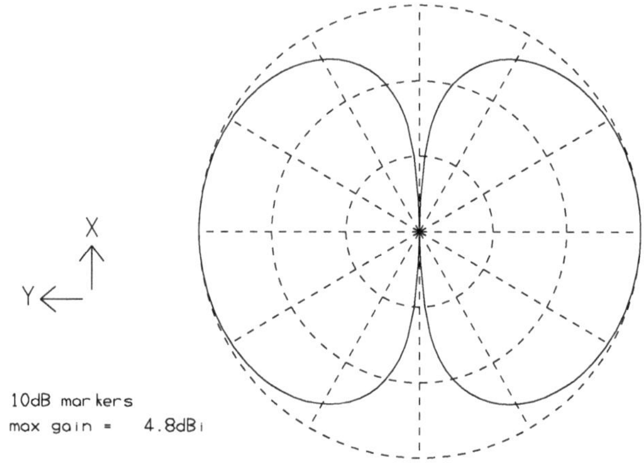

Fig. 8.25 Gain pattern of a two-dipole endfire array with $\lambda/2$ spacing.

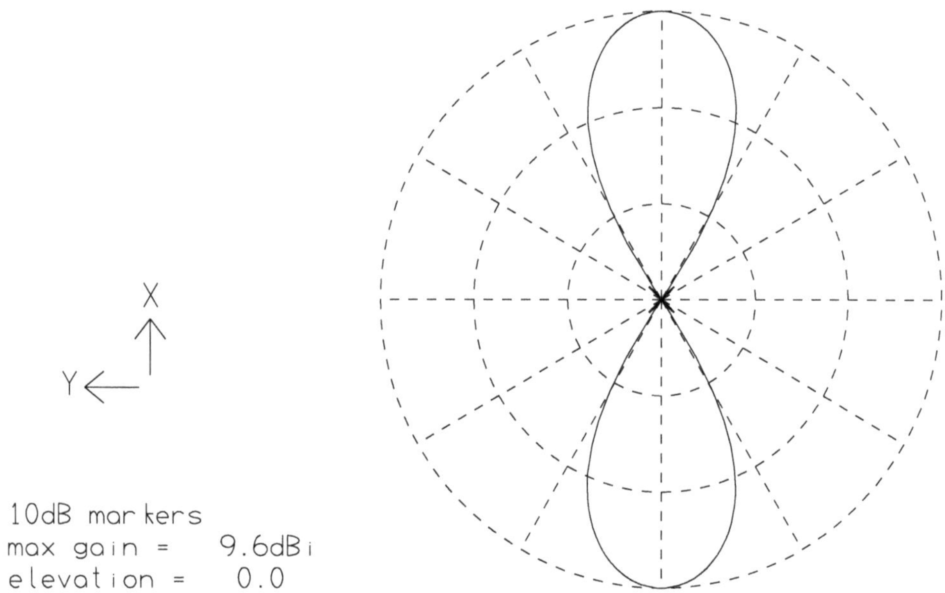

Fig. 8.26 Gain pattern of a five-element broadside array with a Dolph-Chebyshev taper.

We have already seen that there are problems with element spacing greater than $\lambda/2$. However, there turns out to be some utility in making the spacing less than $\lambda/2$, even though there will be some loss in gain and an increase in beamwidth. Consider the two element receive array and make a separation of a quarter wavelength between the array elements (i.e. $D = \lambda/4$). We also make the feeds to the combiner differ in length by a quarter wavelength (i.e. $S = \lambda/4$) to obtain a $\pi/2$ phase increment between the array elements. Then, from (8.44) to (8.46), we obtain that $V_r = 0$, $V_f = -2jh_{\text{eff}}E_f \exp(-\mathrm{j}\beta L)$ and $V_s = (1 - \mathrm{j})h_{\text{eff}}E_s \exp(-\mathrm{j}\beta L)$, i.e. we have a zero response to the rear, an enhanced response to the fore and a reduced response to the sides. Essentially, the array has only one main lobe and this can be seen in Figure 8.27 which shows the gain pattern derived from a numerical simulation. Unlike the above simple theory, the simulations exhibit a non-zero gain to the rear which is, nevertheless, small in comparison to the main lobe. This effect arises due to inter-element coupling that is taken into account by the simulations, but not by the simple theory.

As we have seen above, inter-element interactions can have a significant effect upon the radiation pattern of arrays and must be given serious consideration, especially when the elements are close in comparison to a wavelength. The effect of element interaction can be seen from the simple case of two interacting half-wave dipoles. Figure 8.28 details the interaction of two parallel dipoles through their mutual impedance Z_{12} (itself equal to Z_{21} by symmetry). The mutual impedance is shown as a function of distance between the dipole centres and it will be noted that Z_{12} can be of a similar order to the self impedances Z_{11} (itself equal to Z_{22} by symmetry). It can be seen that the interaction can be quite strong for separations less than half a wavelength. The strong mutual interaction can be used to create arrays in which only one element is directly driven and several *parasitic elements* are driven through their mutual interaction with the driven element. Figure 8.30a shows an example of such an array, known as a Yagi-Uda array after its inventors. The array that is shown has three elements, known as the reflector, driven and director elements respectively. We can estimate the currents excited in the reflector and director element through their mutual impedance with the driven element. Assuming the input of the driven element to be port 1, driven by voltage V_1, there will be zero voltage V_2 at the centre of a parasitic element that is taken to be port 2. We will then have that

$$\begin{aligned} V_1 &= Z_{11}I_1 + Z_{12}I_2 \\ 0 &= Z_{21}I_1 + Z_{22}I_2, \end{aligned} \tag{8.47}$$

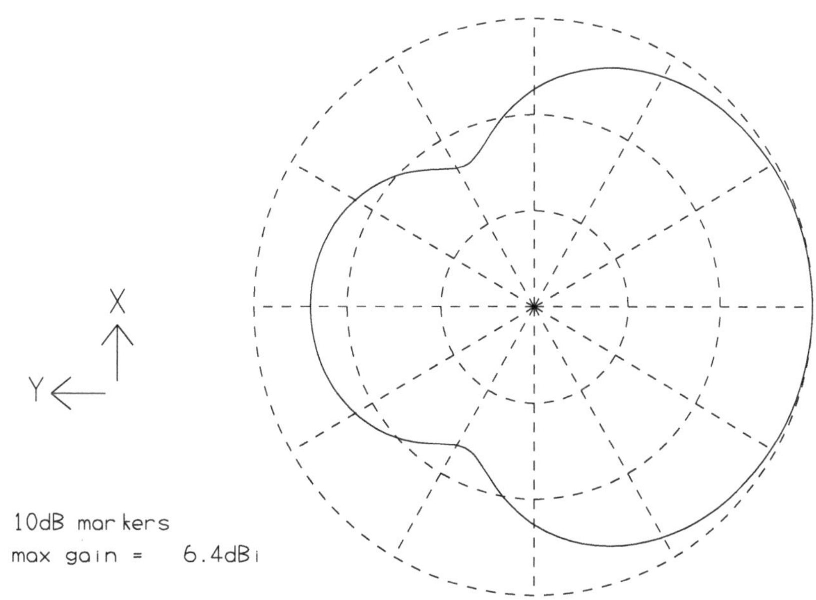

Fig. 8.27 Gain pattern of a two-dipole array with $\lambda/4$ spacing and $\pi/2$ phase increment.

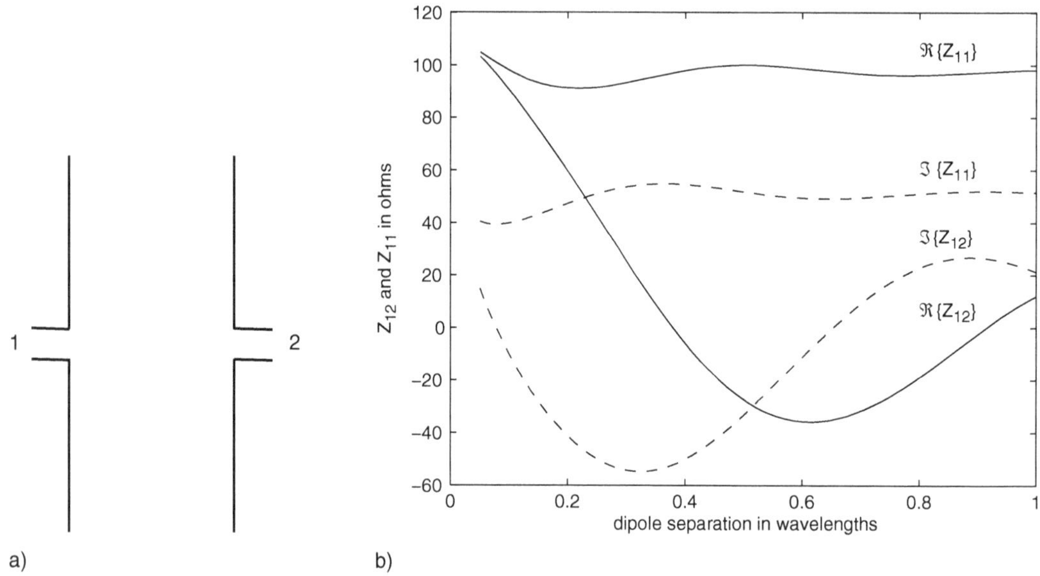

Fig. 8.28 Impedance and mutual impedance of parallel dipoles at various separations.

where I_1 and I_2 are the currents flowing into ports 1 and 2 respectively. Then, from the above equations, we have that $I_2 = -(Z_{12}/Z_{22})I_1$. For element spacings of around a quarter wavelength, we have already seen that we can produce an endfire array with only one main lobe. Further, from Figure 8.28, the mutual impedance will be of the order of self impedance and so the current flowing in the parasitic element will be of the same order of magnitude as that flowing in the driven element. However, to obtain endfire behaviour, we need to ensure the correct phase increment between the driven and parasitic elements. We can achieve this by adjusting the length of the parasitic elements (see Figure 8.29). In order to produce radiation in the direction from the reflector to director elements, the reflector element needs to be larger than the driven element and the director element needs to be shorter.

Figure 8.31 shows the gain pattern in the plane of the antenna for an array with quarter-wavelength spacings between the elements and parasitic element lengths that vary by five per cent from the length of the driven element. It will be noted that, when compared with a dipole, there is substantial narrowing of the main lobe and a consequent increase in gain. This performance can be further enhanced by introducing additional director elements.

Being essentially an extension of the dipole, the Yagi-Uda array is inherently narrowband. To obtain a broadband array, we could create an array of broadband elements such as bicones or discones. However, the Rumsey result of the previous section introduces another possibility. If we consider a structure that is equal to itself upon scaling by $1/\tau$, we will have a structure that has the same properties at frequencies $\tau\omega, \tau^2\omega, \tau^3\omega, \ldots$ (the antenna is said to be a *log-periodic* structure). Providing that τ is close to 1, this structure will have almost the same properties on all frequencies. A

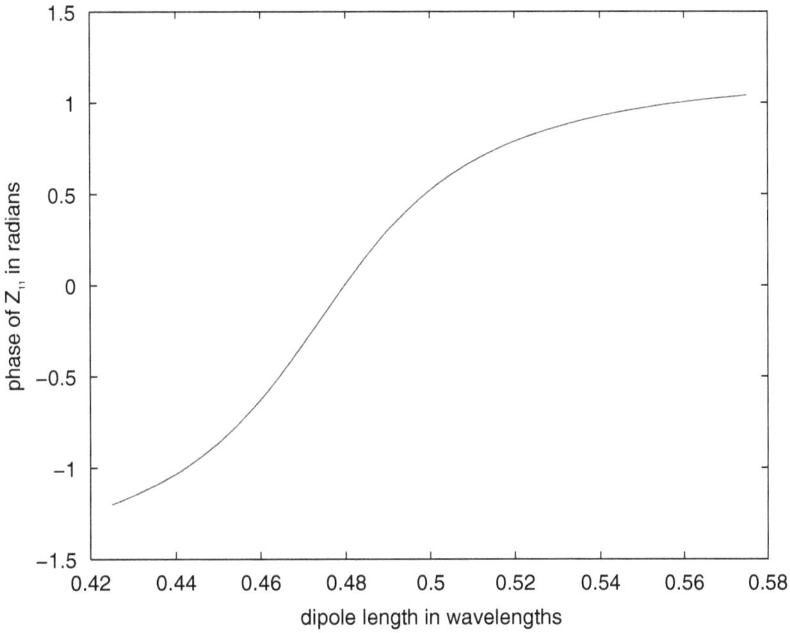

Fig. 8.29 Phase of dipole input impedance as a function of antenna length.

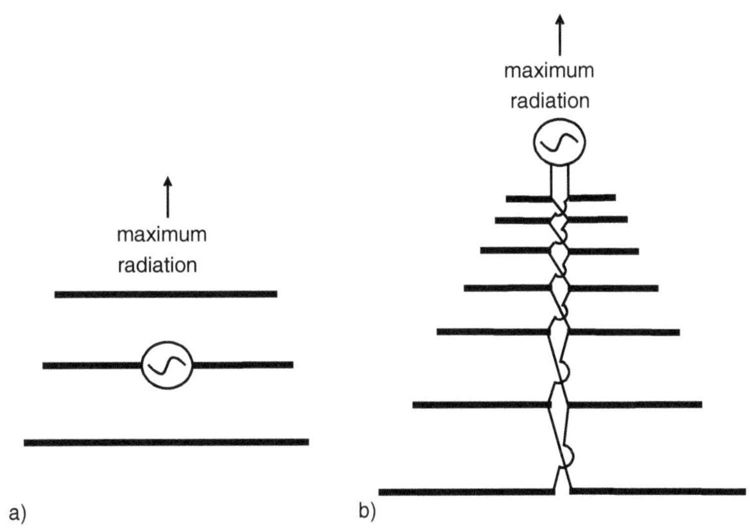

Fig. 8.30 Yagi-Uda and log-periodic arrays.

practical antenna based on this principle is shown in Figure 8.30b. This antenna consists of a series of interconnected dipoles for which the dipole lengths, and their spacings, are scaled by $1/\tau$ as we move between elements along the antenna axis from the longest element to shortest. This is an endfire array that is known as a *log-periodic dipole*

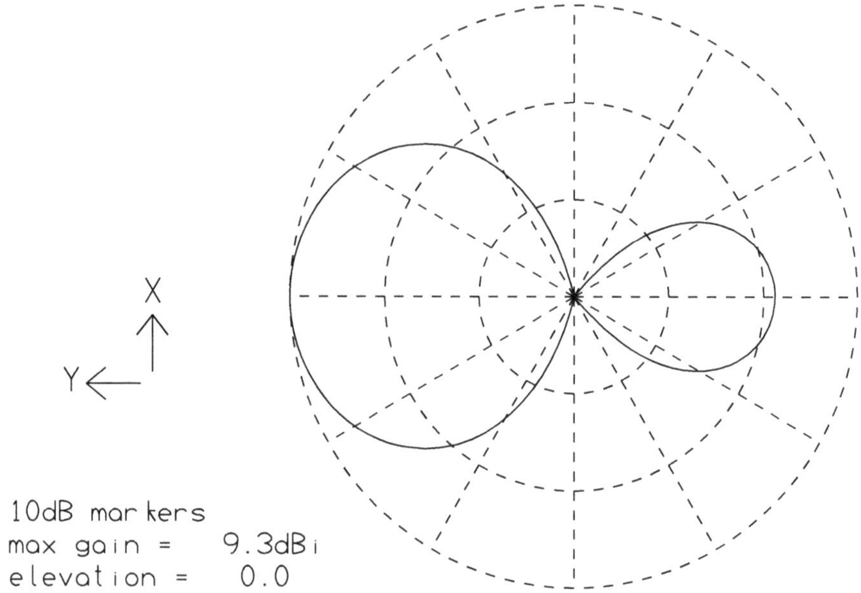

Fig. 8.31 Gain pattern of a three-element Yagi-Uda array.

array (LPDA). Obviously, the Rumsey result requires an infinite number of dipoles and the finite number in the practical implementation could have a serious impact upon performance. Fortunately, it turns out that when the LPDA is driven in the manner shown, only the dipoles that are close to resonance will be active and so the truncation can be chosen in line with a desired frequency range. The antenna can operate effectively between the resonant frequencies of the shortest and longest dipoles and will produce a gain pattern that is similar in nature to that of a three to four element Yagi-Uda antenna.

8.9 Aperture Antennas

A means of transferring energy between two points is through what is known as a *hollow waveguide*. In fact, transmission lines are themselves waveguides, but they always have two conductors and can therefore be analysed in terms of traditional circuit ideas. This is not the case for a hollow waveguide. We consider a waveguide consisting of a metal duct with rectangular cross section (see Figure 8.32a) and analyse wave propagation inside this duct. We first note that, on the surface of a perfect conductor (an effective model of a metal), the tangential electric field will be zero. We have seen earlier that, due to the skin effect, RF currents travel close to the surface of a conductor and are therefore tangential to it. Consequently, since the resistance of a perfect conductor is zero, Ohm's law implies that the voltage drop between any two points on the surface of the conductor must also be zero, i.e. the tangential electric field is zero. This has important implications for waves travelling down our duct.

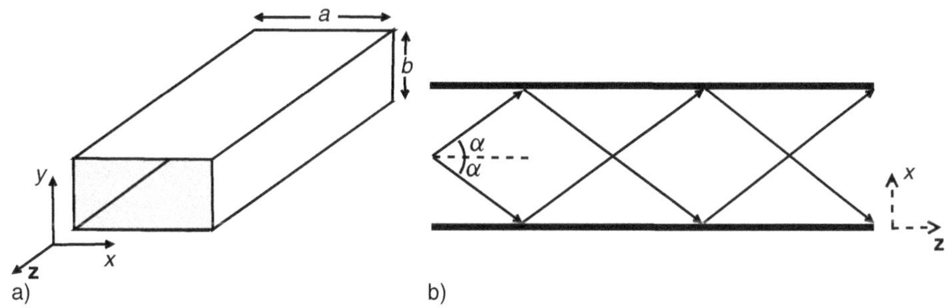

Fig. 8.32 A rectangular waveguide.

We first ask whether a plane EM wave can travel down the waveguide and consider the time-harmonic wave $\mathcal{E} = \Re\{\exp(j\omega t)\mathbf{E}\}$ where

$$\mathbf{E} = E_0 \exp(-j\beta z)\hat{\mathbf{y}} \tag{8.48}$$

where β is the propagation constant in free space. This field is orientated in the $\hat{\mathbf{y}}$ direction and therefore has no component tangential to the horizontal faces of the waveguide. There is, however, a non-zero tangential component at the vertical faces. As a consequence, a plane wave cannot travel down a rectangular waveguide. Consider now an electric field that consists of a combination of two plane waves that travel at angle α to the horizontal, i.e.

$$\mathbf{E} = (\exp(-j\beta(-x\sin\alpha + z\cos\alpha) - \exp(-j\beta(x\sin\alpha + z\cos\alpha))E_0\hat{\mathbf{y}} \tag{8.49}$$

These waves are reflected back and forth between the vertical sides of the waveguide (see Figure 8.32b). Field $\mathbf{E}$ will have zero tangential component on the horizontal sides of the waveguide and on the vertical side at $x = 0$. On the vertical side at $x = a$, the field will have zero tangential component if $\exp(j\beta a \sin\alpha) = \exp(-j\beta a \sin\alpha)$, i.e. if $\sin(\beta a \sin\alpha) = 0$. Consequently, $\mathbf{E}$ will have tangential components that are zero on all surfaces of the waveguide if $\beta a \sin\alpha = n\pi$ where n is an arbitrary integer. There is obviously an infinite set of wave modes that can travel down a rectangular waveguide and, from (8.49), the electric field will have the form

$$\mathbf{E} = 2jE_0 \sin\left(\frac{n\pi x}{a}\right)\exp(-j\beta' z)\hat{\mathbf{y}}, \tag{8.50}$$

where $\beta' = \beta\cos\alpha = \sqrt{\beta^2 - \pi^2 n^2/a^2}$ is the propagation constant in the waveguide. In fact, the above modes are a subset of a doubly infinite set of possible propagation modes with $\beta' = \sqrt{\beta^2 - \pi^2(n^2/a^2 + m^2/b^2)}$ the propagation constant for the nm mode. The nm mode is denoted by TE_{nm} where TE is used to indicate the fact that the electric field is transverse to the propagation direction (note that the magnetic field is not transverse). It will be noted that β' is frequency dependent and this dependence increases with mode number. Consequently, to reduce the problems of dispersion, it is usual to design a waveguide so that it only carries the TE_{10} mode and this can be achieved in the following fashion. Since $\beta = \omega/c$, it is clear that for mode TE_{nm} there is a frequency ω_{nm} below which the propagation constant becomes imaginary and no wave will propagate (in fact, the field decays exponentially and is said to be *evanescent*). Now consider frequencies ω such that

$\omega > \omega_{10}, \omega < \omega_{20}$ and $\omega < \omega_{01}$ (we have assumed that $a > b$). Then, for these frequencies, the TE_{01} mode is the only one that can propagate. A wave can be excited in a waveguide by placing a monopole antenna inside the guide (see Figure 8.33a). The monopole is placed a half wavelength from the end of the waveguide so that the reflected wave coincides with the direct wave (it should be noted that a wavelength in this case refers to that of the TE_{10} mode, i.e. $\lambda_{10} = 2\pi/\beta_{10}$ where β_{10} is the propagation constant of the TE_{10} mode). Although the monopole field will have a complex pattern (it will be a mixture of many modes), this will be reduced to the TE_{10} mode alone only a short distance down the waveguide due to the decay of evanescent modes. An open-ended waveguide can be used as an antenna (see Figure 8.33a) when the radiation is emitted into free space from the end of the guide. However, in order derive the radiation field we need to develop a little more theory.

Consider the electric field of an antenna at a point P, i.e.

$$E = \mathrm{j}\omega\mu_0 I h_{\mathrm{eff}} \frac{\exp(-\mathrm{j}\beta D)}{4\pi D}, \tag{8.51}$$

where D is the distance from the antenna to point P. We consider a surface S that is orthogonal to the line joining the antenna to P (i.e. the plane is tangent to the wavefront), located so that it is a large distance D_1 from the antenna (see Figure 8.34a). Consider

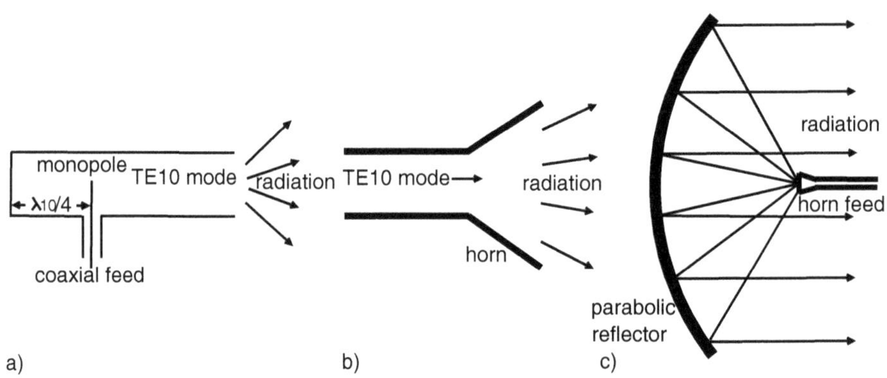

Fig. 8.33 Waveguide, horn and parabolic-reflector antennas.

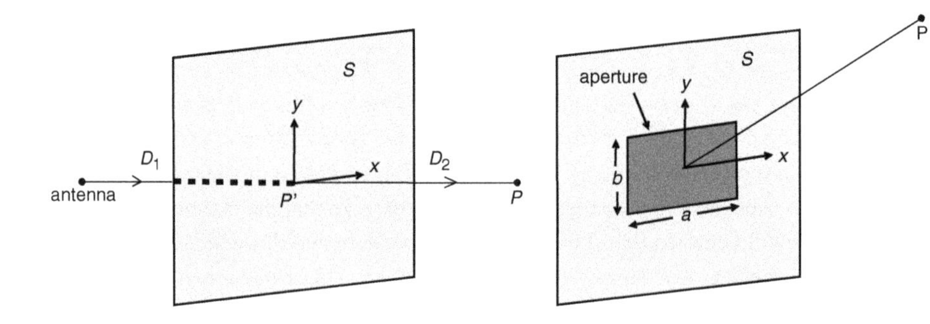

Fig. 8.34 Geometry for aperture theory.

local Cartesian coordinates x and y in the plane, centred on the point of intersection of the line and plane. Then, for a point in the plane, the distance from the antenna is $\sqrt{x^2+y^2+D_1^2} \approx R_1+(x^2+y^2)/2D_1$ and so the field at this point can be approximated as

$$E = \mathrm{j}\omega\mu_0 I h_{\mathrm{eff}} \frac{\exp\left(-\mathrm{j}\beta\left(D_1+\frac{x^2+y^2}{2D_1}\right)\right)}{4\pi D_1}. \tag{8.52}$$

Now define a potential

$$\psi = \mathrm{j}\beta\frac{\exp(-\mathrm{j}\beta\tilde{R})}{2\pi\tilde{R}} \tag{8.53}$$

for a source that is located at P ($\tilde{R}$ is the distance from the source to a general point). For points that lie in the surface S, $\tilde{R} = \sqrt{x^2+y^2+D_2^2} \approx D_2 + (x^2+y^2)/2D_2$, where D_2 is the distance of P from the plane (note that $D = D_1 + D_2$). Consequently, potential ψ can be approximated by

$$\psi = \mathrm{j}\beta\frac{\exp\left(-\mathrm{j}\beta\left(D_2+\frac{x^2+y^2}{2D_2}\right)\right)}{2\pi D_2}. \tag{8.54}$$

Now consider the integral $\int_S \psi E \mathrm{d}S$, then the integrand is given by

$$\psi E = -\beta\omega\mu_0 I h_{\mathrm{eff}} \frac{\exp(-\mathrm{j}\beta\left(D_1+D_2+(x^2+y^2)\left(\frac{1}{2D_1}+\frac{1}{2D_2}\right)\right)}{8\pi^2 D_1 D_2}. \tag{8.55}$$

Consequently,

$$\begin{aligned}\int_S \psi E\mathrm{d}S &= -\beta\omega\mu_0 I h_{\mathrm{eff}}\frac{\exp(-\mathrm{j}\beta(D_1+D_2))}{8\pi^2 D_1 D_2} \\ &\times \int_{-\infty}^{\infty}\int_{-\infty}^{\infty} \exp\left(-\mathrm{j}\beta(x^2+y^2)\left(\frac{1}{2D_1}+\frac{1}{2D_2}\right)\right)\mathrm{d}x\mathrm{d}y. \end{aligned} \tag{8.56}$$

We note that

$$\int_{-\infty}^{\infty} \exp(-\mathrm{j}\alpha x^2)\mathrm{d}x = \sqrt{\frac{\pi}{\mathrm{j}\alpha}} \tag{8.57}$$

from which

$$\int_{-\infty}^{\infty}\int_{-\infty}^{\infty} \exp\left(-\mathrm{j}\beta(x^2+y^2)\left(\frac{1}{2D_1}+\frac{1}{2D_2}\right)\right)\mathrm{d}x\mathrm{d}y = \frac{\pi}{\mathrm{j}\beta}\frac{1}{\frac{1}{2D_1}+\frac{1}{2D_2}} = \frac{2\pi}{\mathrm{j}\beta}\frac{D_1 D_2}{D_1+D_2} \tag{8.58}$$

and, as a consequence,

$$\int_S \psi E = \mathrm{j}\omega\mu_0 I h_{\mathrm{eff}}\frac{\exp(-\mathrm{j}\beta(D_1+D_2))}{4\pi(D_1+D_2)}. \tag{8.59}$$

The right-hand side of (8.59) is the field at point P (note that $D = D_1 + D_2$) and so

$$E = \int_S \psi E\mathrm{d}S. \tag{8.60}$$

Although we have only derived (8.60) for a point P located at $D_2\hat{\mathbf{z}}$, (8.60) is still valid for small deviations from this (i.e. points $\mathbf{R} = X\hat{\mathbf{x}} + Y\hat{\mathbf{y}} + D_2\hat{\mathbf{z}}$ for which $|X| \ll D_2$ and $|Y| \ll D_2$).

We now consider the situation where no radiation passes through plane S, except for an aperture A (this could be the situation where S is an opaque screen that contains a hole). E is now the field of the radiation emanating from this aperture and its value at point P will be given by (8.60) with the integral restricted to A, i.e.

$$E = \int_A \psi E \mathrm{d}S. \tag{8.61}$$

This expression is sometimes known as the *Kirchhoff integral* (Kirchhoff, 1883). Assuming that the aperture is small in comparison to the distance from point P, the distance from a point $\mathbf{r} = x\hat{\mathbf{x}} + y\hat{\mathbf{y}}$ within the aperture to point P can be approximated as $\sqrt{(\mathbf{R}-\mathbf{r})\cdot(\mathbf{R}-\mathbf{r})} = R\sqrt{1 - 2\mathbf{r}\cdot\mathbf{R}/R^2 + r^2/R^2} \approx R - (xX + yY)/R$ where R is the distance of P from the origin and $r = |\mathbf{r}|$. Consequently, $\psi \approx \mathrm{j}\beta \exp(-\mathrm{j}\beta R)\exp(\mathrm{j}\beta(xX + yY)/R)/\pi R$ and equation (8.61) will now reduce to

$$E(\mathbf{R}) = \mathrm{j}\beta \frac{\exp(-\mathrm{j}\beta R)}{2\pi R} \int_A E(x,y,0) \exp\left(\mathrm{j}\beta \frac{xX + yY}{R}\right) \mathrm{d}x\mathrm{d}y. \tag{8.62}$$

Consider a rectangular aperture (see Figure 8.34b) with sides length a and b, then the above expression will imply that

$$E(\mathbf{R}) = \mathrm{j}\beta \frac{\exp(-\mathrm{j}\beta R)}{2\pi R} \int_{-b/2}^{b/2} \int_{-a/2}^{a/2} E(x,y,0) \exp(j\beta(x\hat{X} + y\hat{Y})\mathrm{d}x\mathrm{d}y, \tag{8.63}$$

where $\hat{\mathbf{R}} = \hat{X}\hat{\mathbf{x}} + \hat{Y}\hat{\mathbf{y}} + \hat{Z}\hat{\mathbf{z}}$ is a unit vector in the direction of P. The simplest example is a plane wave $E = E_0\hat{\mathbf{y}}$ that is incident upon the rectangular aperture, then the integral in the above expression is easily evaluated and yields

$$E(\mathbf{R}) = \mathrm{j}\beta \frac{\exp(-\mathrm{j}\beta R)}{2\pi R} E_0 \frac{\sin(\frac{\beta a\hat{X}}{2})}{\frac{\beta\hat{X}}{2}} \frac{\sin(\frac{\beta b\hat{Y}}{2})}{\frac{\beta\hat{Y}}{2}}. \tag{8.64}$$

For a direction that is orthogonal to the aperture (i.e. $\hat{X} \to 0$ and $\hat{Y} \to 0$) the radiation field will be maximum in amplitude, with $E = \mathrm{j}\beta(\int_A E\mathrm{d}S)\exp(-\mathrm{j}\beta R)/2\pi R$, where A is the area of the aperture. Further, the directivity will also be maximum and its value can be calculated from

$$D_{\text{max}} = \frac{4\pi}{\lambda^2} \frac{|\int_A E\mathrm{d}S|^2}{\int_A |E|^2 \mathrm{d}S} \tag{8.65}$$

on noting that the total radiation will actually be that which emerges from the aperture (i.e. $P_{\text{rad}} = (1/2\eta_0)\int_A |E|^2\mathrm{d}S$). For a plane-wave excitation of the aperture, we obtain that $D_{\text{max}} = 4\pi A/\lambda^2$ since $D_{\text{max}} = G_{\text{max}}$. This brings us to the important concept of *effective aperture*, a quantity that is defined as $A_e = \lambda^2 G/4\pi$ ($A_e = A$ in the case of a lossless aperture excited by a plane wave). In the case of an aperture that consists of the end of a rectangular waveguide, the above expression for directivity yields that $D_{\text{max}} = 32A/\pi\lambda^2$ and from which $A_e = 8A/\pi^2$ on assuming the aperture to be lossless.

magnetic fields on the interface ($z = 0$) will imply

$$\exp(-j\beta_1 y \sin\theta_i)\frac{B_i}{\mu_1} + \exp(-j\beta_1 y \sin\theta_r)\frac{B_r}{\mu_1} = \exp(-j\beta_2 y \sin\theta_t)\frac{B_t}{\mu_2} \tag{9.4}$$

and

$$\exp(-j\beta_1 y \sin\theta_i)E_i\cos\theta_i - \exp(-j\beta_1 y \sin\theta_r)E_r\cos\theta_r = \exp(-j\beta_2 y \sin\theta_t)E_t\cos\theta_t. \tag{9.5}$$

We note that, in (9.3), $E_0 = E_i$ and $B_0 = B_i$ for the incident wave, $E_0 = E_r$ and $B_0 = B_r$ for the reflected wave and $E_0 = E_t$ and $B_0 = B_t$ for the transmitted wave (β_1 and β_2 are the propagation constants in media 1 and 2 respectively). In order for Eqs. (9.4) and (9.5) to hold for all y, we will need $\beta_1 \sin\theta_i = \beta_1 \sin\theta_r = \beta_2 \sin\theta_t$. As a consequence $\theta_i = \theta_r$, i.e. the angle of incidence is equal to the angle of reflection. In addition, $\sin\theta_i/c_1 = \sin\theta_t/c_2$, which is known as *Snell's law* ($c_1 = 1/\sqrt{\mu_1\epsilon_1}$ and $c_2 = 1/\sqrt{\mu_2\epsilon_2}$ are the speeds of propagation in media 1 and 2 respectively). Rewriting the magnetic field in terms of the electric field, we now find that

$$\frac{E_i}{\eta_1} + \frac{E_r}{\eta_1} = \frac{E_t}{\eta_2} \text{ and } E_i\cos\theta_i - E_r\cos\theta_r = E_t\cos\theta_t, \tag{9.6}$$

where η_1 and η_2 are the impedances of media 1 and 2 respectively. If we solve these equations, we obtain that

$$E_r = E_i\frac{\eta_1\cos\theta_i - \eta_2\cos\theta_t}{\eta_1\cos\theta_i - \eta_2\cos\theta_t} \text{ and } E_t = E_i\frac{2\eta_2\cos\theta_i}{\eta_1\cos\theta_i - \eta_2\cos\theta_t}. \tag{9.7}$$

We can use Snell's law to eliminate θ_t and so, from (9.7), $E_r = R_V E_i$ where the reflection coefficient R_V of this vertically polarised field is given by

$$R_V = \frac{\frac{\cos\theta_i}{\eta_r^2} - \sqrt{\frac{1}{\eta_r^2} - \sin^2\theta_i}}{\frac{\cos\theta_i}{\eta_r^2} + \sqrt{\frac{1}{\eta_r^2} - \sin^2\theta_i}}, \tag{9.8}$$

in which $\eta_r = \eta_2/\eta_1$ is the relative impedance.

When the electric field is parallel to the plane of the boundary, we have the situation described by Figure 9.1b. By similar arguments to the above case, we can now show that $\theta_i = \theta_r$, $\sin\theta_i/c_1 = \sin\theta_t/c_2$ and $E_r = R_H E_i$, where

$$R_H = \frac{\cos\theta_i - \sqrt{\frac{1}{\eta_r^2} - \sin^2\theta_i}}{\cos\theta_i + \sqrt{\frac{1}{\eta_r^2} - \sin^2\theta_i}} \tag{9.9}$$

is the reflection coefficient of this horizontally polarised field.

9.2 The Friis Equation

One of the most important calculations in designing a radio system is that of the power budget, i.e. the power that must be transmitted in order that a radio signal be received with sufficient strength for its successful demodulation. The basic tool in this calculation is known as the *Friis equation*. Consider a station that transmits a power P_T through an antenna with gain G_A and a receiver station with an antenna of gain G_R. As the signal propagates, it will spread out and the power flow per unit area will have strength $P = G_T P_T / 4\pi R^2$ where R is the distance travelled by the signal. The power at the receiver will be $P_R = PA_{\text{eff}}$ where A_{eff} is the effective aperture of the receiver antenna and, since $A_{\text{eff}} = G_R \lambda^2 / 4\pi$, we will have basic Friis equation

$$P_R = P_T G_R G_T \left(\frac{\lambda}{4\pi R} \right)^2, \tag{9.10}$$

where λ is the wavelength. The Friis equation is often written as $P_R = P_T G_R G_T / L_{\text{spr}}$ where $L_{\text{spr}} = (4\pi R / \lambda)^2$ is known as the *spreading loss*. Propagation losses are often quoted in terms of decibels, a logarithmic scale in which the loss L has the value $10 \log_{10} L$ and, in the case of spreading loss, this will be $20 \log_{10}(4\pi R/\lambda)$.

The Friis equation makes several assumptions that might not be met in practice. One major assumption is that the receiver antenna is *polarisation-matched* to the polarisation of the incoming wave $\boldsymbol{\mathcal{E}}$, i.e. vector $\boldsymbol{\mathcal{E}}$ is parallel to the effective vector length of the receiver antenna. In effect this means that $\mathbf{h}^R_{\text{eff}}$ and $\mathbf{h}^T_{\text{eff}}$ are parallel, where $\mathbf{h}^R_{\text{eff}}$ and $\mathbf{h}^T_{\text{eff}}$ are the effective vector lengths of the receiver and transmitter antennas respectively. If there is a mismatch, the received signal will be reduced by the *polarisation efficiency* $\eta_P = |\mathbf{h}^R_{\text{eff}} \cdot \mathbf{h}^T_{\text{eff}}|^2 / |\mathbf{h}^R_{\text{eff}}|^2 |\mathbf{h}^T_{\text{eff}}|^2$.

Another assumption is that there is no interaction with the environment. At a minimum, however, there will always be some ground reflections and the transmitter and receiver will need to be located considerably above the ground for this to be negligible. We consider the propagation between transmitter and receiver shown in Figure 9.3. The voltage induced in the receiver antenna will then be

$$V = \mathrm{j}\omega\mu_0 I h^T_{\text{eff}} h^R_{\text{eff}} \left(\frac{\exp(-\mathrm{j}\beta R_1)}{4\pi R_1} + R_g \frac{\exp(-\mathrm{j}\beta R_2)}{4\pi R_2} \right), \tag{9.11}$$

where R_1 and R_2 are the lengths of the direct and reflected paths. It should be noted that R_g is the reflection coefficient of the ground that is appropriate to the polarisation of the

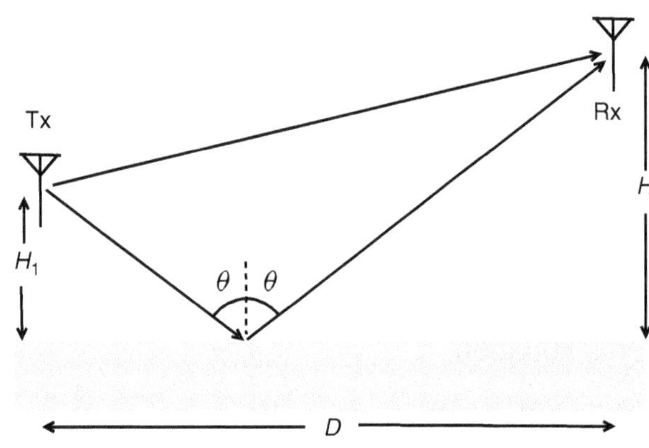

Fig. 9.3 The effect of ground reflections upon propagation.

antennas, h_{eff}^T is the effective length of the transmit antenna and h_{eff}^R is the effective length of the receive antenna. We will assume that the receiver and transmitter are located at heights above the ground that are much smaller than the distance between the stations, then $R_1 = \sqrt{(H_2-H_1)^2+D^2} \approx D+(H_2-H_1)^2/2D$ and $R_1 = \sqrt{(H_2+H_1)^2+D^2} \approx D+(H_2+H_1)^2/2D$. As a consequence,

$$V \approx \mathrm{j}\omega\mu_0 I h_{\text{eff}}^T h_{\text{eff}}^R \frac{\exp(-\mathrm{j}\beta D)}{4\pi D}\left(\exp\left(-j\beta\frac{(H_2-H_1)^2}{2D}\right)+R_g\exp\left(-\mathrm{j}\beta\frac{(H_2+H_1)^2}{2D}\right)\right). \tag{9.12}$$

In other words, the mutual impedance between the antennas is given by

$$Z_{TR} \approx \mathrm{j}\omega\mu_0 h_{\text{eff}}^T h_{\text{eff}}^R \frac{\exp(-\mathrm{j}\beta D)}{4\pi D}\left(\exp\left(-\mathrm{j}\beta\frac{(H_2-H_1)^2}{2D}\right)+R_g\exp\left(-\mathrm{j}\beta\frac{(H_2+H_1)^2}{2D}\right)\right). \tag{9.13}$$

Since the power generated by the transmitter in the receiver will be proportional to $V\bar{V}$, (9.12) will imply that the received power is modified by factor $|1+R_g\exp(-2\mathrm{j}\beta H_1 H_2/D)|^2$ due to reflections, i.e. the Friis equation will now take the form

$$P_R = P_T G_R G_T \left(\frac{\lambda}{4\pi D}\right)^2 \left|1+R_g\exp\left(-2\mathrm{j}\beta\frac{H_1 H_2}{D}\right)\right|^2. \tag{9.14}$$

At frequencies above about 300 MHz, $R_g \approx -1$ and (9.14) will reduce to

$$P_R = 4P_T G_R G_T \left(\frac{\lambda}{4\pi D}\right)^2 \sin^2\left(\beta\frac{H_1 H_2}{D}\right) \tag{9.15}$$

It will be noted that, although there is a general fall in amplitude of order D^{-2}, there are also nulls in the signal strength that are dependent upon H_1, H_2 and D. In particular, these nulls are responsible for the *fading* that is sometimes observed on the signals of mobile stations. In complex environments, reflections from objects other than the ground can contribute to the received signal and exacerbate such problems. Further, the movement of these additional reflectors can cause additional fading and this will also be experienced by fixed stations.

9.3 Huygens' Principle and Propagation by Refraction

We saw in Chapter 7 that the electric field at a point P could be expressed in terms of the electric field on a wavefront S through the integral relation

$$E_P = \int_S \psi E \mathrm{d}S, \tag{9.16}$$

where

$$\psi = \mathrm{j}\beta\frac{\exp(-\mathrm{j}\beta R)}{2\pi R}, \tag{9.17}$$

in which R is the distance from a general point $\mathbf{R}$ on the surface S to point P.

We can approximate the integral over S by dividing the surface into small elements of area δS_i on each of which we approximate the electric field by its value at the centre $\mathbf{R}_i$ of the element, i.e.

$$E_P = \sum_i E_i \delta S_i \mathrm{j}\beta \frac{\exp(-\mathrm{j}\beta R_i^P)}{2\pi R_i^P}, \tag{9.18}$$

where R_i^P is the distance from the point $\mathbf{R}_i$ to point P. From (9.18), we see that we have approximated the integral by a collection of point sources on the wavefront (see Figure 9.4a) and this will mean that, after a short time δt, the new wavefront will be the envelope of spherical wavefronts (each with radius $c\delta t$ where $c = 1/\sqrt{\mu\epsilon}$) emanating from these point sources. This leads us to what is known as *Huygens' principle* (Huygens, 1912):

Each point on a wavefront of a general wave can be considered as the source of a secondary spherical wave. A subsequent wavefront of the general wave can then be constructed as the envelope of secondary wavefronts.

Huygens' principle can be used to study the propagation of a radio wave at the interface between media with differing properties. We consider a wavefront that is incident upon a plane interface with propagation direction at angle θ_i to the normal (see Figure 9.4b). Consider the wavefront at times t and $t + \delta t$. It will intersect the interface at points A and B that will then become the sources of secondary waves. The secondary waves that propagate upwards will travel at speed c_1 and those that propagate downwards will travel at speed c_2. Figure 9.4b shows the incident wave, at times t and $t + \delta t$, and the secondary waves, at time $t + \delta t + \Delta t$. The envelope of the upper secondary waves will be a wavefront of the reflected wave and that of the lower secondary waves will be a wavefront of the transmitted wave. In terms of the upward secondary waves, $AB = c_1\delta t/\cos\theta_r$ and, in terms of the downward secondary wave, $AB = c_2\delta t/\sin\theta_t$. For the incident wave $AB = c_1\delta t/\sin\theta_i$. As expected, these relations imply that $\theta_r = \theta_i$ and $\sin\theta_t/c_2 = \sin\theta_i/c_1$.

The mechanism of refraction does not need there to be a sudden change in the medium's properties and will also occur when there is a continuous variation in these properties. Consider the situation shown in Figure 9.5a where, across the wavefront, the variation in material properties causes a variation in the wave propagation speed. At two points on the wavefront, separated by distance δR, the speeds of the wave will be c and $c + \delta c$ respectively. After a time δt, the wavefront will have moved on, but the differential in wave speeds will cause a tilt in the wavefront, i.e. the direction of propagation will change as the wave propagates (see Figure 9.4a). Consequently, a given point on the wavefront will move through space on a curve that can be nonlinear and this is known as a *ray*. Over a short distance, a ray can be regarded as a section of a circle of radius R and the centre of this circle will be the intersection of the wavefronts. As a consequence,

$$\frac{(c + \delta c)\delta t}{c\delta t} = \frac{R + \delta R}{R} \tag{9.19}$$

and then, in the limit that $\delta R \to 0$, we will have that

$$\frac{\mathrm{d}c}{\mathrm{d}R} = \frac{c}{R}, \tag{9.20}$$

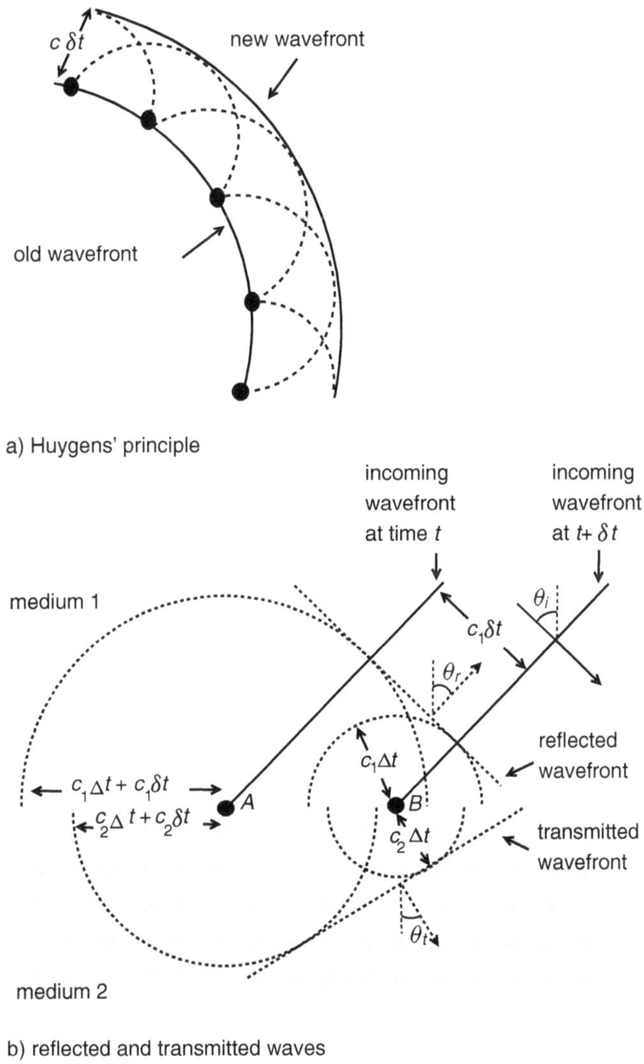

Fig. 9.4 Huygens' principle and refraction.

where $\mathrm{d}c/\mathrm{d}R$ is the gradient of the wave speed across the wavefront. Rays are an important concept in propagation since, at any point on a ray, the ray tangent will be the direction of propagation at that point (the electric and magnetic fields will be orthogonal to this direction). Rays will spread out from a source and, since energy will flow along a ray, any energy that is initially within a bundle of rays will remain within that bundle. Consequently, when rays converge the power density will rise and, when they diverge, it will fall (for example, power density will fall when rays move out from an antenna).

We will now consider how the ray paths depend upon the distribution of propagation speed c in the medium. Consider a ray path in the xy plane. We can parameterise this path in terms of the distance s along the path and then $x = x(s)$ and $y = y(s)$. The tangent to the

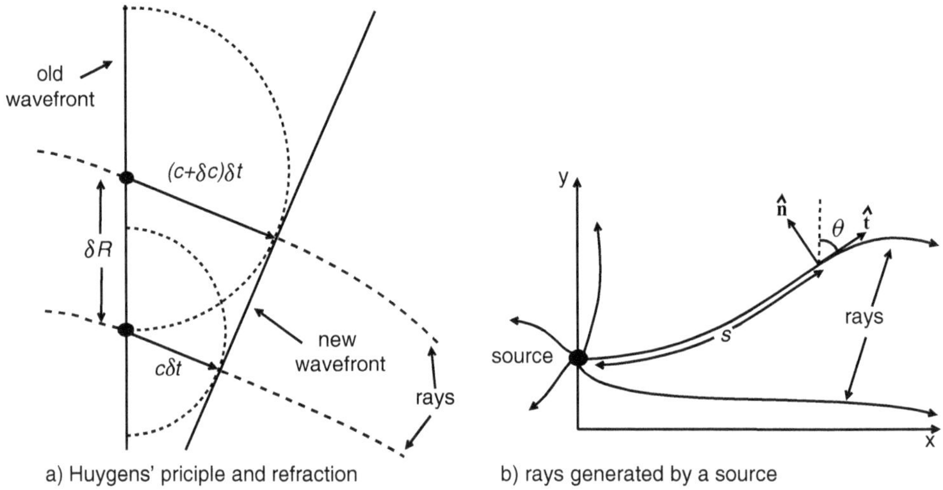

Fig. 9.5 Huygens' principle and ray tracing.

ray is given $\hat{\mathbf{t}} = (\mathrm{d}x/\mathrm{d}s)\hat{\mathbf{x}} + (\mathrm{d}y/\mathrm{d}s)\hat{\mathbf{y}}$ and the normal by $\hat{\mathbf{n}} = -(\mathrm{d}y/\mathrm{d}s)\hat{\mathbf{x}} + (\mathrm{d}x/\mathrm{d}s)\hat{\mathbf{y}}$. From elementary calculus, the gradient of c in the direction $\hat{\mathbf{n}}$ (the gradient along the wavefront) will be given by $\hat{n}_x \partial c/\partial x + \hat{n}_y \partial c/\partial y = -(\mathrm{d}y/\mathrm{d}s)\partial c/\partial x + (\mathrm{d}x/\mathrm{d}s)\partial c/\partial y$. Further, for a curve in the xy plane, the curvature (i.e. $1/R$) is given by

$$\frac{1}{R} = \frac{\left|\frac{\mathrm{d}^2 y}{\mathrm{d}x^2}\right|}{\left(\left(\frac{\mathrm{d}y}{\mathrm{d}x}\right)^2 + 1\right)^{\frac{3}{2}}}, \tag{9.21}$$

where, for the conventions we have adopted, $|\mathrm{d}^2y/\mathrm{d}x^2| = -\mathrm{d}^2y/\mathrm{d}x^2$. We will consider the case where the wave speed c only varies in the y direction and then the gradient across the wavefront will be given by $(\mathrm{d}x/\mathrm{d}s)\partial c/\partial y = (\partial c/\partial y)/\sqrt{(\mathrm{d}y/\mathrm{d}x)^2 + 1}$ (noting that $\mathrm{d}s = \sqrt{\mathrm{d}x^2 + \mathrm{d}y^2}$). From (9.20), we will now find that

$$\frac{\partial c}{\partial y} = -c \frac{\frac{\mathrm{d}^2 y}{\mathrm{d}x^2}}{\left(\frac{\mathrm{d}y}{\mathrm{d}x}\right)^2 + 1} \tag{9.22}$$

and, if we multiply both sides of this equation by $\mathrm{d}y/\mathrm{d}x$, we can rearrange this into the form

$$-\frac{\mathrm{d}\ln c}{\mathrm{d}x} = \frac{1}{2} \frac{1}{\left(\frac{\mathrm{d}y}{\mathrm{d}x}\right)^2 + 1} \frac{\mathrm{d}\left(\left(\frac{\mathrm{d}y}{\mathrm{d}x}\right)^2\right)}{\mathrm{d}x}. \tag{9.23}$$

This can be integrated to yield

$$\frac{1}{c} = C\sqrt{\left(\frac{\mathrm{d}y}{\mathrm{d}x}\right)^2 + 1}, \tag{9.24}$$

where C is a constant of integration. Since $\sqrt{(\mathrm{d}y/\mathrm{d}x)^2+1} = \mathrm{d}s/\mathrm{d}x$ and $\sin\theta = \mathrm{d}x/\mathrm{d}s$ (see Figure 9.5b), we find that

$$\frac{\sin\theta}{c} = \text{constant}, \tag{9.25}$$

i.e. Snell's law is satisfied along a ray.

An important refracting medium, known as the *ionosphere*, is a set of ionised layers of the atmosphere for which the main layer lies at an altitude of about 300 km. This layer is generated by the ionising effect of solar radiation which causes some of the electrons of the gas molecules to be disassociated and to flow freely. Importantly, this medium can cause sufficient refraction to make radio waves, generated at the surface of the Earth, return to the ground at a distance of many thousands of kilometres, hence enabling over-the-horizon communication. To understand how this layer can cause refraction, we need to go back to Maxwell's equations. We will assume that a time-harmonic wave of frequency ω propagates through the ionosphere and then the Maxwell equation (1.35) becomes

$$\int_C \frac{\mathbf{B}}{\mu}\cdot \mathrm{d}\mathbf{r} = \mathrm{j}\omega \int_S \epsilon_0 \mathbf{E}\cdot \mathbf{n}\mathrm{d}S + I, \tag{9.26}$$

where we have assumed that the atmosphere is approximately a vacuum (i.e. $\epsilon = \epsilon_0$). The current I will now consist of disassociated electrons flowing across the surface S. If we neglect the effect of magnetic fields, the Lorentz force on the electrons (see (1.18)) will reduce to $\mathbf{F} = q\boldsymbol{\mathcal{E}}$, where q is the charge on an electron. By Newton's law of motion, $m_e\mathrm{d}\mathbf{v}/\mathrm{d}t = q\boldsymbol{\mathcal{E}}$, where m_e is the mass of an electron and $\mathbf{v}$ is its velocity. Since the fields will be time-harmonic, we can rewrite this as $\mathrm{j}\omega m_e\mathbf{v} = q\mathbf{E}$. If we consider a small area

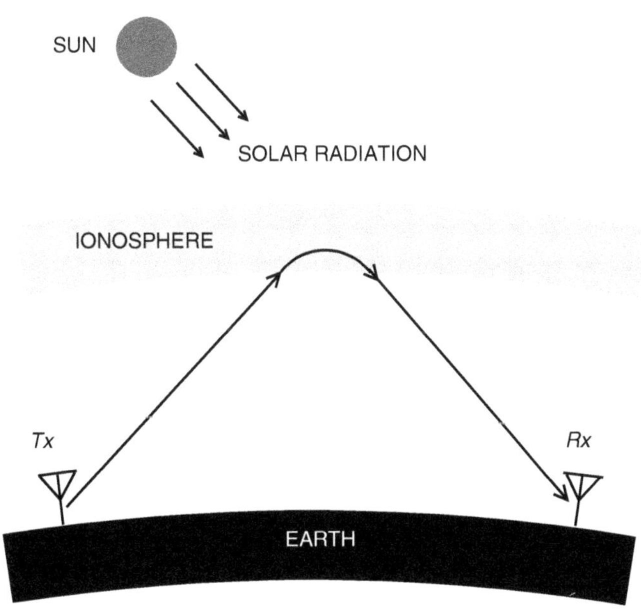

Fig. 9.6 The ionosphere and the refraction of radio waves.

$\mathrm{d}S$ of the surface S, with electron density N_e, the total current flowing through this area will be $qN_e\mathbf{v}\cdot\mathbf{n}\mathrm{d}S$ and the total current I flowing through S will be

$$I=\int_S qN_e\frac{q\mathbf{E}\cdot\mathbf{n}}{\mathrm{j}\omega m_e}\mathrm{d}S. \tag{9.27}$$

Substituting this into (9.26), we obtain

$$\int_C\frac{\mathbf{B}}{\mu}\cdot\mathrm{d}\mathbf{r}=\mathrm{j}\omega\int_S\epsilon_{\mathrm{eff}}\mathbf{E}\cdot\mathbf{n}\mathrm{d}S, \tag{9.28}$$

where the effective permittivity is given by

$$\epsilon_{\mathrm{eff}}=\epsilon_0\left(1-\frac{q^2N_e}{m_e\epsilon\omega^2}\right). \tag{9.29}$$

The speed of wave propagation in the plasma will be $c=1/\mu_0\epsilon_{\mathrm{eff}}$ which can be rewritten as $c=c_0/N$ where $N=\sqrt{\mu_0\epsilon_0/\mu_0\epsilon_{\mathrm{eff}}}$ is known as the *refractive index* of the medium. Snell's law can now be rewritten as $N\sin\theta=$ constant. From (9.29), the refractive index for a plasma can be written as

$$N=\frac{1}{\sqrt{1-\frac{\omega_p^2}{\omega^2}}}, \tag{9.30}$$

where $\omega_p=\sqrt{q^2N_e/m_e\epsilon_0}$ is known as the *plasma frequency*. It will be noted that this refractive index, and hence the propagation speed, will be frequency-dependent and hence, as with tranmission lines, propagation in the ionosphere will suffer from the problem of dispersion.

The simplest model of the ionosphere is known as a *parabolic layer*, and for which

$$\begin{aligned}\omega_p^2&=\omega_m^2\left(1-\frac{(y-h_m)^2}{y_m^2}\right)\quad\text{for}\quad|y-h_m|<y_m\\&=0\quad\text{otherwise,}\end{aligned} \tag{9.31}$$

where h_m is the layer peak height, y_m is its thickness and ω_m is the peak plasma frequency. This is an effective approximation to the dominant ionospheric layer, the *F2 layer*. The peak plasma frequency ω_m of the ionosphere varies with the intensity of solar radiation, peaking around midday (typical values of the parameters at this time are $h_m=300\,\mathrm{km}$, $y_m=100\,\mathrm{km}$ and $\omega_m=2\pi f_m$ where $f_m=10\,\mathrm{MHz}$). After dusk, although the ionisation mechanism of solar radiation has disappeared, the plasma frequency still remains relatively high for an appreciable portion of the night due to the fairly slow decay of the ionisation.

We now consider propagation through a parabolic ionosphere for a ray that has been launched from the ground. From Snell's law we have that

$$N(y)\sin\theta=N(0)\sin\theta_0, \tag{9.32}$$

where θ_0 is the value of θ at the start of the ray ($x=y=0$). For a parabolic layer, Snell's law implies that

$$\frac{\sin^2\theta - \sin^2\theta_0}{\sin^2\theta} = \frac{\omega_m^2}{\omega^2}\left(1 - \frac{(y-h_m)^2}{y_m^2}\right) \quad \text{for} \quad |y-h_m| < y_m$$
$$= 0 \quad \text{otherwise.} \tag{9.33}$$

Then, for the ray to return to the ground, there will need to be a height $h_{\max}$ at which $\theta = \pi/2$ and, from (9.33), this will be given by

$$h_{\max} = h_m - y_m\sqrt{1 - \frac{\omega^2}{\omega_m^2}\cos^2\theta_0}. \tag{9.34}$$

It will be noted that the height will become imaginary for θ_0 less than $\theta_m = \cos^{-1}(\omega_m/\omega)$, i.e. no rays will return to the ground for $\theta < \theta_m$. Since $\mathrm{d}x/\mathrm{d}y = \tan\theta = \sin\theta/\sqrt{1-\sin^2\theta}$, we have from (9.32) that

$$\frac{\mathrm{d}x}{\mathrm{d}y} = \frac{\sin\theta_0}{\sqrt{N^2(y) - \sin^2\theta_0}} \tag{9.35}$$

on noting that $N(0) = 1$. Then, integrating with respect to y,

$$x(y) = \int_0^y \frac{\sin\theta_0}{\sqrt{N^2(y) - \sin^2\theta_0}}\mathrm{d}y. \tag{9.36}$$

If the ray returns to the ground at distance D, $x(h_{\max})$ will be $D/2$ since the ray must be symmetric about the mid point. Consequently, the total ground range D will be

$$D = 2(h_m - y_m)\tan\theta_0 + 2\int_{h_m-y_m}^{h_{\max}} \frac{\sin\theta_0}{\sqrt{\cos^2\theta_0 - \frac{\omega_m^2}{\omega^2}(1 - \frac{(y-h_m)^2}{y_m^2})}}\mathrm{d}y$$
$$= 2(h_m - y_m)\tan\theta_0 +$$
$$+ \frac{y_m\omega\sin\theta_0}{\omega_m}\ln\left(\frac{1 + \frac{\omega}{\omega_m}\cos\theta_0}{1 - \frac{\omega}{\omega_m}\cos\theta_0}\right). \tag{9.37}$$

For $\omega < \omega_m$, D will be a monotonic function of θ_0 and there will only be one ray for each range. However, when $\omega > \omega_m$, the minimum value of D is greater than zero and there is a region around the source (known as the *skip zone*) which rays cannot reach. Figure 9.7 shows examples of the variation of range D with initial angle θ_0 for these two cases. In the case that $\omega > \omega_m$, whilst there are ranges to which there is no propagation, there are also ranges for which there are two propagation paths (i.e. two initial angles θ_0). Up to now we have made the simplifying assumption that the ionosphere is flat (a reasonable assumption for short ranges), but the same conclusions follow for the more realistic case of an ionosphere that is a spherical shell surrounding the Earth. Figure 9.8 shows the rays for the case of a spherical ionosphere with the same ionospheric parameters as for the results of Figure 9.7. In the case that $f > f_m$ (f and f_m denote ω and ω_m in terms of hertz) the skip zone can be clearly seen, as can the dual propagation paths for ranges that

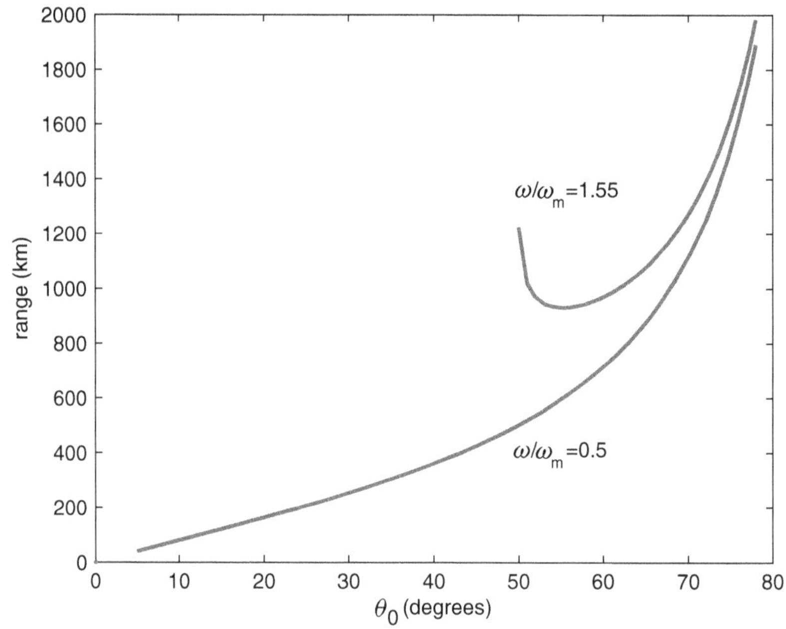

Fig. 9.7 Variation of range with initial angle.

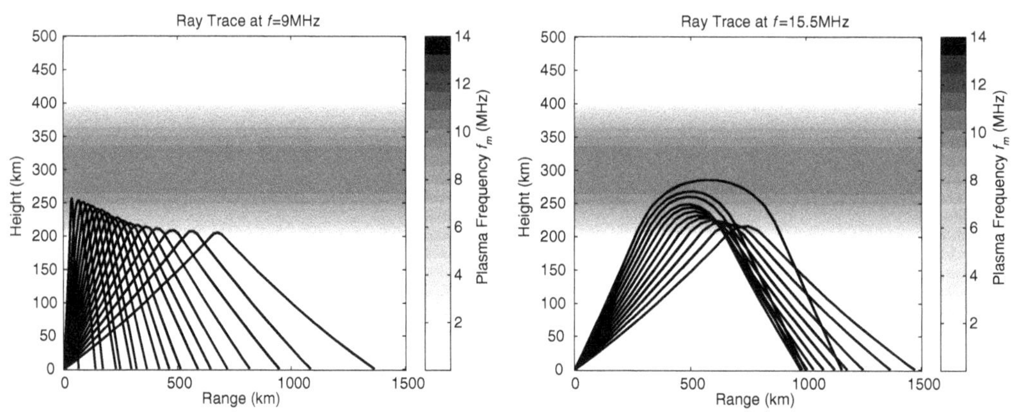

a) $f < f_m$ b) $f > f_m$

Fig. 9.8 Rays for $f < f_m$ and $f > f_m$ (NB $\omega = 2\pi f$ and $\omega_m = 2\pi f_m$).

can be reached. For the dual propagation paths, it will be noted that one ray returns low in the ionosphere and the other high. As a consequence, these rays are known as a *low ray* and *high ray* respectively.

From the above considerations, it is clear that the ionosphere can provide long-range communications through the process of refraction. The idea of the ionosphere was first suggested in 1938 by Carl Friedrich Gauss as a way of explaining anomalous variations in Earth's magnetic field. However, it was with the transatlantic communication experiments of Guglielmo Marconi in 1901 that the importance of the ionosphere was recognised. In 1902 Oliver Heaviside suggested that an ionised layer above the Earth could be responsible for the transatlantic communications achieved by Marconi through a process of reflection from this layer (the layer is sometimes known as the *Heaviside layer*). Marconi performed his experiments at a frequency of 300 kHz and, for a long time, it was believed that a frequency of 1.5 MHz was the highest for which such communications were possible. Indeed, in 1912 the US Congress confined amateur radio enthusiasts to frequencies above 1.5 MHz in the belief that these frequencies were of no use. Consequently, it was left to radio amateurs to show the utility of communications at frequencies above 1.5 MHz. During the 1920s, amateurs had increasing success at these higher frequencies over ranges of thousands of kilometres and with only very modest power. Further, it was clear that such communications were only possible through the refraction caused by the ionosphere. In 1923, the success of the amateurs led Marconi, and his assistant Charles Franklin, to investigate radio propagation at high frequencies (frequencies between 3 and 30 MHz) and this soon led to their use as a major means of long-distance communication.

A major problem for communications via the ionosphere turned out to be the diurnal variation of this medium due to the diurnal variation of its driving mechanism. Thus far, we have only talked in terms of a single ionospheric layer, but during the daytime there can be several layers. The various layers are generated by the ionising effect of solar radiation, their differing nature being a result of the different mixtures of gases at their respective altitudes. The highest layer (known as the F2 layer) occurs at an altitude of around 320 km or more; it has a thickness of about 200 km and a peak plasma frequency of about 10 MHz during the daytime. Just below the F2 layer is the F1 layer at a height of around 170 km; it has a thickness of about 50 km and a peak plasma frequency of about 5 MHz. Below this, at a height of around 110 km, there is the E layer with a thickness of about 50 km and a peak plasma frequency about 3 MHz. Finally, at a height of around 80 km, there is the D layer with a thickness of about 40 km and a peak plasma frequency of about 0.5 MHz.

Although all of these layers can sustain propagation, the major useful propagation comes from the F2 layer. During the day, the D and E layers cause severe attenuation of the radio waves through collisions in the plasma, the attenuation being proportional to the plasma density in these layers and the inverse square of the wave frequency. For the strongest signals, this means that communications should occur as close as possible to the highest frequency that will sustain the desired propagation path. Operating at this frequency will also mean that the apex of the propagation path will be the highest possible (Figure 9.9 shows the variation of propagation path and propagation loss with frequency), which means that the path will travel less distance in the D layer and hence suffer even less attenuation. By using the highest frequency, we therefore ensure that propagation occurs via the F2 layer. At night, the D, E and F1 layers rapidly disappear due to fast recombination between the electrons and ions. For the F2 layer, however, the recombination process is much slower, and this layer continues through the night, although the peak plasma frequency will be down to a few MHz by dawn. Consequently, the F2 layer provides the only useful ionospheric propagation mechanism at night.

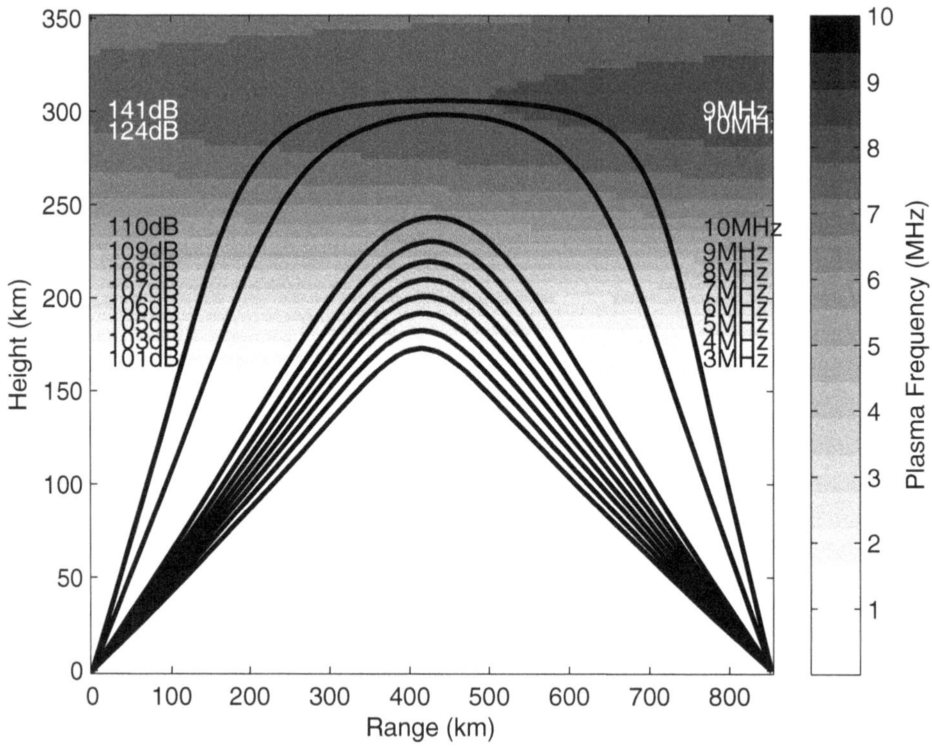

Fig. 9.9 Variation of propagation paths with frequency.

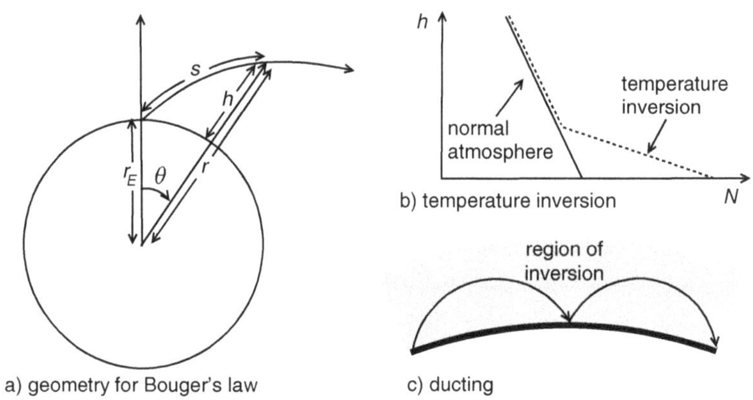

Fig. 9.10 Bouger's law and atmospheric ducting.

For communications over long ranges, the curvature of Earth must be taken into account. If we consider a polar coordinate system that is based upon the centre of the Earth (see Figure 9.10a), the equivalent of Snell's law is Bouger's Law, i.e

$$N(r)\, r^2 \frac{\mathrm{d}\theta}{\mathrm{d}s} = C, \tag{9.38}$$

where $\mathrm{d}s = \sqrt{r^2 \mathrm{d}\theta^2 + \mathrm{d}r^2}$. Referring to Figure 9.10a, θ is the angle between the radial vector through the start of the ray and that through the current position on the ray, s is the distance along this section of the ray and r is the radial distance to the current point. If γ is the initial elevation of the ray (the angle between the ground and the ray tangent), we have that $C = r_E N_0 \cos\gamma$ where $N_0 = N(r_E)$ and r_E (≈ 6378.135 km) is the radius of the Earth. Equation (9.38) can now be rearranged into the form

$$\frac{\mathrm{d}r}{\mathrm{d}\theta} = \frac{r\sqrt{N^2r^2 - N_0^2 r_E^2 \cos^2\gamma}}{r_E N_0 \cos\gamma}. \tag{9.39}$$

For the ionosphere defined by (9.31), this equation can be solved numerically to produce the rays shown in Figure 9.8.

We will now consider propagation in that part of the atmosphere much closer to Earth. There is no ionosphere close to the surface of the Earth and refraction in this layer will be caused by the atmospheric gases alone. Under normal conditions the refractive index in this layer can be taken to be linear in altitude $h = r - r_E$, i.e. for the first few kilometres above the ground $N(h) \approx N_0(1+\alpha h)$ where $\alpha \approx -0.0000425\,\mathrm{km}^{-1}$ and $N_0 = 1.000315$. The effect of the atmosphere is very small and the rays will normally escape to higher altitudes where, if the wave frequency is low enough, they will be refracted back to Earth by the ionosphere. However, it turns out that, under abnormal weather conditions, the value of α can be modified to a value that causes a ray to return to Earth. For this to be possible, there will need to be a point on the ray trajectory at which $\mathrm{d}r/\mathrm{d}\theta$ changes sign, i.e. $\mathrm{d}r/\mathrm{d}\theta = 0$. From (9.39), this will imply that $Nr = N_0 r_E \cos\gamma$ or $(1+\alpha h)r = r_E \cos\gamma$. For propagation close to the ground $h \ll r_E$ and so this condition will reduce to $h(\alpha r_E + 1) \approx r_E(\cos\gamma - 1)$. Consequently, since h must be positive, we will need $r_E\alpha \ll -1$, i.e. $\alpha < -0.000136$, for there to be a suitable h at which propagation turns back towards ground. For the lower atmosphere, the nature of the refractive index, and hence α, is related to meteorological conditions through the Debye formula

$$N = 1 + \frac{7.76\times 10^{-5}}{T}(P + 4810\frac{e}{T}), \tag{9.40}$$

where T is the temperature (in kelvin), P is the atmospheric pressure (in millibars) and e is the water vapour pressure (in millibars). Under normal circumstances, pressure will decrease with height ($P \approx P_0 \exp(-y/H)$ where $H \approx 8.5$ km and $P_0 \approx 1010$ mb) and likewise temperature. The rate at which temperature decreases is known as the *lapse rate* and is typically about 6.5°C per kilometer. Occasionally, however, a layer of warm air can form above a layer of cold air (i.e. a *temperature inversion*) and this can cause a sufficiently negative value of α (see Figure 9.10b) and hence enable over-the-horizon propagation. Temperature inversions can arise in a number of ways. If air at a higher altitude sinks, the air can be warmed by adiabatic heating and result in an inversion. This is a subsidence inversion and is usually associated with a zone of high pressure. Another form of inversion arises at a weather front, where a flow of cold air can undercut warm air and form what is known as a frontal inversion. A third kind of inversion is caused when air is cooled from underneath by a cold surface. This can occur on a clear night when the ground rapidly cools by the mechanism of radiation and then itself cools the air directly above it. Alternatively, in coastal regions dry warm air can be blown from land over the cool humid air just above the sea. All of these mechanisms can give rise to conditions that significantly extend radio coverage through a process known as *atmospheric ducting* (see Figure 9.10c). In such a duct the propagation is refracted back to the ground where it suffers reflection followed by further refractions and reflections.

9.4 Scatter and Diffraction

Consider a wavefront that is incident upon a screen that forms a partial barrier to the wave. Huygens' principle suggests that we picture the wavefront as a series of point sources (see Figure 9.11a). After a time δt, the envelope of the wavefronts generated by the point sources along the screen will constitute the reflected wavefront and the envelope from the sources above the screen will constitute the transmitted wave. Just above the screen, the envelope of the reflected wave will be circular in the backward direction since the source at the top of the screen is the highest to contribute to the reflected wave. Just below the top of the screen, the envelope will also be circular in the forward direction since this will be the bottommost to contribute to the transmitted wave. The reflected wave above the screen is known as back scatter and the transmitted wave behind the screen is known as forward scatter. However, a more common name for forward scatter is *diffraction*.

We consider a screen that is located in the xy plane of a Cartesian coordinate system (see Figure 9.11) and will calculate the diffraction for this screen by means of the Kirchhoff integral (Blaunstein, 2000). As we have seen in the previous chapter, we can calculate the field at the receiver from the field of the transmitter using the Kirchhoff integral relation (8.53), i.e.

$$E_R = \int_A \psi E_T \, \mathrm{d}S, \tag{9.41}$$

where aperture A is now that part of the xy plane above the screen (the integral is essentially the mathematical expression of Huygens' principle). Potential ψ is given by

$$\psi = \mathrm{j}\beta \frac{\exp(-\mathrm{j}\beta r_R)}{2\pi r_R}, \tag{9.42}$$

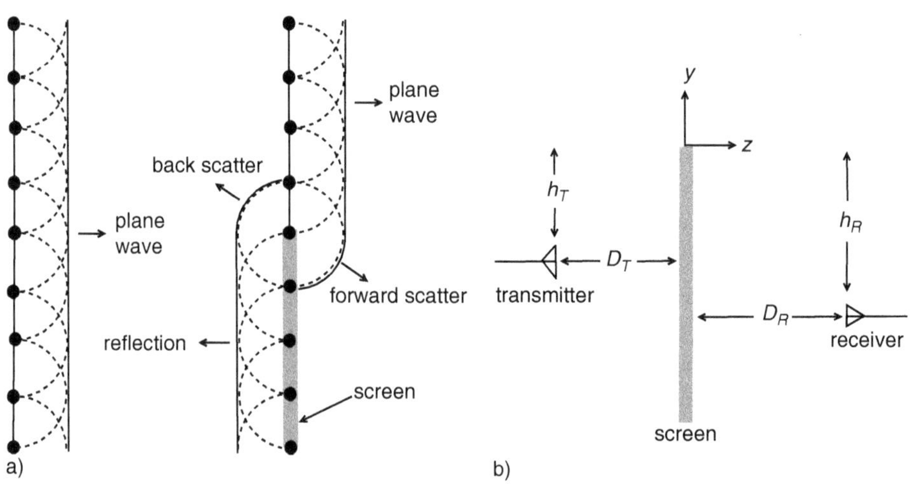

Fig. 9.11 Huygens' principle and diffraction over a screen.

in which r_R is the distance from the receiver to a general point in the aperture A. The field of the transmitter in A has the form

$$E_T = \frac{\mathrm{j}\omega\mu_0 I}{4\pi} h_{\mathrm{eff}}^T \frac{\exp(-\mathrm{j}\beta r_T)}{r_T}, \tag{9.43}$$

where r_T is the distance of the transmitter from a general point in the aperture A. When both receiver and transmitter are well separated from the screen, $r_T \approx D_T + (x^2+(y+h_T)^2)/2D_T$ and $r_R \approx D_R + (x^2+(y+h_R)^2)/2D_R$. The field at the receiver is then given by

$$E_R = \frac{-\beta\omega\mu_0 I}{8\pi^2} h_{\mathrm{eff}}^T \exp(-\mathrm{j}\beta D) \tag{9.44}$$

$$\times \int_0^\infty \int_{-\infty}^\infty \frac{\exp\left(-\frac{\mathrm{j}\beta}{2}\left(\frac{x^2}{D_T}+\frac{x^2}{D_R}+\frac{(y+h_T)^2}{D_T}+\frac{(y+h_R)^2}{D_R}\right)\right)}{D_T D_R} \mathrm{d}x\mathrm{d}y,$$

where $D = D_T + D_R$ is the horizontal distance between the antennas. On noting that

$$\int_{-\infty}^\infty \exp(-\mathrm{j}\alpha x^2)\mathrm{d}x = \sqrt{\frac{\pi}{\mathrm{j}\alpha}} \tag{9.45}$$

it is possible to perform the x integral analytically and then

$$E_R = \frac{-\beta\omega\mu_0 I}{8\pi^2}\sqrt{\frac{2\pi}{\mathrm{j}\beta D D_T D_R}} h_{\mathrm{eff}}^T \exp(-\mathrm{j}\beta D) \tag{9.46}$$

$$\times \int_0^\infty \exp\left(-\frac{\mathrm{j}\beta}{2}\left(\frac{y^2}{D_T}+\frac{y^2}{D_R}+\frac{2yh_T}{D_T}+\frac{2yh_R}{D_R}+\frac{h_T^2}{D_T}+\frac{h_R^2}{D_R}\right)\right)\mathrm{d}y.$$

We can then rearrange (9.46) into the form

$$E_R = \frac{-\beta\omega\mu_0 I}{8\pi^2}\sqrt{\frac{2\pi}{\mathrm{j}\beta D D_T D_R}} h_{\mathrm{eff}}^T \exp(-\mathrm{j}\beta D) \tag{9.47}$$

$$\times \exp\left(-\frac{\mathrm{j}\beta}{2}\left(\frac{h_T^2}{D_T}+\frac{h_R^2}{D_R}-\frac{D_T D_R}{D}\left(\frac{h_T}{D_T}+\frac{h_R}{D_R}\right)^2\right)\right)$$

$$\times \int_0^\infty \exp\left(-\frac{\mathrm{j}D\beta}{2D_T D_R}\left(y+\left(\frac{D_R h_T}{D}+\frac{D_T h_R}{D}\right)\right)^2\right)\mathrm{d}y.$$

If we now introduce the new variable $Y = \sqrt{D\beta/2D_T D_R}\,(y + (D_R h_T + D_T h_R)/D)$, we can simplify (9.47) into

$$E_R = \frac{-\omega\mu_0 I}{4\pi^2 D}\sqrt{\frac{\pi}{\mathrm{j}}} h_{\mathrm{eff}}^T \tag{9.48}$$

$$\times \exp\left(-\mathrm{j}\beta\left(D+\frac{(h_T-h_R)^2}{2D}\right)\right)\int_v^\infty \exp\left(-\mathrm{j}Y^2\right)\mathrm{d}Y,$$

where $v = \sqrt{\beta/2D_T D_R D}\,(D_T h_R + D_R h_T)$. The behaviour of the integral in the above expression is illustrated in Figure 9.12 where the solid line is the real part and the

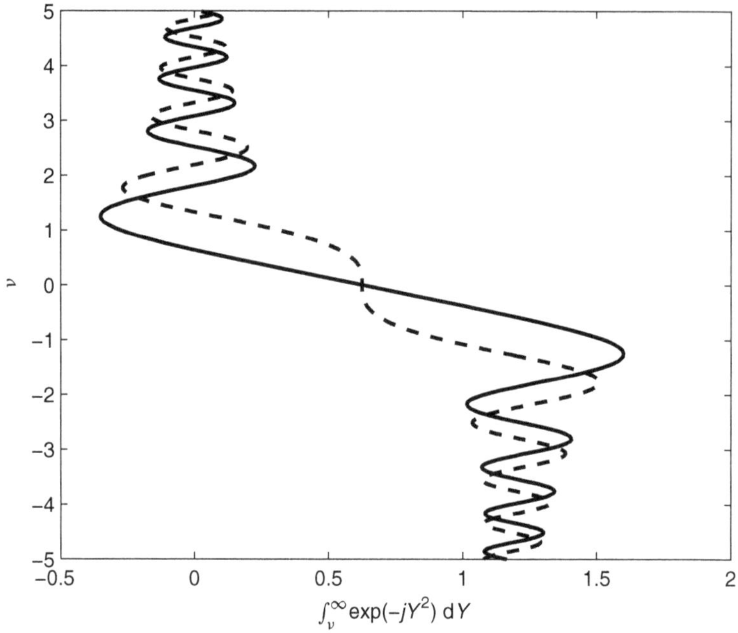

Fig. 9.12 The integral $\int_\nu^\infty \exp\left(-jY^2\right) \mathrm{d}Y$ (the solid line is the real part and broken line the imaginary part).

broken line the imaginary part. (Note that, if the receive or transmit antenna is above the screen, the corresponding value of h_R or h_T will need to be negative.) A positive value of ν will correspond to optical obscuration by the screen and a negative value will correspond to line of sight between receiver and transmitter. From the behaviour of the integral, the field increases quite dramatically as we move from positive to negative values of ν. This behaviour is to be expected, but it is also clear that there can still be substantial propagation by diffraction when the path is obscured. The voltage induced in the receive antenna will be $h^R_{\text{eff}} E_R$ where h^R_{eff} is the effective length of the receive antenna. Consequently, the mutual impedance between the receive and transmit antennas will be

$$Z_{TR} = \frac{\mathrm{j}\omega\mu_0}{4\pi D}\sqrt{\frac{\mathrm{j}}{\pi}} h^T_{\text{eff}} h^R_{\text{eff}} \times \exp\left(-\mathrm{j}\beta R_{TR}\right) \int_\nu^\infty \exp\left(-\mathrm{j}Y^2\right) \mathrm{d}Y, \tag{9.49}$$

where $R_{TR} \approx D + (h_T - h_R)^2/2D$ is the distance between the antennas.

Propagation over a more complex obstacle can be estimated from the expression for propagation over a single screen using a method devised by Bullington (1947). The lowest unobscured paths are drawn from both receiver and transmitter and their point of intersection (see Figure 9.13). The obstacle is then replaced by a single screen with its top at the point of intersection and the propagation calculated using this screen (the result of such a calculation is usually within a few dB of the true result). This process can only yield

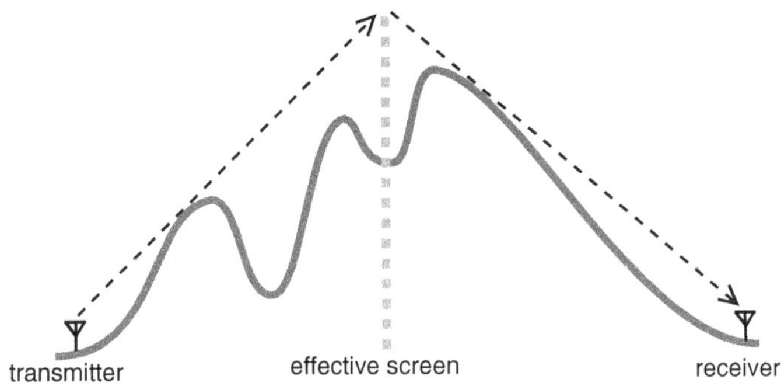

Fig. 9.13 Diffraction over a complex obstacle.

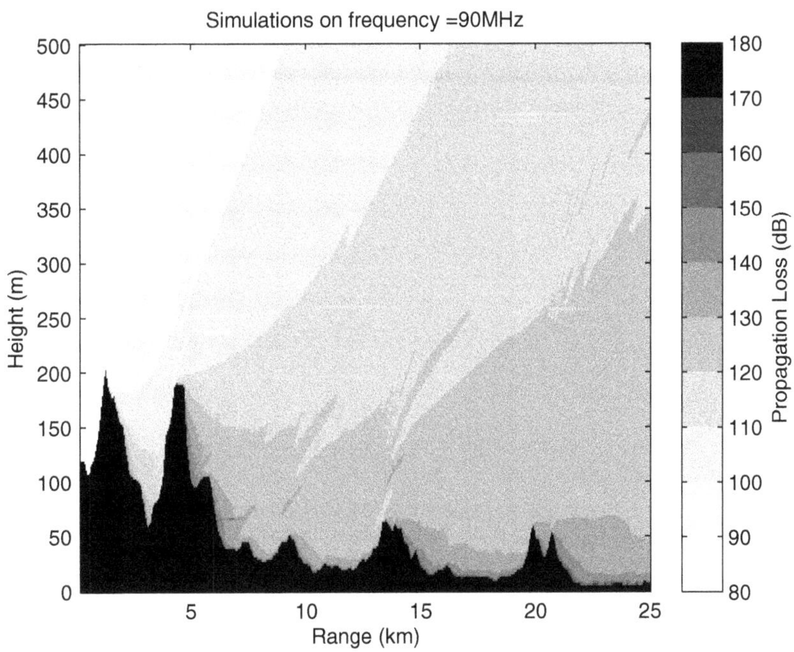

Fig. 9.14 Propagation by diffraction over a complex terrain.

a first approximation to propagation since, in reality, there can be repeated diffraction and reflection processes that are missed by such an approximation. A more detailed analysis will usually involve a numerical solution of Maxwell's equations. Figure 9.14 shows an example of such an analysis for a complex terrain, the propagation being described in terms of the propagation loss. It will be noted that the signal at a distant point is the combination of waves that have arrived by a multitude of diffraction and reflection processes.

Another situation where we expect there to be little energy transfer between receiver and transmitter is when the antennas are both located close to a conducting ground. In this case, since the rays connecting the antennas only graze the ground, the angle θ_i in

(9.8) and (9.9) will be almost $\pi/2$ and the reflection coefficient will have a value close to -1. Consequently, from (9.13), we expect the mutual impedance between the antennas to be zero. There is, however, a small amount of power that arrives through what is known as a *surface wave*. Consider the situation shown in Figure 9.15. The Kirchhoff integral (9.41) will still be appropriate (Monteath, 1973), but the field E_T, and the potential ψ, will now need to take into account reflections from the ground. In this case

$$E_T \approx \frac{\mathrm{j}\omega\mu_0 I}{4\pi} h_{\mathrm{eff}}^T (1+R) \frac{\exp\left(-\mathrm{j}\beta\left(Z + \frac{x^2+y^2}{2Z}\right)\right)}{Z} \tag{9.50}$$

and

$$\psi \approx \mathrm{j}\beta(1+R) \frac{\exp\left(-\mathrm{j}\beta\left(Z + \frac{x^2+y^2}{2Z}\right)\right)}{2\pi Z}, \tag{9.51}$$

where R is the reflection coefficient. For propagation close to the ground, the propagation direction of the reflected wave will make only a small angle α with the horizontal and, in the case of vertical polarisation, we will have from (9.8) that

$$\begin{aligned} R &= \frac{\sin\alpha\, \eta_r{}^{-2} - \sqrt{\eta_r{}^{-2} - \cos^2\alpha}}{\sin\alpha\, \eta_r{}^{-2} + \sqrt{\eta_r{}^{-2} - \cos^2\alpha}} \\ &\approx \frac{2\alpha}{\tilde{\eta}_r} - 1, \end{aligned} \tag{9.52}$$

where $\tilde{\eta}_r = \eta_r\sqrt{1-\eta_r^2}$. Since the antennas are close to the ground, $\alpha \approx y/Z$ and therefore $1 + R \approx 2y/D\tilde{\eta}_r$. From (9.41), we now find that

$$\begin{aligned} E_R = &\frac{2I}{\eta_0}\left(\frac{-\omega^2\mu_0^2}{16\pi^2}\right) h_{\mathrm{eff}}^T \frac{\exp(-\mathrm{j}\beta D)}{Z^2} \times \\ &\times \int_0^\infty \int_{-\infty}^\infty \frac{4y^2}{\tilde{\eta}_r{}^2 Z^2} \exp\left(\frac{-\mathrm{j}\beta(x^2+y^2)D}{2Z^2}\right) \mathrm{d}x\mathrm{d}y, \end{aligned} \tag{9.53}$$

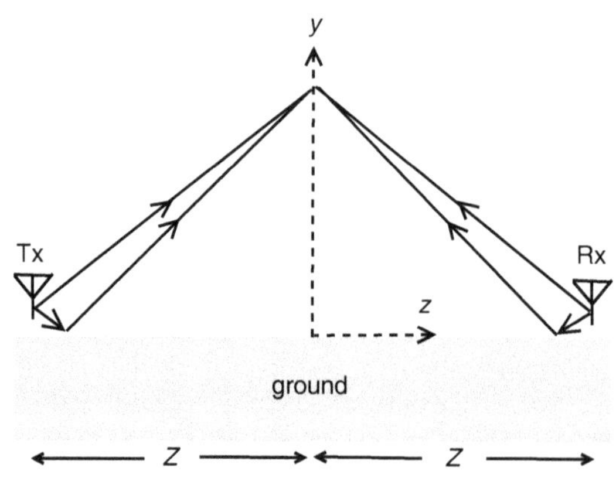

Fig. 9.15 Surface-wave propagation.

where $D = 2Z$. From the relations $\int_0^\infty t^2 \exp(-jqt^2)\,dt = (1/2jq)\int_0^\infty \exp(-jqt^2)\,dt$ and $\int_0^\infty \exp(-jqt^2)\,dt = \frac{1}{2}\sqrt{\pi/jq}$, we now find that

$$E_R \approx \frac{\eta_0 I h^T_{\text{eff}}}{2\pi \tilde{\eta}_r^{\,2} D^2} \exp(-j\beta D) \tag{9.54}$$

and, from which, the mutual impedance is given by

$$Z_{TR} \approx \frac{\eta_0 h^R_{\text{eff}} h^T_{\text{eff}}}{2\pi \tilde{\eta}_r^{\,2} D^2} \exp(-j\beta D). \tag{9.55}$$

The important thing to note is that the mutual impedance will fall away as D^{-2} rather than as D^{-1}, as would be the case for free space. For the strongest ground waves, it is clear that we need η_r to be small. This turns out to be the case for strongly conducting media such as sea water or damp ground (both of these media contain molecules with ionic bonds that disassociate in the presence of water and therefore conduct). Equation (9.55) represents the mutual impedance between antennas that are vertically polarised. In the case that the antennas are horizontally polarised, however, it turns out that the surface wave is negligible in comparison to the vertical case. Surface-wave propagation will still occur when the ground is non-flat (e.g. as a result of the curvature of the Earth) and is another mechanism for over-the-horizon propagation. Surface-wave propagation is essentially a diffraction process to which ground reflections make a major contribution. In the case of propagation over the surface of the Earth, we can regard this as diffraction over the surface of a sphere (radius $R_e = 6378$ km). However, the radius of the Earth is so large that (9.55) still provides an effective approximation for propagation over distances of hundreds of kilometres.

To understand the effect of conductivity, we need to go back to Maxwell's equations. We will assume time-harmonic waves of frequency ω and then the Maxwell equation (1.35) becomes

$$\int_C \frac{\mathbf{B}}{\mu} \cdot d\mathbf{r} = j\omega \int_S \epsilon \mathbf{E} \cdot \mathbf{n} dS + I. \tag{9.56}$$

The current I will consist of all conduction electrons flowing across the surface S, but we will first consider the current dI flowing across a small area dS. If we consider a unit length cylinder orthogonal to dS, we can imagine this to be a wire and then Ohm's law will imply that $dI = \sigma \mathbf{E} \cdot \mathbf{n} dS$, where σ is the conductivity. Consequently, $I = \int_S dI = \int_S \sigma \mathbf{E} \cdot \mathbf{n} dS$ and, substituting this into (9.56), we obtain

$$\int_C \frac{\mathbf{B}}{\mu} \cdot d\mathbf{r} = j\omega \int_S \epsilon_{\text{eff}} \mathbf{E} \cdot \mathbf{n} dS, \tag{9.57}$$

where the *effective permittivity* is given by

$$\epsilon_{\text{eff}} = \epsilon + \frac{\sigma}{j\omega}. \tag{9.58}$$

The *effective impedance* of the medium will now be given by

$$\eta_{\text{eff}} = \sqrt{\frac{\mu}{\epsilon_{\text{eff}}}} = \sqrt{\frac{\mu}{\epsilon}} \frac{1}{\sqrt{1 - j\frac{\sigma}{\omega\epsilon}}} \tag{9.59}$$

For low frequencies, and/or high conductivity, the effective impedance is low and so surface-wave propagation over such a medium will be strong. The benefit of low frequencies for long-range propagation was observed in the early days of radio and, as a consequence, much early broadcasting took place at low frequencies in order to achieve wide coverage. Indeed, it was believed that 1.5 MHz was the limit of useful propagation. This was based on the extrapolation of experimental observations, but we now know that the experimenters were only observing surface waves for which such a conclusion is valid. At this time the experimenters were oblivious to the strong waves that were reaching the ground many hundreds of kilometres away by ionospheric refraction. It was not until the 1920s that it was realised that considerably greater ranges could be achieved at frequencies above 1.5 MHz through this mechanism.

Over-the-horizon propagation can also occur through scatter by small anomalies in the refractive index, usually caused by turbulence in the ionosphere or neutral atmosphere. If a wave of amplitude E_{inc} is incident upon a small dielectric anomaly of volume V, a field (see Coleman (2017) for example)

$$E_{scat} = E_{inc}\frac{\beta^2 V}{4\pi r}\left(\frac{3\epsilon_r - 3}{\epsilon_r + 2}\right)\exp(-\mathrm{j}\beta r)\sin\theta \tag{9.60}$$

will be scattered in all directions where r is the distance from the scatterer and θ is the angle between the polarisation vector of the incident field and the direction of scatter (ϵ_r is the relative permittivity of the dielectric). Essentially, the incoming wave polarises the material in the direction of its electric field and the resulting oscillating dipoles then re-radiate some of the energy that they have absorbed. We consider a rectangular lattice of such scatterers with spacing d and a wave incident from below at angle ϕ to the horizontal. Let the waves be scattered downwards from anomalies in a horizontal line and in a direction at angle ψ to the horizontal (see Figure 9.16a). The total scattered field will be strongest when the scattered waves from the different anomalies are in phase. This will mean that $X'Y = XY'$ from which $d\cos(\psi) = d\cos(\phi)$ and hence $\phi = \psi$. Now consider waves that are scattered downwards from anomalies in a vertical line (Figure 9.16b). It will be noted that waves scattered from the upper anomaly will have travelled a further distance to those scattered by the lower anomaly. Consequently, to be in phase, we will

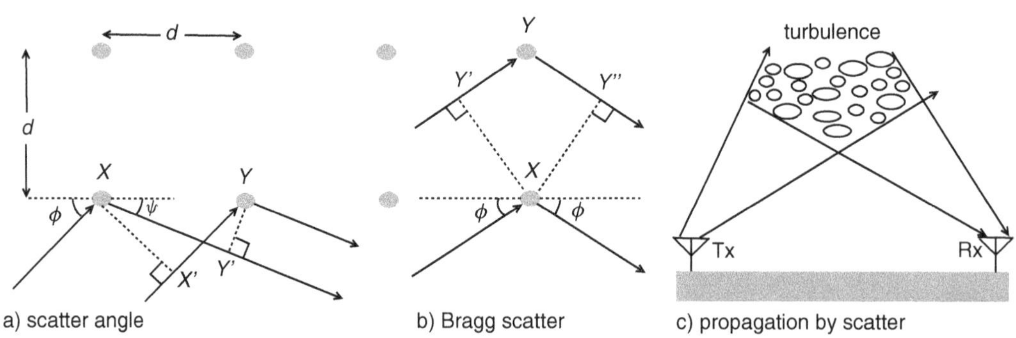

Fig. 9.16 Bragg scatter and propagation by scatter.

need $Y'Y + YY'' = \lambda$ (or a multiple of λ), i.e. $2d\sin(\phi) = \lambda$. For a given frequency, the angle of scatter will depend on the spacing of the scatterers. Such scatter is known as *Bragg scattering* and was first used to explain the effect of crystal structure upon the propagation of X-rays. Turbulence in the propagation medium can cause irregular fluctuations in the refractivity and this can be modelled as a set of dielectric anomaly lattices with different spacings. The strength of the refractivity fluctuations at these different spacings is known as the spectrum of the irregularity and its nature depends on the process that caused the turbulence. Turbulence in the troposphere causes a radio wave to scatter in a multitude of directions and this can be used to produce over-the-horizon propagation that is known as *tropospheric scatter* (see Figure 9.16). Likewise, turbulence in the ionosphere, especially in the auroral regions, can cause scatter that will also result in over-the-horizon propagation.

9.5 Conclusion

In the current chapter we have studied the propagation of radio waves in detail. Whilst radio waves might at first appear to be restricted to line of sight, we have found that there are mechanisms, such as refraction and diffraction, that can provide over-the-horizon propagation. Indeed, ionospheric propagation through refraction can provide global communications at HF frequencies. The ability to propagate radio waves over vast distances has led to an explosion in radio usage and there is now great pressure upon the available spectrum. Consequently, modern radio systems make use of techniques that optimise spectral usage and many of these techniques depend upon the careful management of propagation. In the next chapter we will consider some of the more important of these techniques.

10 Modern Radio Systems

In the previous chapters, we have studied the basic elements of radio. This has included the generation and detection of RF signals, their launch and capture as radio waves and the propagation of these waves. Thus far, we have concentrated on radio signals that consist of a carrier with simple analogue modulation. However, modern communication is increasingly accomplished by means of digital modulation, even when conveying analogue information. This has allowed coding that has decreased the bandwidth requirement of analogue signals such as television and audio. However, there is now considerable pressure on the available spectrum and this has necessitated the development of novel approaches that can optimise spectral usage. In the current chapter we consider some of these approaches, including techniques such as spread spectrum, cellular radio and MIMO. Additionally, we consider some of the non-communication uses of radio including surveillance, navigation and astronomy.

10.1 Digital Communication Systems

In digital communications, information is transferred as a stream of distinct modulation states and such communications, in the form of Morse code, were in frequent use well before the age of radio. Morse code was developed as a way of sending complex messages down telegraph wires in an efficient manner, the modulation states consisting of a series of long and short pulses (dashes and dots) that conveyed the information. In the early days of the telegraph, when clicks were the information sent, the dots and dashes were differentiated by the time between them. Table 10.1 shows the Morse code for some basic characters (the letters and numbers), but there is a Morse code for most of the important characters (e.g. · — · — ·— for a period). The basic idea of the code was developed by Samuel Morse in 1834 and further refined by Alfred Vail. In later usage, tones were transmitted with the dots and dashes represented by short and long tones with the dash being 3 times the length of the dot. The dots and dashes were separated by the length of a dot, the letters by the length of 3 dots and words by the length of 7 dots. The complexity of the code for a particular character was chosen on the basis of its frequency of usage. Consequently, the most frequently used character 'e' was designated a single dot. Modern analysis of the code shows it to be fairly close to the theoretical ideal in terms of efficiency. For reasons explained in Chapter 2, Morse code became a convenient means of conveying complex messages in radio and continued to be used for long-distance communications well into the 1960s.

Table 10.1 Some basic Morse code.

A	·—	M	——	Y	—·——
B	—···	N	—·	Z	——··
C	—·—·	O	———	1	·————
D	—··	P	·——·	2	··———
E	·	Q	——·—	3	···——
F	··—·	R	·—·	4	····—
G	——·	S	···	5	·····
H	····	T	—	6	—····
I	··	U	··—	7	——···
J	·———	V	···—	8	———··
K	—·—	W	·——	9	————·
L	·—··	X	—··—	0	—————

Whilst digital radio communications have always been with us, the development of efficient means of digitising data has resulted in nearly all radio communications becoming digital. Most broadcasting has now switched over to digital forms and mobile phones have been digital for quite some time. A major issue, however, is the way in which this digital data is modulated onto the radio carrier. Binary communications use two modulation states and these are represented by phase states in the case of *phase-shift keying* (PSK), by amplitude states in the case of *amplitude-shift keying* (ASK) and two frequency states in the case of *frequency-shift keying* (FSK). The modulating sequence $g_0g_1g_2g_3g_4\ldots$ for binary communications consists of single-bit symbols (these can take the values 0 or 1). Consider a modulated signal of the form

$$E(t) = A(t) * \cos(\omega(t)t + \phi(t)), \tag{10.1}$$

then, in the case of ASK,

$$A(t) = A_0 g_{[\frac{t}{T}]}, \quad \omega(t) = \omega_c \quad \text{and} \quad \phi(t) = 0, \tag{10.2}$$

where A_0 is a constant amplitude, $[\frac{t}{T}]$ is the integer part of $\frac{t}{T}$, T is the duration of the pulse representing a bit and ω_c is the carrier frequency. In the case of PSK,

$$A(t) = A_0, \quad \omega(t) = \omega_c \quad \text{and} \quad \phi(t) = \pi g_{[\frac{t}{T}]} \tag{10.3}$$

and, in the case of FSK,

$$A(t) = A_0, \quad \omega(t) = \omega_c + \Delta f g_{[\frac{t}{T}]} \quad \text{and} \quad \phi(t) = 0, \tag{10.4}$$

where Δf is the frequency deviation. PSK exhibits the smallest bit-error rate against noise (10^{-5} for 10 dB of SNR) with FSK and ASK far behind (about 10^{-3} and 10^{-2} of bit-error rate for 10 dB of SNR, respectively). It is possible to shorten the modulating sequence ($g_0g_1g_2g_3g_4\ldots$) by using a quaternary base (symbols can take the values 0, 1, 2 or 3), but this requires a modulation system with four states. The most important example of this is quadrature phase-shift keying (QPSK) for which

$$A(t) = A_0, \quad \omega(t) = \omega_c \quad \text{and} \quad \phi(t) = \frac{\pi}{4}(2g_{[\frac{t}{T}]} + 1). \tag{10.5}$$

QSPK has the same symbol error rate as PSK, but it has the advantage that it conveys more information in a symbol and hence is more efficient. The key to effective digital communications is the efficient representation of the data to be communicated in terms of the modulation states, i.e. the *coding*. The high efficiency of modern codes has meant that most forms of data (voice, television, etc.) can now be transferred digitally with a much lower bandwidth than could be achieved with analogue modulation. However, the continual pressure on radio spectrum has meant that even greater efficiency is required. Consequently, more imaginative ways of utilising the spectrum have needed to be developed and some of these are the subject of the next few sections.

10.2 Spread-Spectrum Systems

The Shannon–Hartley theorem

$$C = B\log_2(1 + \mathrm{SNR}) \tag{10.6}$$

provides a relationship between channel capacity C (bits per second), the channel bandwidth B (Hz) and the SNR. The relationship suggests that an increase in bandwidth will improve capacity. Indeed, it even suggests that there can even be transmission of data when the SNR is less than 1 ($C \approx 1.433 \times B \times \mathrm{SNR}$). Systems that increase capacity by increasing bandwidth are known as *spread-spectrum* systems. Surprisingly, the idea of spread-spectrum systems was first suggested in 1941 by the Hollywood actress Hedy Lamarr and the pianist George Antheil. They donated their idea to the US government to help the war effort, but the idea was not taken up until the 1980s. Since then, it has become the basis of many important technologies, 3G mobile communications being an important example. Two major examples of spread-spectrum systems are the frequency-hopping (FH) and the direct-sequence spread-spectrum (DSSS) varieties.

In FH spread-spectrum systems, the signal will hop around a set of frequencies in a pseudo-random fashion and this will be simultaneously tracked by a receiver that has knowledge of this sequence. In theory, the signal will only occupy a channel containing interference for a limited time and this will result in an overall increase in SNR. In addition, many users can use the same frequencies by each using a different pseudo-random sequence of frequencies since, with different random sequences, there will be a low probability of collision amongst these users. An important example of this approach is the Bluetooth system for interconnecting electronic devices.

In the DSSS variety of spread spectrum, a digital baseband sequence $a(t)$ is modulated by a much higher bit rate pseudo-random sequence $p(t)$ to form a new sequence $a(t)p(t)$ that is now *spread* in frequency. The new sequence is then used to modulate a carrier and this results in a wide bandwidth RF signal. At the receiver, the process is then reversed. The RF signal is first demodulated and then the spread baseband sequence is multiplied by a replica of the pseudo-random sequence in order to reconstruct the original baseband sequence.

There are two major advantages to the DSSS approach. Firstly, if we were to reduce the overall power level we could then maintain the same channel capacity by increasing the bandwidth. Secondly, although an increase in bandwidth might seem wasteful of spectrum, this is compensated for by the fact that many users can occupy the same channel by using uncorrelated pseudo-random sequences (multiplication by the wrong sequence will simply produce noise).

A realisation of DSSS is illustrated by Figures 10.2 and 10.3 with the delay τ in the receiver system used to compensate for the delay caused by propagation. In mobile communications one of the major problems is that of multiple propagation paths with the differential delays causing *inter-symbol interference*. For communication channels that suffer from multi-path propagation, DSSS offers a solution by adding together copies of a signal that have been demodulated using a range of delay parameter τ. Due to the pseudo-random nature of the code, signals that do not match the delay will only produce noise. However, signals that match the delay will coherently add. Importantly, this approach does not require synchronisation of the receiver and transmitter.

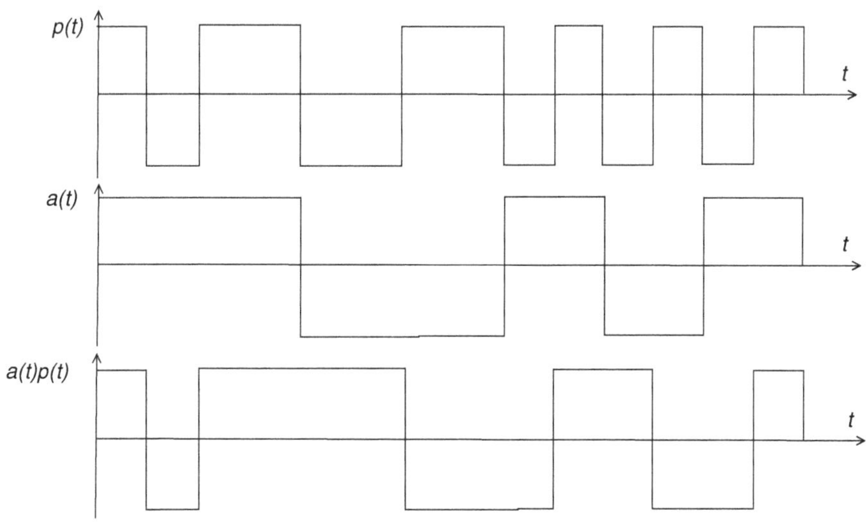

Fig. 10.1 Formation of a spread baseband signal.

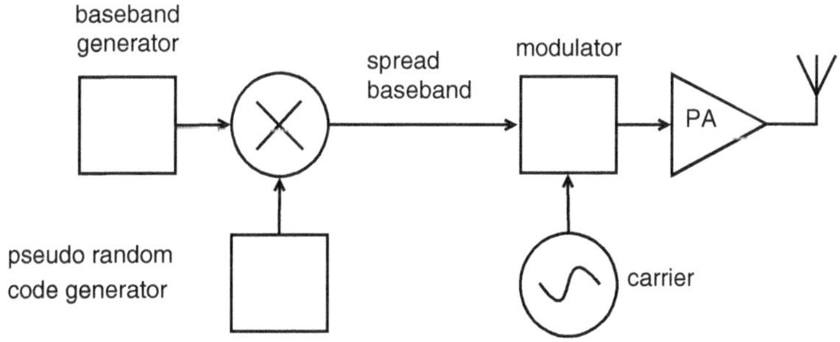

Fig. 10.2 A simple DSSS transmitter.

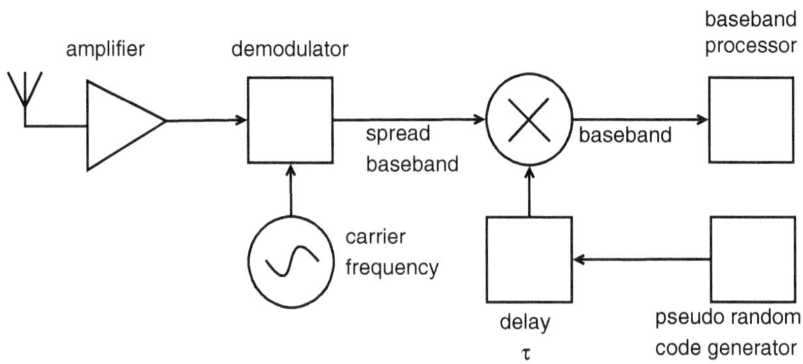

Fig. 10.3 A simple DSSS receiver.

For broadcast radio, an important spread-spectrum system is known as *digital audio broadcasting* or DAB for short. This system provides multiple broadcast services, together with text information. The data is split up into a total of 1,536 parallel data streams (also known as multiplexes) that are transmitted using DQPSK (differential quadrature phase-shift keying) on 1,536 sub-carriers that are separated by 1 kHz. (In DQPSK the changes in phase carry the information rather than absolute phase states.) The multiplexes are orthogonal since their carriers are harmonically related (i.e. they are easily separated from each other using fast DFT techniques) and the data rate of each multiplex is kept low enough so that the multiplexes do not interfere with each other. Such a system is known as OFDM or *orthogonal frequency-division multiplexing*. As with users of mobile communications, mobile users of broadcast services will also suffer from multi-path problems. With DAB, however, we essentially have many low-data-rate channels for which the delays of multi-path are insufficient to cause such significant overlap of the now very much longer symbols, i.e. inter-symbol interference will be reduced. As mentioned in Chapter 9, fading is another problem caused by multi-path. However, the phase shift that causes the fading is frequency-dependent (see Eq. (9.15) of Chapter 9) and so, at any one time, only a limited amount of the multiplexes will be affected by the fading (i.e. we have frequency-selective fading). As a consequence, providing that each service is suitably spread across the carriers, the effect of fading will be minimised.

10.3 Cellular Radio

Although digital and spread spectrum techniques have led to a far more efficient usage of the spectrum, the explosion in personal communications (including video and internet) has forced even more dramatic solutions. The most important of these is *cellular radio*, a system that limits the coverage of a channel so that its frequency can be reused at other locations. The total network area is divided into small cells, each of which contains a radio base station (RBF). The users within a cell will then have their transmit power limited so that they can only communicate with the RBF of that cell. As a consequence, the same set

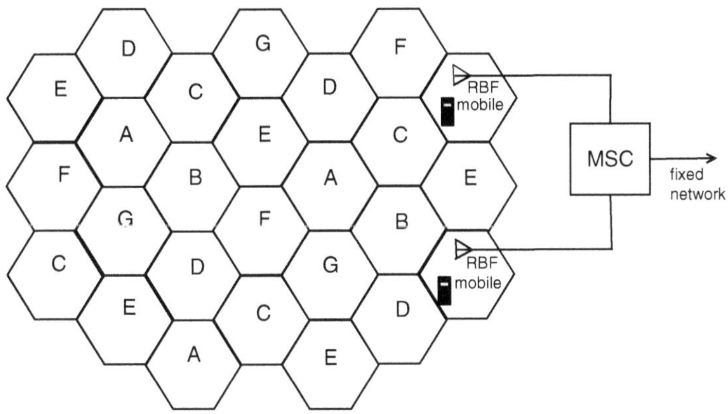

Fig. 10.4 A cellular radio system.

of channels can be used in other cells that are sufficiently isolated. The RBFs within the network are all interconnected through a mobile switching centre (MSC) which passes control from one RBF to another as a user passes from one cell to another, appropriately changing the channels in the process.

The design of a cellular system can be affected by many factors and the cells can vary in size according to the topography and location of the RBF. To ensure that adjacent cells do not share frequencies, the system needs to be designed around a *cluster* of cells that ensures this. A typical cluster consists of seven cells and Figure 10.4 shows a cellular system that is based on such a cluster (frequency sets are labelled A to G). It will be noted this cluster topology allows reuse of the frequency sets in adjacent clusters. However, it is inevitable that there will be some interference between the cells and this will be the dominant source of interference. From Figure 10.4, it will be noted that the minimum distance between cells with the same frequency set is approximately $4.583R$ where R is the cell radius. Then, assuming that all transmitters have the same power level, and that power decays as $(1/\text{distance})^n$ ($n \approx 4$ in most cases), the *signal-to-interference ratio (SIR)* is given by

$$\mathrm{SIR} = \frac{\text{minimum power within a cell}}{6 \times \text{maximum power between cells}} = \frac{R^{-n}}{6(4.583R)^{-n}} = \frac{4.583^n}{6}. \tag{10.7}$$

For the configuration of Figure 10.4 this would suggest an SIR of greater than 10 dB.

10.4 MIMO

The problem of communications capacity has pushed us into ever-more exotic techniques, one of the more recent being MIMO (multiple input multiple output). The systems that we have so far considered are SISO (single input single output) systems for which, as we have seen, there are several techniques for increasing capacity. MIMO, however, opens up further possibilities. The architecture for a typical MIMO system is

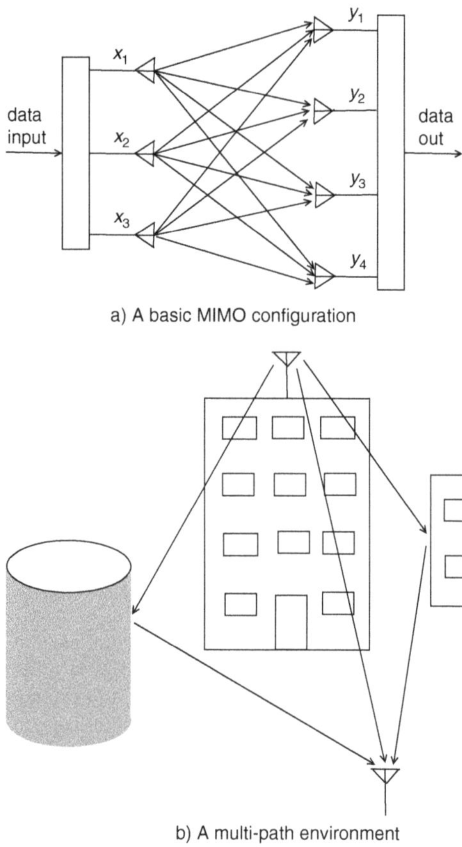

Fig. 10.5 MIMO and multi-path.

shown in Figure 10.5a and it will be noted that there are several transmit antennas (N_T say) and several receive antennas (N_R say). Our study of array antennas has already shown us that we can use multiple antennas on transmit, and receive, to increase the strength of the received signal (typically by a factor $N_T N_R$) and hence the SNR. By the Shannon theorem, this means that we have an increase in the channel capacity, i.e. we have what is known as *array gain*. As we have mentioned with respect to DSSS, it is possible to distinguish signals in the same channel through modulation of the baseband by different pseudo-random codes. If we modulate the baseband on separate transmit antennas by different pseudo-random codes, it is now possible to distinguish the contributions from different transmitters at the receive end. Further, if the baseband signals at each transmit antenna are the same, it would then be possible to beamform the transmitter at the receive end. In a multi-receiver system, such as a mobile telephone system, this would allow transmit beamforming at the receivers. Consequently, there are obvious gains to be made with MIMO systems. It will be noted that all of the above advantages also apply to MISO (multiple input single output) systems, but we will now see is that there are tremendous gains to be made when there are multiple receive antennas.

We have previously noted that multi-path can be a problem for communications systems by causing both inter-symbol interference and fading. These problems can be overcome by means of diversity techniques in which copies of the signal are transmitted at different frequencies (*frequency diversity*) or at different times (*time diversity*). This can be expensive in resource and so, as we have seen in the previous section, the alternative is to spread the signal over many channels so that, at any one time, only a small part of the signal is in a fading channel. Through multiple antennas, MIMO systems can allow what is commonly called *space diversity*. The simplest form of spatial diversity is for the receiver to have two, or more, antennas that are suitably spaced (half a wavelength or more). For sufficient spacing, it is likely that at least one antenna will receive signals that do not meet the conditions for fading. MIMO systems, however, have the potential to increase data capacity by turning multi-path from a problem into an advantage. Figure 10.5b shows a typical multi-path situation. It will be noted that each path effectively represents a separate communication channel and so has the potential to carry different data from the other paths, provided the system can separate out these paths. In the simplest approach, the signal to be transmitted is split into N_T data streams and these are transmitted through the propagation medium by the N_T separate antennas. At the receive end there will now be N_R data streams, but each of these will contain contributions from all of the transmitted streams. The trick is to separate out the individual transmitted streams. If x_1 to x_{N_T} are the transmitted signal, and y_1 to y_{N_R} are the received signals, then

$$
\begin{aligned}
y_1 &= h_{11}x_1 + h_{12}x_2 + \cdots + h_{1N_T}x_{N_T} \\
y_2 &= h_{21}x_1 + h_{22}x_2 + \cdots + h_{2N_T}x_{N_T} \\
&\vdots \\
y_{N_R} &= h_{N_R1}x_1 + h_{N_R2}x_2 + \cdots + h_{N_RN_T}x_{N_T}.
\end{aligned} \tag{10.8}
$$

The coefficient h_{ij} describes the propagation between the jth transmitter and the ith receiver. These coefficients are found by transmitting pilot signals from the transmitter, a process that is repeated at intervals short enough to account for time variations in the channel. In order extract the transmitted data streams from the received data streams, we need to solve (10.8). Whether this can be done depends on the number of effective communication channels. The number of transmitted data streams will need to be limited by this number and the minimum of N_T and N_R. For an environment rich in multi-path, however, MIMO has the potential to greatly increase capacity without additional power or bandwidth. MIMO can be combined with other techniques for increasing capacity and the combination MIMO with OFDM is one promising avenue.

10.5 Radar Systems

One of the major non-communications applications of radio waves is *radar* (radio detection and ranging). In a classical radar, the transmitted signal is interrupted by a *target* from which a small amount of energy is reradiated back to a receiver. The receiver will normally ascertain the direction of the target using a steerable array (mechanical or electronic steering) and the time of flight of the signal will then provide the target range. As far back as 1904, the German inventor Christian Hulsmeyer invented a system that detected ships by the reflection of radio waves. However, it was not until the mid-1930s that radar, in its currently recognisable form, came into existence. In 1935, a team led by the British scientist Robert Watson-Watt developed what is often credited as being the first radar system. This development was crucial to the defence of Britain in the second world war and big strides were made in its further development during this war. Radar systems can be regarded as radio systems in which the environment modulates the signal, hence allowing an operator to glean information about the environment. Whilst their primary use has been for detecting ships and aircraft, they are now increasingly used for gleaning information about the natural environment (wind profiling radars for example).

A typical radar configuration is shown in Figure 10.6. As mentioned earlier, the antennas will normally be steerable arrays whose steer direction will provide the direction of the target, with accuracy depending on the size of array. The range, however, is obtained from the time of flight and the accuracy with which this can be measured depends on the nature of the radar signal. For a signal consisting of pulses separated by time T, ie. *pulse-repetition frequency* $1/T$, the accuracy will be $cT/2$ (c is the propagation speed). The power P_R returned from the target is related to the transmitted power P_T through the *radar equation*,

$$P_R = P_T G_R G_T \left(\frac{\lambda}{4\pi R_T}\right)^2 \left(\frac{\lambda}{4\pi R_R}\right)^2 \frac{4\pi\sigma}{\lambda^2}, \tag{10.9}$$

where R_T and R_R are the ranges of the target from the transmitter and receiver respectively and G_T and G_R are the gains of the transmit and receive antennas respectively. σ is the *radar cross section* of the target and represents the amount of power reflected when a field with unit power per unit area is incident. Typically, a light aircraft has a cross section of about 2 m^2 and a jumbo jet a cross section of about 100 m^2. In essence the radar equation is the double application of the Friss equation with the target acting as both a receiver and transmitter. Radar cross sections can be quite complex, often depending on both the direction of the illumination and the direction of the reception.

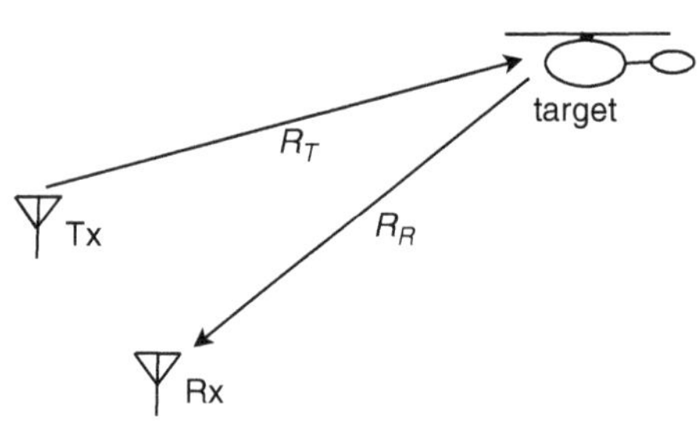

Fig. 10.6 A general radar configuration.

Besides signal returns from a desired target (an aircraft for example) there will often be returns from unwanted targets such as rough sea and rough ground. Sea in particular can return a considerable amount of power and has the potential to mask the desired target. Such interference, termed *clutter*, will be in addition to the noise that is experienced by communication systems. In radar, the signal-to-clutter ratio (SCR) can be just as important as SNR in determining radar performance. Fortunately, the motion of the target will itself cause a frequency shift in the radar return and this will normally allow the radar return to be distinguished from the clutter. The frequency shift is known as the *Doppler shift* and is related to the target dynamics through

$$\Delta f = -\frac{f}{c}\left(\frac{\mathrm{d}R_T}{\mathrm{d}t} + \frac{\mathrm{d}R_R}{\mathrm{d}t}\right). \tag{10.10}$$

To see this, consider the time δt between the crests of the radar carrier signal at the transmitter. When the first crest arrives at the receiver it will have travelled distance $R_T(t+R_T/c)+R_R(t+R_T/c)$ and when the second crest arrives it will have travelled distance $R_T(t+R_T/c+\delta t)+R_r(t+R_T/c+\delta t)$ i.e. the second crest will have travelled the extra distance $(\mathrm{d}R_T/\mathrm{d}t+\mathrm{d}R_R/\mathrm{d}t)\delta t$. The time between crests will have increased by $(\mathrm{d}R_T/\mathrm{d}t+\mathrm{d}R_R/\mathrm{d}t)\delta t/c$ and so the frequency of the wave will have changed according to (10.10). The Doppler of a target will not always ensure that it can be distinguished from clutter since targets travelling transverse to the look direction will generate a low Doppler. Further, the dynamic nature of the sea will mean that its radar returns will also exhibit Doppler and could serve to mask slow-moving targets.

For many radars the transmit and receive antennas are closely located and we have what is known as a *monostatic radar* ($R_T = R_R$). This is obviously necessary for compact platforms such as ships and aircraft. Consequently, in such radars, care must be taken to ensure that the transmit and receive phases do not overlap, i.e. the receive phase will need to occur between the transmit pulses. However, a radar signal consisting of pulses is not always convenient and, in the case of continuous radar signal, the transmitter and receiver will need to be well separated in order to avoid what is known as direct signal interference or DSI for short.

A radar for which the transmitter and receiver are well separated is known as *bistatic radar* (see Figure 10.7a) and a particularly important example is the passive bistatic

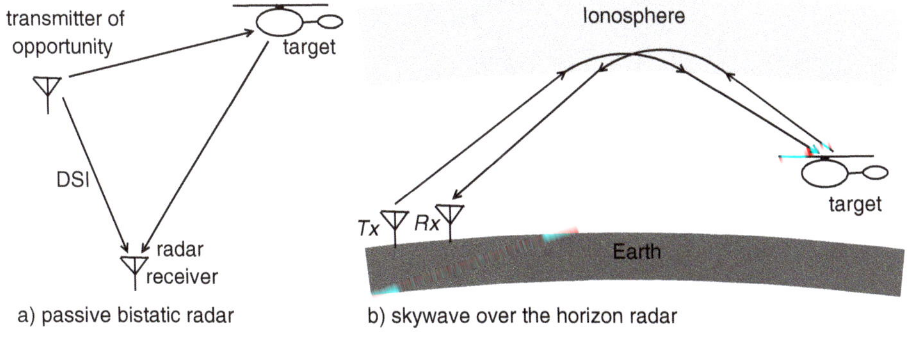

Fig. 10.7 Bistatic radar.

radar. In this case, the radar illuminator is a transmitter of opportunity, a broadcast transmitter for example. The signal from such a transmitter is usually continuous and so there are no quiet periods for reception, necessitating good isolation of the receiver from the transmitter. An important example of such an illuminator is that provided by the DAB system. In the UK, each transmitter has a power of the order of 10 kW, an operating frequency of around 222 MHz and a bandwidth of 1.536 MHz. For a continuous radar signal, the range resolution ΔR is related to the bandwidth B of the illuminator by $\Delta R = c/2B$ and so this example of a passive radar will have a range resolution of approximately 98 metres. However, as mentioned above, DSI is a major issue for such radars and the receive antennas will need good nulling in the direction of the transmitting antenna. However, a small amount of DSI is necessary for cross-correlation with the target returns for the purpose of target detection. Given the pressure on radio spectrum, passive radar offers a way of optimising the usage of spectrum by reusing radio signals created for other purposes.

Another form of bistatic radar is known as *over-the-horizon radar* (OTHR). These radars operate on frequencies in the HF band (3 – 30 MHz) and use sky-wave, or surface-wave, propagation in order to achieve over-the-horizon reach. Such radars usually transmit a *frequency-modulated continuous wave* (FMCW) signal in order to avoid the interference that pulse modulation can produce and typically have bandwidths between 10 and 50 kHz. To reduce the DSI that occurs due to ground-wave propagation, the transmitter and receiver will normally need to be separated by 100 km or more and the radar will need to operate at frequencies for which the receiver is inside the skip zone of the transmitter. Further, due to the dispersion caused by the ionosphere, sky-wave radars need to operate with relatively narrow bandwidths and consequently have limited range resolution. Nevertheless, skywave OTHR radar has been found to be an invaluable tool for long-range large surveillance.

For airborne radar, there is a further option known as *synthetic aperture radar* or SAR for short. This radar uses the flight path to simulate a long array antenna. A radar with relatively low resolution is located on the aircraft which, during flight, records radar returns (phase and amplitude) at regular intervals (see Figure 10.8). After a suitably large number of samples, these are combined in the fashion of an array. Returns from

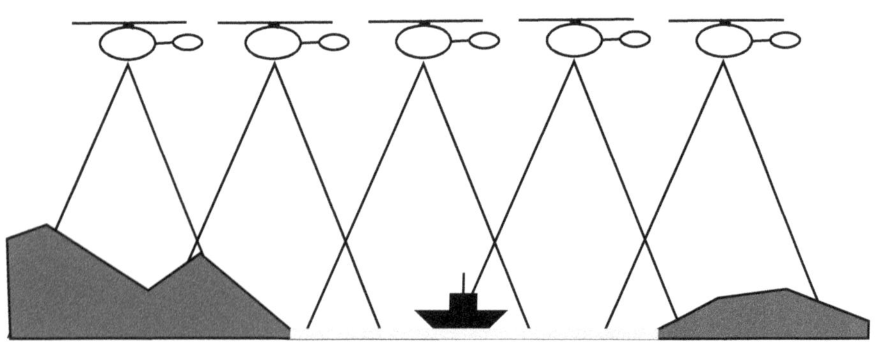

Fig. 10.8 A typical SAR scenario.

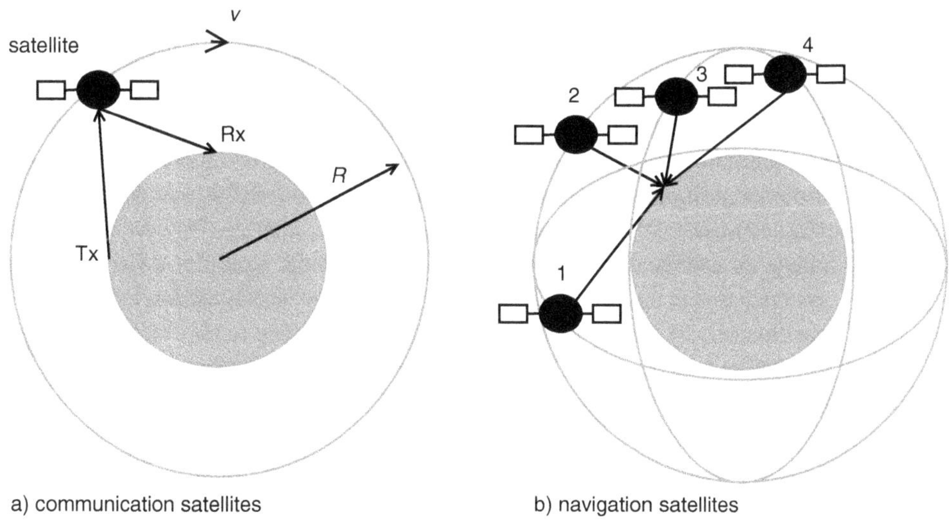

Fig. 10.9 Communication and navigation satellites.

a particular point on the ground are combined, with phase and amplitude corrections to account for the change in aircraft position during the observation period, to form a high-resolution image. It is clear that such radars work best with static, or at least slowly moving targets. However, moving targets can be handled with more sophisticated processing.

10.6 Satellite Systems

The first artificial satellite to be put into orbit around the Earth was Sputnik 1, launched by the Soviet Union in 1957. This satellite merely carried radio beacons (at frequencies of 20,005 MHz and 40.002 Hz) which only lasted for 21 days, but it started the era of radio in space. Satellites offered the possibility of very long-range over-the-horizon communication that did not depend on the ionosphere. Consequently, in 1962, the USA launched the first orbiting communications satellite known as *Telstar* and a new era of global communications started. Unfortunately, because of their orbital motion, satellites such as Telstar could only communicate for a limited time. Further, as we have seen with radar targets, a Doppler shift is imposed upon the signals and this needs to be compensated for. For a circular orbit, the speed v of the satellite is related to the radius of orbit R through $v = \sqrt{GM/R}$ where G is the gravitational constant ($G = 6.67 \times 10^{-11}$ N·m^2/kg^2) and M is the mass of the Earth (5.98×10^{24} kg). If a satellite is placed in an orbit at an altitude of 35,785 km above the Earth, the angular speeds of the Earth and satellite will match and the satellite will be fixed relative to Earth's surface. This is known as a geostationary orbit. Geostationary satellites are able to provide continuous communication without any Doppler effect. However, since the satellites are at a great distance from Earth, they require powerful transmitters and suffer a significant time delay in communication.

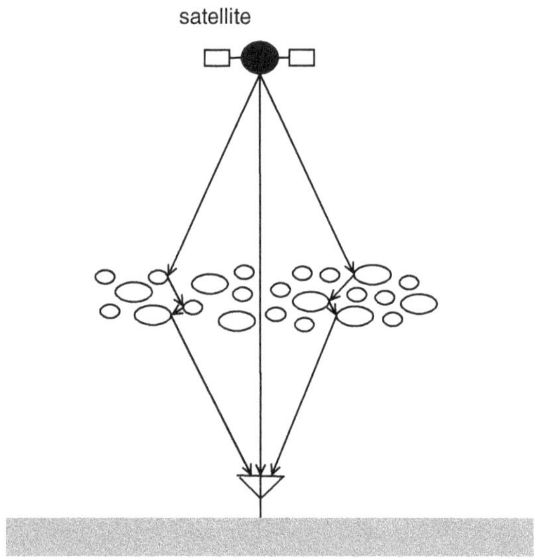

Fig. 10.11 The effect of turbulence upon satellite signals.

phase shift, the dipoles are fed from a power splitter, one by a transmission line of length L and the other by a line of length $L + \lambda/4$.

As we have mentioned in the previous chapter, both the neutral atmosphere, and the ionosphere, can be subject to turbulence that can cause irregular fluctuations in the refractive index. Because of this, there can be multiple propagation paths between the transmitter and the receiver (see Figure 10.11). This will cause there to be a spread of ranges between the receiver and transmitter, rather than one single range. Furthermore, if the random fluctuations are in motion, there will be a spread of Doppler shifts. In the equatorial, and auroral, regions the ionosphere is prone to turbulence that can, at times, be severe enough to cause appreciable range and Doppler spread. This phenomenon is known as *scintillation* and can severely compromise the operation of satellite systems. In the case of communication systems it can lower the capacity of a channel and in the case of navigation systems it can reduce the attainable accuracy. HF propagation can also suffer from the effects of scintillation and this can compromise the operation of HF communication systems and sky-wave OTHR.

10.7 Noise and Radio Astronomy

Although the last century has seen the technology of radio progress by leaps and bounds, factors such as propagation and noise still largely remain outside the control of the engineer. Our understanding of propagation has allowed us to choose the frequencies most appropriate to a radio system and modern electronics has allowed us to keep equipment noise down to an almost negligible level. However, external noise still remains a severe constraint. Figure 10.12 illustrates the various contributions to this

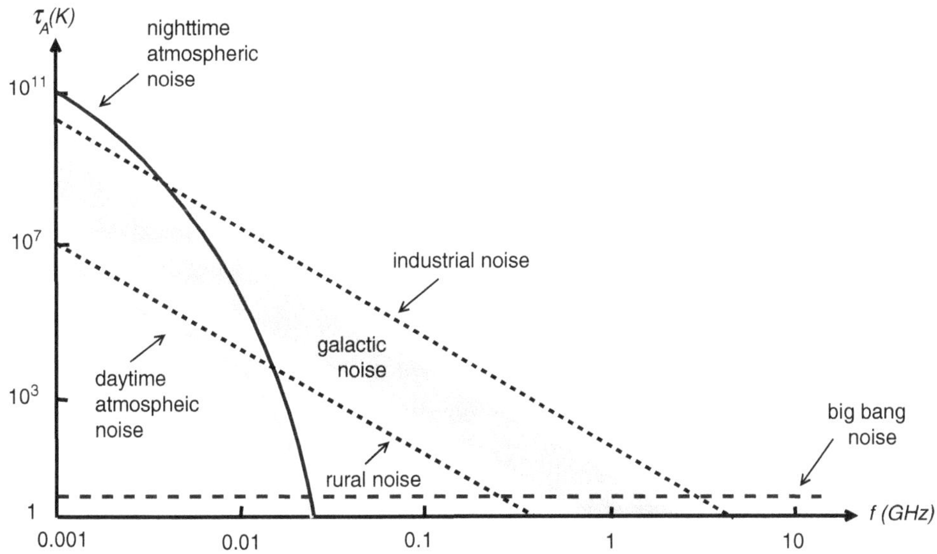

Fig. 10.12 Antenna temperature for external noise sources.

noise. Firstly, we live in an environment that is rich in man-made noise (car ignitions, plasma televisions and motors, to name but a few sources), and Figure 10.12 shows the typical variation of industrial noise with frequency. We might think we could do a lot better in the countryside, but Figure 10.12 also shows some typical man-made rural noise and it will be noted that its contribution is still appreciable. Besides man-made noise, there is also considerable noise from natural phenomena such as thunderstorms, known as atmospheric noise. These storms occur frequently and worldwide and their lightning flashes will generate considerable radio frequency energy across a broad spectrum. Such interference will propagate locally through the mechanism of surface waves and globally through refraction by the ionosphere. Consequently, at any location, there will be noise contributions from points across the planet. Furthermore, there is a considerable difference between the day and night thunderstorm contributions due to the difference between the daytime and nighttime ionospheric propagation. Due to its means of propagation, thunderstorm noise is highly directional and, as a consequence, its effect on a communications system can be highly dependent upon the type of receive antenna that is employed. Figure 10.12 also shows what is known as cosmic noise, i.e. noise coming from outside the Earth and its atmosphere. Here there are two major contributions: noise originating from the myriad of radio sources in the galaxy and the background cosmic radiation left over from the Big Bang at the start of the universe. The galactic noise is shown as a band since its strength depends on the view of the galaxy from Earth and this will change with time.

Obviously, the ionosphere blocks out cosmic radiation at lower frequencies, but it is the dominant source of naturally occurring noise at higher frequencies. At frequencies above 10 GHz, there are other problems that arise. Firstly, radiation from the sun can cause ionisation that causes further atmospheric noise due to discharge. Secondly, as where h_{eff} is the effective length of the antenna (note that this can be direction-dependent). The delay $\tau = D\cos\theta/c$ at antenna 2 is caused by the extra distance the wave must travel to this antenna. Consider the product of the two outputs, i.e.

$$V_1V_2 = h_{\text{eff}}^2 E_0^2 (\cos(2\omega t - \omega\tau) + \cos(\omega\tau)). \tag{10.14}$$

If we measure V_1 and V_2 over a number of periods, and average V_1V_2 over this time, we obtain $h_{\text{eff}}^2 E_0^2 \cos(\omega\tau)$. Then, if we divide this by the product of the amplitudes of V_1 and V_2 (i.e. $h_{\text{eff}}^2 E_0^2$), we obtain a quantity $\Phi = \cos(\phi)$ where $\phi = \omega\tau$. Unfortunately, from a given value of Φ, we will only be able to ascertain ϕ up to a multiple of 2π. However, if the source has at least two frequencies (ω_1 and ω_2), and we find ϕ for these two frequencies (i.e. ϕ_1 and ϕ_2), we can obtain an unambiguous value for τ through $\tau = (\phi_2 - \phi_1)/(\omega_2 - \omega_1)$. In the situation that the interferometer is moving with respect to the source (this will happen due to the rotation of Earth), an alternative option is to measure the rate of change $\mathrm{d}\phi/\mathrm{d}t$ since this quantity will not exhibit an ambiguity. Quantity $\mathrm{d}\theta/\mathrm{d}t$ will be known from the motion of the interferometer with respect to the source and so θ can then be calculated from the time derivative of the relation $\phi = \omega D\cos\theta/c$. The major thing to be noted is that both ϕ and $\mathrm{d}\phi/\mathrm{d}t$ are proportional to the length of the baseline D and so the greater the baseline the more sensitive the measurement to changes in θ. This in turn leads to a more accurate estimation of θ. Figure 10.14 shows the kind of Galactic map that can be obtained with such observations. (This shows the distribution of noise temperature across the Galaxy at a frequency of 20 MHz.)

10.8 Conclusion

In the current chapter we have considered modern radio techniques for communications, surveillance and astronomy. The pressure on the radio spectrum has forced an emphasis on the optimisation of spectral usage and we have discussed techniques for achieving this end. For communications this has led to technologies such as spread spectrum, cellular radio and MIMO. In the sphere of radar, technologies such as passive bistatic radar allow the use of existing broadcasts as radar illuminators of opportunity (digital radio and digital television for example). Radio remains a dynamic technology and major developments in hardware have made possible techniques that could not have been contemplated several decades ago. OTHR is an example of a technology that is only possible because of the development of high-speed computing. Radio processing in the digital domain is now changing the face of radio and we are at the stage where both receiver and transmitter are digital almost to the antenna. This has heralded in a new era of software radio in which the function of hardware can be changed by simply changing the software.

Appendix A: Some Useful Mathematics

A.1 Trigonometric Identities

$$\exp(j\alpha) = \cos\alpha + j\sin\alpha \tag{A.1}$$

$$\sin(\theta+\phi) = \sin\theta\cos\phi + \cos\theta\sin\phi \tag{A.2}$$

$$\cos(\theta+\phi) = \cos\theta\cos\phi - \sin\theta\sin\phi \tag{A.3}$$

$$\sin\theta + \sin\phi = 2\sin\left(\frac{\theta+\phi}{2}\right)\cos\left(\frac{\theta-\phi}{2}\right) \tag{A.4}$$

$$\cos\theta + \cos\phi = 2\cos\left(\frac{\theta+\phi}{2}\right)\cos\left(\frac{\theta-\phi}{2}\right) \tag{A.5}$$

$$\sin\theta - \sin\phi = 2\cos\left(\frac{\theta+\phi}{2}\right)\sin\left(\frac{\theta-\phi}{2}\right) \tag{A.6}$$

$$\cos\theta - \cos\phi = 2\sin\left(\frac{\theta+\phi}{2}\right)\sin\left(\frac{\phi-\theta}{2}\right) \tag{A.7}$$

$$\sin\theta\sin\phi = \frac{1}{2}\left(\cos(\theta-\phi) - \cos(\theta+\phi)\right) \tag{A.8}$$

$$\cos\theta\cos\phi = \frac{1}{2}\left(\cos(\theta-\phi) + \cos(\theta+\phi)\right) \tag{A.9}$$

$$\sin\theta\cos\phi = \frac{1}{2}\left(\sin(\theta-\phi) + \sin(\theta+\phi)\right) \tag{A.10}$$

$$\sin\left(\frac{\pi}{2}\pm\theta\right) = \cos\theta \tag{A.11}$$

$$\cos\left(\frac{\pi}{2}\pm\theta\right) = \mp\sin\theta \tag{A.12}$$

$$\sin(2\theta) = 2\sin\theta\cos\theta \tag{A.13}$$

$$\cos(2\theta) = \cos^2\theta - \sin^2\theta \tag{A.14}$$

$$\cos^2(\theta) = \frac{1}{2} + \frac{1}{2}\cos(2\theta) \tag{A.15}$$

$$\cos^3(\theta) = \frac{3}{4}\cos(\theta) + \frac{1}{4}\cos(3\theta) \tag{A.16}$$

$$\cos^2\theta + \sin^2\theta = 1 \tag{A.17}$$